Zoltán Tuba, DSc
Editor

Ecological Responses and Adaptations of Crops to Rising Atmospheric Carbon Dioxide

Ecological Responses and Adaptations of Crops to Rising Atmospheric Carbon Dioxide has been co-published simultaneously as *Journal of Crop Improvement*, Volume 13, Numbers 1/2 (#25/26) 2005.

Pre-publication REVIEWS, COMMENTARIES, EVALUATIONS . . .

"A VALUABLE AND WIDE-RANGING ANALYSIS of the impacts of elevated carbon dioxide concentrations in plants. The book addresses tree species, agricultural and horticultural crops, and pastures. The important issues of interactions with other environmental factors is addressed in studies on rising carbon dioxide levels in combination with drought, elevated temperature, nitrogen, and ozone. Readers are offered a comprehensive overview of the consequences of increased air carbon dioxide concentrations for crop plants and of the mechanisms involved. Among other experts in the plant sciences, this volume will be of interest to ecologists, physiologists, and agronomists. Tuba's work as editor and his selection of subjects and contributors is to be commended."

Rafael Martinez-Carrasco, PhD
Research Scientist
Institute for Natural Resources and Agricultural Biology of Salamanca
CSIC

More pre-publication
REVIEWS, COMMENTARIES, EVALUATIONS . . .

"This book not only helps us gain a deeper insight into the effect of rising carbon dioxide and changing climate, it also poses new questions to stimulate future research that will certainly help to maintain and strengthen world agriculture for the growing population. Experts from America, Europe, Asia, and Africa provide detailed results of research studies with thoroughly prepared figures and ample references."

Claus Buschmann, PhD
Associate Professor of Botany
University of Karlsruhe, Germany

"A VERY USEFUL AND MUCH-NEEDED BOOK. The structure and composition of the chapters is of high quality. The scientific value of the presented papers is on a high level."

Michal V. Marek, DSc
Professor
Institute of Landscape Ecology
Academy of Sciences
Budejovice, Czech Republic

More pre-publication
REVIEWS, COMMENTARIES, EVALUATIONS . . .

"Brings together the critical issues for the improvement of crop productivity under rising atmospheric carbon dioxide. . . . Presents state-of-the-art research results and reviews of the processes involved. The reader is provided with a comprehensive view of current knowledge and research. . . . Each chapter contains a summary and concluding section, amply supported by recent data, tables and graphs, and a list of important references."

Krisztina R. Végh, PhD
Senior Researcher
Research Institute for Soil Science
and Agricultural Chemistry
Hungarian Academy of Sciences
(RISSAC)

FPP

Food Products Press®
An Imprint of The Haworth Press, Inc.

New York • London • Victoria (AU)
www.HaworthPress.com

Ecological Responses and Adaptations of Crops to Rising Atmospheric Carbon Dioxide

Ecological Responses and Adaptations of Crops to Rising Atmospheric Carbon Dioxide has been co-published simultaneously as *Journal of Crop Improvement*, Volume 13, Numbers 1/2 (#25/26) 2005.

Monographic Separates from the *Journal of Crop Improvement*®

For additional information on these and other Haworth Press titles, including descriptions, tables of contents, reviews, and prices, use the QuickSearch catalog at http://www.HaworthPress.com.

The *Journal of Crop Improvement*® is the successor title to the *Journal of Crop Production**, which changed title after Volume 9, Numbers 1/2 (#17/18) 2003. The journal under its new title begins as the "*Journal of Crop Improvement*®," Volume 10, Numbers 1/2 (#19/20) 2004.

Ecological Responses and Adaptations of Crops to Rising Atmospheric Carbon Dioxide, Zoltán Tuba, DSc (Vol. 13, No. 1/2 #25/26, 2005). *Examines in detail the ecophysiological responses of crops to elevated air carbon dioxide and the economic significance of these changes.*

New Dimensions in Agroecology, David Clements, PhD and Anil Shrestha, PhD (Vol. 11, No. 1/2 #21/22 and Vol. 12, No. 1/2 #23/24, 2004). *Provides extensive information on current innovative agroecological research and education as well as emerging issues in the field.*

Adaptations and Responses of Woody Plants to Environmental Stresses, edited by Rajeev Arora, PhD (Vol. 10, No. 1/2 #19/20, 2004). *Focuses on low-temperature stress biology of woody plants that are of horticultural importance.*

Cropping Systems: Trends and Advances, edited by Anil Shrestha, PhD* (Vol. 8, No. 1/2 #15/16 and Vol. 9, No. 1/2 #17/18, 2003). *"Useful for all agricultural scientists and especially crop and soil scientists. Students, professors, researchers, and administrators will all benefit. . . . THE CHAPTER AUTHORS INCLUDE PRESENT AND FUTURE LEADERS IN THE FIELD with broad international perspectives. I am planning to use this book in my own teaching" (Gary W. Fick, PhD, Professor of Agronomy, Cornell University)*

Crop Production in Saline Environments: Global and Integrative Perspectives, edited by Sham S. Goyal, PhD, Surinder K. Sharma, PhD, and D. William Rains, PhD* (Vol. 7, No. 1/2 #13/14, 2003). *"TIMELY. . . . COMPREHENSIVE. . . . The authors have considerable experience in this field. I hope this book will be read widely and used for promoting soil health and sustainable advances in crop production." (M. S. Swaminathan, PhD, UNESCO Chair in Ecotechnology, M. S. Swaminathan Research Foundation, Chennai, Tami Nadu, India)*

Food Systems for Improved Human Nutrition: Linking Agriculture, Nutrition, and Productivity, edited by Palit K. Kataki, PhD and Suresh Chandra Babu, PhD* (Vol. 6, No. 1/2 #11/12, 2002). *Discusses the concepts and analyzes the results of food based approaches designed to reduce malnutrition and to improve human nutrition.*

Quality Improvement in Field Crops, edited by A. S. Basra, PhD, and L. S. Randhawa, PhD* (Vol. 5, No. 1/2 #9/10, 2002). *Examines ways to increase nutritional quality as well as volume in field crops.*

Allelopathy in Agroecosystems, edited by Ravinder K. Kohli, PhD, Harminder Pal Singh, PhD, and Daizy R. Batish, PhD* (Vol. 4, No. 2 #8, 2001). *Explains how the natural biochemical interactions among plants and microbes can be used as an environmentally safe method of weed and pest management.*

The Rice-Wheat Cropping System of South Asia: Efficient Production Management, edited by Palit K. Kataki, PhD* (Vol. 4, No. 1 #7, 2001). *This book critically analyzes and discusses production issues for the rice-wheat cropping system of South Asia, focusing on the questions of soil depletion, pest control, and irrigation. It compiles information gathered from research institutions, government organizations, and farmer surveys to analyze the condition of this regional system, suggest policy changes, and predict directions for future growth.*

The Rice-Wheat Cropping System of South Asia: Trends, Constraints, Productivity and Policy, edited by Palit K. Kataki, PhD* (Vol. 3, No. 2 #6, 2001). *This book critically analyzes and discusses available options for all aspects of the rice-wheat cropping system of South Asia, addressing the question, "Are the sustainability and productivity of this system in a state of decline/stagnation?" This volume compiles information gathered from research institutions, government organizations, and farmer surveys to analyze the impact of this regional system.*

Nature Farming and Microbial Applications, edited by Hui-lian Xu, PhD, James F. Parr, PhD, and Hiroshi Umemura, PhD* (Vol. 3, No. 1 #5, 2000). *"Of great interest to agriculture specialists, plant physiologists, microbiologists, and entomologists as well as soil scientists and evnironmentalists. . . . very original and innovative data on organic farming." (Dr. André Gosselin, Professor, Department of Phytology, Center for Research in Horticulture, Université Laval, Quebec, Canada)*

Water Use in Crop Production, edited by M.B. Kirkham, BA, MS, PhD* (Vol. 2, No. 2 #4, 1999). *Provides scientists and graduate students with an understanding of the advancements in the understanding of water use in crop production around the world. You will discover that by utilizing good management, such as avoiding excessive deep percolation or reducing runoff by increased infiltration, that even under dryland or irrigated conditions you can achieve improved use of water for greater crop production. Through this informative book, you will discover how to make the most efficient use of water for crops to help feed the earth's expanding population.*

Expanding the Context of Weed Management, edited by Douglas D. Buhler, PhD* (Vol. 2, No. 1 #3, 1999). *Presents innovative approaches to weeds and weed management.*

Nutrient Use in Crop Production, edited by Zdenko Rengel, PhD* (Vol. 1, No. 2 #2, 1998). *"Raises immensely important issues and makes sensible suggestions about where research and agricultural extension work needs to be focused." (Professor David Clarkson, Department of Agricultural Sciences, AFRC Institute Arable Crops Research, University of Bristol, United Kingdom)*

Crop Sciences: Recent Advances, Amarjit S. Basra, PhD* (Vol. 1, No. 1 #1, 1997). *Presents relevant research findings and practical guidance to help improve crop yield and stability, product quality, and environmental sustainability.*

∞ ALL FOOD PRODUCTS PRESS BOOKS
AND JOURNALS ARE PRINTED
ON CERTIFIED ACID-FREE PAPER

Ecological Responses and Adaptations of Crops to Rising Atmospheric Carbon Dioxide

Zoltán Tuba, DSc
Editor

Ecological Responses and Adaptations of Crops to Rising Atmospheric Carbon Dioxide has been co-published simultaneously as *Journal of Crop Improvement*, Volume 13, Numbers 1/2 (#25/26) 2005.

Food Products Press®
An Imprint of The Haworth Press, Inc.

New York • London • Victoria (AU)
www.HaworthPress.com

Published by

Food Products Press®, 10 Alice Street, Binghamton, NY 13904-1580 USA

Food Products Press® is an imprint of The Haworth Press, Inc., 10 Alice Street, Binghamton, NY 13904-1580 USA.

Ecological Responses and Adaptations of Crops to Rising Atmospheric Carbon Dioxide has been co-published simultaneously as *Journal of Crop Improvement*, Volume 13, Numbers 1/2 (#25/26) 2005.

Cover design by Jennifer M. Gaska.

Cover photograph courtesy of Zoltán Tuba.

Library of Congress Cataloging-in-Publication Data

Ecological responses and adaptations of crops to rising atmospheric carbon dioxide / Zoltán Tuba, editor.

p. cm.

"Co-published simultaneously as Journal of crop improvement, volume 13, numbers 1/2 2005."

Includes bibliographical references and index.

ISBN-10: 1-56022-120-8 (hard cover : alk. paper)

ISBN-10: 1-56022-121-6 (soft cover : alk. paper)

ISBN-13: 978-1-56022-120-3 (hard cover : alk. paper)

ISBN-13: 978-1-56022-121-0 (soft cover : alk. paper)

1. Crops–Physiology. 2. Crops–Effect of atmospheric carbon dioxide on. I. Tuba, Zoltán.

SB112.5.E29 2005

632′.1–dc22

2004026112

Indexing, Abstracting & Website/Internet Coverage

This section provides you with a list of major indexing & abstracting services and other tools for bibliographic access. That is to say, each service began covering this periodical during the year noted in the right column. Most Websites which are listed below have indicated that they will either post, disseminate, compile, archive, cite or alert their own Website users with research-based content from this work. (This list is as current as the copyright date of this publication.)

<u>Abstracting, Website/Indexing Coverage</u> . . . . <u>Year When Coverage Began</u>

- *AGRICOLA Database <http://www.natl.usda.gov/ag98>* **1998**
- *AGRIS <http://www.fao.org/agris>* **1998**
- *BIOBASE (Current Awareness in Biological Science) <http://www.elsevier.nl>* **1998**
- *Cambridge Scientific Abstracts (Water Resources Abstracts/ Agricultural & Environmental Biotechnology Abstracts) <http://www.csa.com>* **2001**
- *Chemical Abstracts Service <http://www.cas.org>* **1998**
- *Derwent Crop Production File* **1998**
- *EBSCOhost Electronic Journals Service (EJS) <http://ejournals.ebsco.com>* **2004**
- *Environment Abstracts. Available in print–CD-ROM– on Magnetic Tape. For more information check: <http://www.cispubs.com>* **1998**
- *Field Crop Abstracts (c/o CAB Intl/CAB ACCESS) <http://www.cabi.org/>* **1998**
- *FISHLIT <http://www.nisc.co.za>* **2004**
- *Foods Adlibra* **1998**

(continued)

- ***Food Science and Technology Abstracts (FSTA) Scanned, abstracted and indexed by the International Food Information Service (IFIS) for inclusion in Food Science and Technology Abstracts (FSTA) <http://www.foodsciencecentral.com>*** **2002**
- ***Google <http://www.google.com>*** **2004**
- ***Google Scholar <http://scholar.google.com>*** **2004**
- ***Haworth Document Delivery Center <http://www.HaworthPress.com/journals/dds.asp>*** **2004**
- ***HORT CABWEB (c/o CAB Intl/CAB ACCESS)*** **1998**
- ***Plant Breeding Abstracts (c/o CAB Intl/CAB ACCESS) <http://www.cabi.org/>*** **1998**
- ***Referativnyi Zhurnal (Abstracts Journal of the All-Russian Institute of Scientific and Technical Information– in Russian) <http://www.viniti.ru>*** **1998**
- ***TROPAG & RURAL (Agriculture and Environment for Developing Regions), available from SilverPlatter, both via the Internet and on CD-ROM <http://www.kit.nl>*** **2003**

Special Bibliographic Notes related to special journal issues (separates) and indexing/abstracting:

- indexing/abstracting services in this list will also cover material in any "separate" that is co-published simultaneously with Haworth's special thematic journal issue or DocuSerial. Indexing/abstracting usually covers material at the article/chapter level.
- monographic co-editions are intended for either non-subscribers or libraries which intend to purchase a second copy for their circulating collections.
- monographic co-editions are reported to all jobbers/wholesalers/approval plans. The source journal is listed as the "series" to assist the prevention of duplicate purchasing in the same manner utilized for books-in-series.
- to facilitate user/access services all indexing/abstracting services are encouraged to utilize the co-indexing entry note indicated at the bottom of the first page of each article/chapter/contribution.
- this is intended to assist a library user of any reference tool (whether print, electronic, online, or CD-ROM) to locate the monographic version if the library has purchased this version but not a subscription to the source journal.
- individual articles/chapters in any Haworth publication are also available through the Haworth Document Delivery Service (HDDS).

Ecological Responses and Adaptations of Crops to Rising Atmospheric Carbon Dioxide

CONTENTS

ABOUT THE EDITOR

Zoltán Tuba, DSc, is Full Professor and Head of Department of Biology PhD School and Plant Ecological Research Group of Hungarian Academy of Sciences at the Department of Botany and Plant Physiology, Szent István University, Gödöllö, Hungary. He has been chairman or invited key speaker at numerous international scientific conferences and has authored and co-authored dozens of papers in peer-reviewed scientific journals, book chapters, and books. He is on the editorial boards of *Photosynthetica*, *Journal of Crop Improvement*, *Community Ecology*, *Acta Botanica Hungarica*, and *Botanicai Kozlemenyek*. Dr. Tuba's main research interests include the ecology of desiccation tolerance in plants, the likely impact of global climate change on plant ecology and physiology, ecology and ecophysiology of natural vegetation, stress and production ecology of crops, weeds, and grasslands, and the bioindication of environmental pollution.

Preface

The atmospheric CO_2 values of the Earth have changed from the first days to the end of the 20th century from 278 to 367 ppm. During the past century the average temperature of the Earth's surface has risen with about 0.5-0.7°C. The fact of climate change has been accepted, however there is a broad discussion concerning the nature of that, whether it is an irreversible or cyclic phenomenon or not.

Documents of the UN Conference on Human Environment have suggested profound examinations concerning the impact of atmospheric gases and also that of CO_2 emissions on climate, human health, and the life of plants and animals.

Climate change has been labelled by the 1987 Report of the World Commission on Environment (Our Common Future) as a global problem. Suggestions upon that were mainly focusing on the reduction of emissions.

The Framework Convention on Climate Change was one of the major results of the UN Conference on Environment and Development. Heads of states have expressed their goodwill and considerations in favour of reducing the emissions of greenhouse gases. Exact obligations and deadlines were not described yet. CO_2 absorbing abilities of plant vegetation was mentioned first in this document as possible means of balance regulations.

In 1997 the Kyoto Protocol has been undersigned, a document which determined exact figures for the reduction of CO_2 emissions in the developed World. Unfortunately some states have avoided to ratify the Protocol. Later, the UN World Summit on Sustainable Development has made attempts to urge all countries to accept the Kyoto Protocol.

Elevated carbon dioxide concentrations in air may result in new parametres influencing natural ecosystems, agriculture, and forestry with a chance for beneficial growth of biological production. These facts also belong to the consequences of the climate change. It is a most essential and highly actual scien-

[Haworth co-indexing entry note]: "Preface." Láng, István. Co-published simultaneously in *Journal of Crop Improvement* (Food Products Press, an imprint of The Haworth Press, Inc.) Vol. 13, No. 1/2 (#25/26), 2005, pp. xxiii-xxiv; and: *Ecological Responses and Adaptations of Crops to Rising Atmospheric Carbon Dioxide* (ed: Zoltán Tuba) Food Products Press, an imprint of The Haworth Press, Inc., 2005, pp. xiii-xiv. Single or multiple copies of this article are available for a fee from The Haworth Document Delivery Service [1-800-HAWORTH, 9:00 a.m. - 5:00 p.m. (EST). E-mail address: docdelivery@haworthpress.com].

tific challenge to gain more information concerning the magnitude and the zonal variability of climate change as well as to learn more about the behaviour of plants. The present volume of the *Journal of Crop Improvement* is dedicated to this endeavour.

I have no doubts that information carried by the chapters published may have a positive response in the scientific world.

István Láng
Member of the Hungarian Academy of Sciences
Member of the World Commission on Environment and Development

Foreword

The carbon dioxide concentration in the Earth's atmosphere is increasing and a lively discussion among scientists, politicians, and the lay public exists about the effects of this phenomenon. The increase is attributed to fossil fuel burning and land use changes, so it is closely linked to the increase of human population, agriculture, and the increasing demand for energy. The increased CO_2 concentration is expected to influence climate, but it is also well accepted that elevated CO_2 will directly affect the process of photosynthesis.

The implications of elevated CO_2 and temperature for photosynthesis biochemistry are considered, as the important entry point for carbon and energy into the system. This indicates that more carbon is expected to flow into crop ecosystems with an associated increase in productivity. These, in turn, will also influence the processes related to soil formation, and soil composition itself.

Crops, including grasslands contribute a significant range of goods and services for the Earth. These include food for the expanding human population directly or through livestock production, rough materials of goods, sources and sinks of carbon, water, nitrogen, and other elements and habitants for biodiversity (consult with Campbell, B., Abstr. Bot., 21:219-220, 1997). Accelerating changes in the environment are expected to impact on this role of crop ecosystems in global ecology. A better understanding is needed for how these changes in land use, atmospheric CO_2 concentrations, climate, and other aspects of global change will alter biological processes and the functioning of crop ecosystems.

The known changes in plant responses given to increased air CO_2 concentration raise important questions about the consequences of these changes for human and livestock diets. For example, it is still quite unclear how much the predicted lowering of protein contents in crops and herbage and presence of greater amounts on non-structural carbohydrate in crop and forage diets will

[Haworth co-indexing entry note]: "Foreword." Tuba, Zoltán. Co-published simultaneously in *Journal of Crop Improvement* (Food Products Press, an imprint of The Haworth Press, Inc.) Vol. 13, No. 1/2 (#25/26), 2005, pp. xxv-xxvi; and: *Ecological Responses and Adaptations of Crops to Rising Atmospheric Carbon Dioxide* (ed: Zoltán Tuba) Food Products Press, an imprint of The Haworth Press, Inc., 2005, pp. xv-xvi. Single or multiple copies of this article are available for a fee from The Haworth Document Delivery Service [1-800-HAWORTH, 9:00 a.m. - 5:00 p.m. (EST). E-mail address: docdelivery@haworthpress.com].

http://www.haworthpress.com/web/JCRIP

alter the animal and human nutrition physiology. Elevated CO_2 may act at multiple levels in crop and grassland ecosystems. Mathematical modeling allows us an analysis of the key drivers in these complex systems and will assist in refining future predictions (Campbell, B., Abstr. Bot., 21:219-220, 1997).

This volume of *Journal of Crop Improvement* contains a welcome contribution to the scientific assessment of the future consequences of the increased air CO_2 concentration for crops (including all major crops of the Earth and grassland species, citrus, pulp tree and grapevine). Concerning their production ecology both the quantity and quality of their dry matter production are concerned. The papers cover most of the climatic regions of the Earth (from the North to the Equator). The closing paper is devoted to the role of agricultural soil carbon in the global carbon cycling suggesting that only limited increase of agricultural soil carbon may be expected due to the increased atmospheric CO_2 concentrations. Therefore in the context of the United Nations Framework Convention on Climate Change (UNFCCC), the importance of CO_2 impacts on carbon storage in agricultural soils appears to be relatively limited. Contributions from the experts of the area were invited. All manuscripts were reviewed by at least two reviewers in the usual evaluation procedure of scientific journals. We wish to thank the many colleagues who carried out the careful reviewing of the submitted manuscripts.

I am convinced, that the volume can and will serve as a reference book for experts as well as for newcomers and it will stimulate further research in this field. I can recommend the volume the most warmly not only to those explicitly involved in this research area and agricultural practice, but also to the researchers working in ecology, plant sciences, and other areas of biology. It is also a useful tool for the graduate and PhD students interested in this scientific area. Because of its aspects of practical importance, it also provides useful information to those involved in environmental sciences and practice.

I am very grateful to the *Journal of Crop Improvement* Editor-in-Chief, Dr. Amarjit Basra, for his invitation to print this volume in his journal and as a separate book, both published by The Haworth Press. Finally, I thank Mr. Zsolt Ürmös, MSc, for his essential help in the editorial work and all the authors for their contributions and co-operation.

Zoltán Tuba

Is the Long-Term Elevated Air CO_2 Environment Beneficial for Plants, Crops and Vegetation?

Zoltán Tuba

SUMMARY. Rising atmospheric carbon dioxide concentration and air temperature are of major concern when considering the possible effects of global climate change on crops and vegetation and their magnitudes. Although production has been found to increase in many cases, other experiments have also indicated increased hazards for plant growth because of the increased frequency of weather extremes, such as droughts, floods, and extreme temperatures. Thus at the same time, elevated CO_2 can be foreseen as an indirect constrain and as a stress alleviator. In the shorter term, the effect of CO_2 is direct and is mediated by the photosynthesis.

KEYWORDS. Limitation, stress alleviator, acclimation

Prof. Dr. Zoltán Tuba is affiliated with Plant Ecology Departmental Research Group of Hungarian Academy of Sciences and the Department of Botany and Plant Physiology, Szent István University, Gödöllõ, H-2103, Páter K. u.1., Hungary (E-mail: tuba@fau.gau.hu).

The present work was supported by the Hungarian Széchenyi Agroecology (OM-3B/0057/2002) project.

[Haworth co-indexing entry note]: "Is the Long-Term Elevated Air CO_2 Environment Beneficial for Plants, Crops and Vegetation?" Tuba, Zoltán. Co-published simultaneously in *Journal of Crop Improvement* (Food Products Press, an imprint of The Haworth Press, Inc.) Vol. 13, No. 1/2 (#25/26), 2005, pp. 1-6; and: *Ecological Responses and Adaptations of Crops to Rising Atmospheric Carbon Dioxide* (ed: Zoltán Tuba) Food Products Press, an imprint of The Haworth Press, Inc., 2005, pp. 1-6. Single or multiple copies of this article are available for a fee from The Haworth Document Delivery Service [1-800-HAWORTH, 9:00 a.m. - 5:00 p.m. (EST). E-mail address: docdelivery@haworthpress.com].

http://www.haworthpress.com/web/JCRIP

Digital Object Identifier: 10.1300/J411v13n01_01

However, in the longer term, the effects of elevated CO_2 became more and more indirect and its effects are mediated by source-sink interactions within plants, resources (nutrients, water) temperature, microbes, herbivores, and land-use management practice. As we must cultivate our crops and other agronomic plants in the fields under the naturally occurring unavoidable environmental constrains from the North to the Equator, therefore we need to gain the possible deepest and detailed scientific knowledge on the responses of all field growing agronomic plants. This demand requires more and more intense scientific research on this field.

Global climate change appears to be the greatest ecological problem of the future. When considering prospects of mankind, sound and detailed information about the unavoidable ecological effects of the global climatic change are inevitably needed. Unfortunately, this statement would be correct even if all the factors causing this change could be brought to a halt at once. This is because these factors will continue to exert their effect for a long time. For example, the carbon dioxide emitted into the atmosphere will have a long-term effect and this is one of the most influential ecological factors at global scale. The concentration of CO_2 in the atmosphere has been increasing in the two last centuries from about 280 ppm to a present value of 360 ppm, and is expected to reach more than twice the pre-industrial concentration by the end of this century (Houghton et al., 1990). The increasing use of fossil fuels and deforestation causes severe impacts on climatic components, such as temperature and rainfall. There is general agreement among different climate models that the accumulation of CO_2 and other greenhouse gases (CH_4, N_2O, and chlorofluorocarbons) will cause an increase in mean global temperature at the surface of earth. Although debate has continued about this topic since the end of the 19th century, it is only in the recent years that it has become possible to model the process in a more realistic way (Amthor, 1995).

Rising atmospheric carbon dioxide concentration and air temperature are of major concern when considering the possible effects of global climate change on crops and vegetation and their magnitudes. The predicted rise in global surface temperature together with all its ecological consequences are mainly attributed to the increased level of CO_2 in the atmosphere.

The increasing concentration of CO_2 affects the plants directly, causing changes in their chemical composition, physiological processes, production and fitness (Drake et al., 1997). Consequently, the tolerance, reproduction, distribution and abundance of plants will be altered. Species composition and diversity of plant communities, the vegetation dynamics and the structure of vegetation itself are all expected to change, too. All these changes in individual plants and in the vegetation will affect the subsistence of every living creature, so that of mankind.

As a consequence of increased photosynthesis at elevated CO_2, dry matter production of terrestrial C_3 plants is expected to increase. Reviews by Kimball (1983) and Cure and Acock (1986) of experiments done under a wide range of conditions have shown that a doubling of atmospheric CO_2 from 330 to 660 ppm may increase the productivity of C_3 species by an average of 33%. If these observations are extrapolated to the field, this means that the CO_2 enrichment since the start of the industrial revolution could be expected to have increased dry matter production of C_3 herbaceous plants by 7.5%-9.0%. Other studies (Poorter, 1993; Wand et al. 1999) reported that the average growth stimulation of C_3 plants is higher (40-44%) than 33% and the average increase of production in C_4 plants in response to a doubled CO_2 concentration is about 22-33%, which is lower than the average for C_3 plants.

On a global scale grasslands are agriculturally important ecosystems. They contribute a significant range of goods and services for the earth. Grasslands cover 24% of the Earth's vegetated area (Goudriaan, 1995). Considering terrestrial ecosystems, grasslands are the second largest vegetation formation after the forests and their area is further increasing with the land use changes (clearance of forests, urbanization, extensive agricultural practices). They occur over a broad range of climatic and soil conditions vary from intensively managed sown pastures to natural grasslands (e.g., Szemán, 2002). Grasslands similarly to other natural vegetation can largely affect the global cycling of carbon. Despite their importance, the potential effects of climate change on grasslands have received much less attention than effects on other ecosystems such as forests (Eamus and Jarvis, 1989; Ceulemans and Mousseau, 1994).

A prediction of the effects of global change on plants and vegetation, like agriculturally important grasslands may be based on studies of the effects of temperature, radiation, humidity on single species or small plot vegetation (Tuba et al., 1997). However, scaling from these results (Jarvis, 1995), usually obtained on plants or small plots growing in artificial substrates and/or controlled environments, to natural conditions in which competition (Jongen and Jones, 1997) or symbiosis (Simard et al., 1997) may play a major role, is often misleading. An attempt to predict the consequences of increased air CO_2 concentration on crops and vegetation must therefore take account of both the direct effects of elevated CO_2 and the indirect effects mediated by climatic change, as well as their interaction, bearing in mind the complexity of the natural ecosystems in which plants are growing.

Central to many studies are the questions of future productivity of agricultural crops as well as that of agriculturally important natural or semi-natural grassland ecosystems and the potential carbon storage capacity of crop and grassland ecosystems (e.g., Sala et al., 1996). Increased atmospheric concentrations of CO_2 may lead to increases in soil carbon storage, but this is uncertain due to limitations in other (e.g., water, nutrients) resources and interactions with climatic changes. Since only a small percentage of carbon added to the soil becomes stabilized, the impact of CO_2 fertilization of crops is considered to be very small (Smith et al., 1997). Therefore in the context of the United Nations Framework Convention on Climate Change (UNFCCC), the importance of CO_2 impacts on carbon storage in agricultural soils appears to be relatively limited (Smith et al., 2001). It is still unclear whether the world's grassland ecosystems can sequester a significant fraction of the increasing atmospheric CO_2.

Although production has been found to increase in many cases, other experiments have also indicated increased hazards for plant growth because of the increased frequency of weather extremes, such as droughts, floods and extreme temperatures.

In long-term elevated atmospheric CO_2 concentration can not be considered as unambiguously beneficial for natural vegetation, like grasslands and field grown agronomic plants but rather a profound change in a source the vegetation and outdoor crops have to acclimate to and by indirect way in interaction with other environmental factors it can appear as a stress effect. Further under current atmospheric CO_2 concentration vegetation and field grown agronomic plants continuously have to face environmental constrains (limited resources, extreme temperatures, herbivores, pollution, fire, etc.), which occur as naturally induced stress effects (Geider et al., 2001; Lichtenthaler, 1996). On the other hand, there are many indications that plants may generally cope better with various environmental stresses at higher CO_2 concentration. It can help plants to cope, e.g., with drought (Arp et al., 1998), heat and frost (Wayne et al., 1998) and salinity (Reuveni et al., 1997) stress. Thus at the same time elevated CO_2 can be foreseen as an indirect constrain and as a stress alleviator.

Many of experimental evidences support the prediction that the effects of elevated CO_2 will occur as not a direct beneficial effects of CO_2 on plants growing under natural environmental conditions. In the shorter term, the effect of CO_2 is direct and is mediated by the photosynthesis. However, in the longer term, the effects of elevated CO_2 became more and more indirect and its effects are mediated by source-sink interactions within plants, resources (nutrients, water), temperature, microbes, herbivores and land use/management practice (Hunt et al., 1997). In agricultural plant production systems, only few but not all of the environmental constraints can be controlled or their limiting effects can be reduced. Horticultural greenhouses are the only plant production systems where one can exclude or minimize the effects of all limiting constrains using radiation, temperature, nutrient, water, weed, pest control, etc. (e.g., Helyes and Pék, 2001). Therefore, greenhouse plants production represents the system which can provide useful information on plants growth under elevated air CO_2 concentration and as much as possible unlimited environmental conditions. Indeed, it is true that historically the first information about plants under high CO_2 concentrations were obtained from greenhouse cultivation (e.g., in Moe, 1984). Although these data are not really relevant for global climate change viewpoint, they can be useful in the studying and understanding of the longer-term direct effects of elevated CO_2 under less stressed conditions. However, we must to cultivate our crops and other agronomic plants in the fields under the naturally occurring unavoidable environmental constrains from the North to the Equator, therefore we need to gain the possible deepest and detailed scientific knowledge on the responses of all field growing agronomic plants. This demand requires more and more intense scientific research in this field.

REFERENCES

Amthor, J.S. (1995). Terrestrial higher-plant response to increasing atmospheric [CO_2] in relation to the global carbon cycle. *Global Change Biol.* 1:243-274.

Arp, W.J., J.E.M. Van Mierlo, F. Berendse and W. Snijders. (1998). Interactions between elevated CO_2 concentration, nitrogen and water: effects on growth and water use of six perennial plant species. *Plant Cell Environ.* 21: 1-11.

Ceulemans, R. and M. Mousseau. (1994). Effects of elevated atmospheric CO_2 on woody plants. *New Phytol.* 127: 425-446.

Cure, J.D. and B. Acock. (1986). Crop responses to carbon dioxide doubling: a literature survey. *Agriculture and Forest Meteorol.* 38: 127-145.

Drake, B.G., M.A. González-Meler and S.P. Long. (1997). More efficient plants: A consequence of rising atmospheric CO_2? *Annual Review of Plant Physiology and Plant Molecular Biol.* 48: 609-639.

Eamus, D. and P.G. Jarvis. (1989). Direct effects of increase in the global atmospheric CO_2 concentration on natural and commercial trees and forests. *Advances in Ecological Research* 19: 1-55.

Geider, R.J., E.H. Delucia, P.G. Falkowski, A.C. Finzi, J.P. Grime, J. Grace, T.M. Kana, J. LaRoche, S.P. Long, B.A. Osborne, T. Platt, I.C. Prentice, J.A. Raven, W.H. Schlesinger, V. Smetacek, V. Stuart, S. Sathyendranath, R.B. Thomas, T.C. Vogelman, P. Williams and F.I. Woodward. (2001) Primary productivity of planet earth: biological determinants and physical constraints in terrestrial and aquatic habitats. *Global Change Biol.* 7:849-882.

Goudriaan, J. (1995). Global carbon cycle and carbon sequestration. In: *Carbon sequestration in the biosphere*, ed. A.M. Beran, Berlin, Heidelberg, Springer Verlag. NATO ASI series, 133. pp. 3-18.

Helyes, L. and Pék, Z. (2001): The simultaneous effect of water supply and radiation on tomato flowering and setting. *Acta Horticulturae* 542: 227-233.

Houghton, J.T., G.J. Jenkins and J.J. Ephraums (eds.) (1990). *Climate change: the IPCC scientific assessment*, Cambridge University Press.

Hunt, R., J.P. Grime, S. Díaz, G.M. Spring, J.H.C. Cornelissen and R.L. Colasanti. (1997). Effects of elevated carbon dioxide on British native grassland species and communities. *Abstracta Botanica 21: 275-288.*

Jarvis, P.G. (1993). Global change and plant water relations. In: *Water transport in plants under climatic stress*, eds. Borghetti, M., J. Grace, and A. Raschi, Cambridge University Press, pp. 1-13.

Jarvis, P.G. (1995). Scaling processes and problems. *Plant, Cell and Environment* 18:1079-1089.

Jongen, M. and M.B. Jones. (1997). Effects of elevated CO_2 on species co-occurring in a neutral grassland. *Abstracta Botanica* 21: 249-260.

Kimball, B.A. (1983). Carbon dioxide and agricultural yield: an assemblage and analysis of 430 prior observations. *Agronomy Journal* 75: 779-788.

Körner, Ch., F.A. Bazzaz and C.B. Field. (1996). The significance of biological variation, organism interactions and life histories in CO_2 research. In: *Carbon dioxide, populations and communities*, eds. Ch. Körner and F.A. Bazzaz, San Diego, Academic Press. pp. 443-456.

Lichtenthaler, H.K. (1996). Vegetation stress: an introduction to the stress concept in plants. *Journal of Plant Physiol.* 148: 4-14.

Moe, R. (1984). CO_2 enrichment in Scandinavia. *Acta Horticulturae* 162: 217-225.

Poorter, H. (1993). Interspecific variation in the growth response of plants to an elevated ambient CO_2 concentration. *Vegetatio* 104/105: 77-97.

Reuveni, J., J. Gale and M. Zeroni. (1997), Differentiating day from night effects of high ambient CO_2 on the gas exchange and growth of *Xanthium strumarium* L. exposed to salinity stress. *Annals Bot.* 79: 191-196.

Sala, O.E., W.K. Laurenroth and I.C. Burke. (1996). Carbon budgets of temperate grasslands and the effects of global change. In: *Global change: Effects on coniferous forests and grasslands*, eds. A.I. Breymeyer, D.O. Hall, J.M. Melillo and G.I. Agren, Chichester, New York, Brisbane, Toronto, Singapore, John Wiley & Sons, pp.101-120.

Simard, S.W., D.A. Perry, M.D. Jones, D.D. Myrold, D.M. Durall and R. Molina. (1997). Net transfer of carbon between ectomycorrhizal tree species in the field. *Nature*. 388: 579-582.

Smith, P., D.S. Powlson, M.J. Glendining and J.U. Smith. (1997). Potential for carbon sequestration in European soils: preliminary estimates for five scenarios using results from long-term experiments. *Global Change Biol.* 3: 67-79.

Smith, P., P.D. Falloon, D.S. Powlson and J.U. Smith (2001). Carbon mitigation options in agriculture: Improving our estimates for Kyoto. In: *Sustainable management of soil organic matter*, eds. R.M. Rees, B.C. Ball, C.D. Campbell and C.A. Watson, CABI, Wallingford, UK., pp. 324-329.

Szemán, L. (2002). Effect of seed mixture components on the diversity of grassland. In: *Multi-function grasslands*. eds. Durand, J.L., Emile, J.C., Huyaghe, C., Lemaire, G., EGF. Grassland Science In Europe Volume 7. LaRochelle, France. p. 848-849.

Tuba, Z., Z. Nagy, K. Szente and A. Raschi (eds.). (1998). Grassland ecology and ecophysiology under elevated air CO_2 and temperature. *Abstracta Botanica* 21: 217-336.

Wand, S.J.E., G.F. Midgley, M.H. Jone and P.S. Curtis. (1999). Responses of wild C_4 and C_3 grass (*Poaceae*) species to elevated atmospheric CO_2 concentration: a test of current theories and perceptions. *Global Change Biol.* 5: 723-741.

Wayne, P.M., E.G. Reekie and F.A. Bazzaz. (1998). Elevated CO_2 ameliorates birch response to high temperature and frost stress: Implications for modelling climate-induced geographic range shifts. *Oecologia* 114: 335-342.

Rice Growth, Yield and Photosynthetic Responses to Elevated Atmospheric Carbon Dioxide Concentration and Drought

J. T. Baker
L. H. Allen, Jr.

SUMMARY. Rice is a major food crop that should respond favorably to expected future increases in atmospheric carbon dioxide concentration. Due to uncertainties in the timing and amounts of monsoonal rainfall, drought is common in some rainfed rice production systems. In this paper, we summarize results of experiments conducted by the University of Florida and USDA-ARS at Gainesville, FL, USA, where the effects

J. T. Baker is Plant Physiologist and works for the Alternate Crops and Systems Laboratory, USDA-ARS, Building 001, Room 342, 10300 Baltimore Avenue, Beltsville, MD 20705-2350 USA.

L. H. Allen, Jr. is Soil Scientist, USDA-ARS, Building 350, and affiliated with the University of Florida, Gainesville, FL 32611-0965 USA.

The technical assistance of James Brown, Serge Edme, and Steve Covell is gratefully acknowledged.

The authors acknowledge the support provided by the National Research Initiative Competitive Grants Program, USDA-CSRS, project number 9401541, the US Environmental Protection Agency through Interagency Agreement NO. DW12934099, and US Department of Energy, Carbon Dioxide Research Division through Interagency Agreement NO. DE-AI05-88ER69014 and DE-AI02-93ER61720. This work is a contribution of the Agricultural Research Service, US Department of Agriculture.

[Haworth co-indexing entry note]: "Rice Growth, Yield and Photosynthetic Responses to Elevated Atmospheric Carbon Dioxide Concentration and Drought." Baker, J. T., and L. H. Allen, Jr. Co-published simultaneously in *Journal of Crop Improvement* (Food Products Press, an imprint of The Haworth Press, Inc.) Vol. 13, No. 1/2 (#25/26), 2005, pp. 7-30; and: *Ecological Responses and Adaptations of Crops to Rising Atmospheric Carbon Dioxide* (ed: Zoltán Tuba) Food Products Press, an imprint of The Haworth Press, Inc., 2005, pp. 7-30. Single or multiple copies of this article are available for a fee from The Haworth Document Delivery Service [1-800-HAWORTH, 9:00 a.m. - 5:00 p.m. (EST). E-mail address: docdelivery@haworthpress.com].

http://www.haworthpress.com/web/JCRIP
Digital Object Identifier: 10.1300/J411v13n01_02

and interactions of elevated atmospheric carbon dioxide concentration and periodic drought were examined in relation to grain yield and canopy-scale gas exchanges, specifically photosynthesis, respiration, and evapotranspiration.

Elevated CO_2 increased rice growth, grain yield and canopy photosynthesis while reducing evapotranspiration by about 10%. During drought stress cycles, this water savings under elevated CO_2 allowed photosynthesis to continue for one to two days longer compared with the ambient CO_2 treatment. Rice canopy photosynthesis saturated with respect to CO_2 near 500 μmol mol^{-1} and we found little evidence of photosynthetic acclimation or down-regulation in response to long-term CO_2 enrichment treatments of 350 and 700 μmol mol^{-1}. Under a much broader range of long-term CO_2 treatments (160 to 900 μmol mol^{-1}), a significant degree of photosynthetic down regulation was detected. Daytime CO_2 enrichment resulted in higher canopy dark respiration compared with the ambient grown controls when compared at a common, near ambient nighttime CO_2. We also detected a rapid and reversible, direct inhibition of canopy dark respiration rate with rising chamber CO_2 at an air temperature of 28°C. *[Article copies available for a fee from The Haworth Document Delivery Service: 1-800-HAWORTH. E-mail address: <docdelivery@haworthpress.com> Website: <http://www.HaworthPress.com>]*

KEYWORDS. Climate change, photosynthesis, respiration, water use, acclimation

INTRODUCTION

Rice paddies in Asia play a significant role in the global budgets of greenhouse gasses such as carbon dioxide (CO_2) and methane (IPCC, 2001). The atmospheric concentration of CO_2 was about 358 μmol mol^{-1} in 1995 and is increasing by about 1.6 μmol mol^{-1} per year or nearly 0.5% per year (Keeling et al., 1995). Work by Jones et al. (1994, 1999) and Mann (1999) suggests that there has been a 0.3 to 0.6°C warming of the earths surface since the late 19th century. Compared with pre-industrial levels, atmospheric concentrations of CO_2 have increased by 30% (Houghton et al., 1996). This has led to projections of future increases in global air temperatures as well as potential shifts in regional scale precipitation patterns and the possibility of reduced soil water availability in some regions of the world (Wilson and Mitchell, 1987; Hansen et al., 1988).

Rice is the World's most important grain crop in terms of direct human food consumption. Rice is generally classified into four broad categories depending on soil and water habitat: irrigated or paddy grown rice, lowland rainfed rice, upland rice, and deep-water rice. Of the total of 140 to 145 million hectares of land area planted to rice annually, roughly half or about 53% is grown as irrigated, flooded paddy rice, one-fourth as rainfed lowland rice, 10-14% as upland rice and about 6% as deep-water rice (IRRI, 1975, 1982, 1984). About 75% of the worlds rice harvest comes from irrigated rice where the fields are flooded for much of the growing season.

Drought caused by uncertain monsoonal rainfall frequently reduces rice plant populations and final yield (IRRI, 1975). Excluding irrigated rice, about half of the world's rice land area depends on rainfall and is often subjected to drought stress (O'Toole and Moya, 1978; IRRI, 1982). Since soil water availability currently limits crop growth more than all other environmental factors combined (Mauney et al., 1979; Boyer, 1982), it is important to quantify the effects and possible interactions of water management practices with CO_2 upon major food crops such as rice.

The occurrence (or non-occurrence) of 'photosynthetic acclimation,' or 'down-regulation of photosynthesis,' in response to CO_2 enrichment is a naturally occurring phenomenon that, due to a lack of understanding, limits our ability to predict plant and ecosystem responses to CO_2 enrichment. In some plant species, photosynthesis was initially stimulated by CO_2 enrichment but then subsequently declined with continued CO_2 enrichment (Sage, 1994; Drake et al., 1997). One hypothesis put forward to explain this decline in photosynthesis following CO_2 enrichment is an end-product feedback inhibition of photosynthesis resulting from an imbalance between 'source' and 'sinks' for photoassimilates (Stitt, 1991; Bryant et al., 1998). Long-term CO_2 enrichment for rice and some other species can result in decreases in the content of ribulose-1,5-bisphosphate carboxylase/oxygenase (Rubisco) and other photosynthetic pigments as well as decreases in Rubisco activity (Besford et al., 1990; Rowland-Bamford et al., 1991; Bowes, 1993). There is much evidence that increases in tissue carbohydrate content under CO_2 enrichment modulates gene expression for many photosynthetic enzymes (Koch, 1996; Jang and Sheen, 1997; Moore et al., 1999).

Since 1987, numerous studies have been conducted at the University of Florida, USA on the effects of atmospheric carbon dioxide concentration on crop plants, including rice. For reviews of experiments aimed at determining the effects of CO_2 and air temperature on canopy-level gas exchange, growth and yield of rice, the reader is referred to Baker and Allen (1993a,b) and Allen et al. (1996). Our objective here is to summarize recent results of several papers that examined growth, yield and canopy gas exchanges (photosynthesis, respiration, and evapotranspiration) of rice canopies subjected to ambient and elevated CO_2 as well as periodic drought.

METHODS

Controlled Environment Chambers

Rice (cv. IR-72) plants were grown season-long in eight naturally sunlit *S*oil-*P*lant-*A*tmosphere-*R*esearch (SPAR) chambers at ambient (350 μmol mol^{-1}, 4 chambers) and elevated (700 μmol mol^{-1}, 4 chambers) CO_2 treatments. These chambers were constructed of an aluminum frame, canopy zone tops covered with transparent polyethylene telephtalate "Sixlight" film (Taiyo Kogyo Co., Tokyo[1]). Above ground chamber dimensions were 2.0×1.0 m^2 in cross section by 1.5 m in height. Each chamber top was attached to a lysimeter constructed of aluminum measuring 2.0×1.0 m^2 in cross section and 0.6 m deep, to provide a water tight, flooded root environment for growing rice in paddy culture. Soil filled each lysimeter to a depth of 0.5 m and paddy flood water depth was maintained at 50 mm above the soil surface using a float-actuated water valve to supply deionized water to each soil lysimeter. The main chamber dimensions were 2.0×1.0 m^2 in cross-section with a north-to-south sloping top that averaged 1.4 m in height, for a volume of 2.8 m^3. The 0.05 m space above the flood water provided an additional 0.1 m of aerial volume and the air handling ductwork had a volume of 0.3 m^3. In 1993, the full-wall doors of the south sides of each chamber were replaced with an access compartment that averaged 1.2 m in height with a 2.0×0.5 m^2 base, which provided 1.2 m^3 of additional chamber volume. All together, the total aerial volume of each chamber was 4.4 m^3. The calculated air exchange rate was 2.7 chamber volumes per minute.

Air was circulated from top to bottom within each chamber using fans located in external ducting. Exit duct air velocity was 3 m s^{-1}. Chamber dry bulb air temperature was measured with an aspirated, shielded thermocouple suspended 0.3 m above the canopy. Dew point air temperature was measured in the exit duct air stream with a dew point hygrometer (Dew-10, General Eastman Instruments). The air handling system, located in the external ducting contained cold- and hot-water heat exchangers and an electrical resistive-heat coil to control dew point and dry bulb air temperatures. Dew point was controlled to 18/12°C (day/night) using the cold-water heat exchanger to remove excess humidity. Dry bulb air temperature was controlled to 28/21°C (day/night) by reheating the air, as needed, using the hot-water heat exchanger and an electrical resistive-heat coil.

Rice paddies are a very weak net source of CO_2 to the atmosphere rather than a sink for CO_2 from the atmosphere. The biophysical reasons for this effect are described in detail by Baker et al. (2000a).

---

1. Use of trade or manufacturer's name does not imply endorsement.

Canopy Photosynthesis, Respiration and Evapotranspiration

The chamber system measured CO_2 using a common gas sampling system with infrared gas analyzers (IRGAs) for measuring CO_2 (Siemens, Ultramat 22P). Daytime chamber CO_2 was maintained by supplying pure CO_2 from a compressed CO_2 gas cylinder to mass flow controllers located in the external ducting of each chamber. The CO_2 injection rate was monitored by the mass flow controller which also functioned as a mass flow meter. Daytime canopy carbon exchange or net photosynthesis (P_n) was calculated from mass balance equations averaged every 300 s. Canopy gross photosynthesis (P_g) was calculated by adding daily estimates of canopy dark respiration (R_d) to P_n.

Each day near dusk, the controlled-environment chamber system was automatically switched from measuring photosynthesis to measuring canopy apparent dark respiration rate (R_d) during the night period from 1900 to 0600 h. During this night period, computer-control of chamber CO_2 by injections of pure CO_2 was suspended. Computer-controlled solenoid valves connected to high-pressure air-lines were used to control the opening and closing of air vent gates located in the ductwork of each chamber. At the end of each hour, during the night period, the air vent gates were opened and the chamber atmosphere was purged with ambient air driven by an electrical fan. The R_d was measured by automatically venting and resealing the chambers once an hour at night and monitoring the rise in chamber CO_2 from canopy respiration. The condensate from the cooling coils used to control dew point air temperature (DPT) passed through tipping bucket rain gauges to calculated evapotranspiration rate (ET). All canopy gas exchange measurements were expressed on a ground area basis. Specific methods for controlling chamber environmental set points (hardware, control algorithms) as well as the quality of those environmental controls are given by Pickering et al. (1994).

Tests for Photosynthetic Acclimation to CO_2 Treatment

To test for possible photosynthetic acclimation, a short-term CO_2 cross-switching experiment was conducted from 29 Aug. to 2 Sept. 1994 (45-49 days after planting, DAP), prior to the drought stress treatments (Baker et al., 1997b). Each morning from 0600 to 1200 EST, the CO_2 in each chamber was controlled to 160, 1000, 250, 700, and 350 $\mu mol\ mol^{-1}$ on 45 through 49 DAP, respectively. In this way, P_g light response of the long-term 350 and 700 $\mu mol\ mol^{-1}$ treatments could be compared at a range of common short-term CO_2. After 1200 h, the CO_2 control set-point was returned to its original long-term growth concentration treatment. Canopy P_g light response curves were fit with second order polynomials using the NOINT option of the GLM procedure provided by the SAS Institute (Cary, NC). These results are also compared to results from a previous experiment conducted in 1987 on the rice cultivar IR-30.

In 1987, the rice cultivar IR-30 was grown season-long in subambient (160 and 250) ambient (330) and superambient (500, 660 and 900 μmol mol^{-1}) CO_2 treatments. Tests for photosynthetic acclimation were conducted from 62 to 67 days after planting when canopy net photosynthesis (P_n) was compared by short-term cross-switching of CO_2 treatments to 160, 300, and 660 μmol mol^{-1}. Details of this experiment are given by Baker et al. (1990a).

Plant Culture, Growth Measurements, and Drought Stress Treatments

The rice cultivar used in this experiment (cv. IR-72) is grown throughout Southeast Asia. Rice was direct seeded by hand into 11 rows 0.18 m apart in each chamber on 15 July 1994. The plants were thinned to 200 plants m^{-2}, and flood water was applied on 20 July. Shades made of black, densely-woven, polypropylene fibers were maintained at canopy height along the outside of each chamber to provide a light environment similar to that found created by border rows in a field crop. On 20 July, prior to flooding, the soil in each chamber was fertilized with P and K at a rate of 9.0 g m^{-2}. Nitrogen as urea was applied at 16.5 g N m^{-2} four times during the growing season at 17, 38, 76, and 95 days after planting (DAP). The soil used was the top soil from a Kendrick fine sand (a member of the loamy, siliceous, hyperthermic, family of Arenic Paleudults).

Rice plants were destructively sampled at 19, 32, 52, 80, and 96 DAP. At 19 and 32 DAP, seven plants were sampled while at 52, 80, and 96 DAP 20 plants were sampled from each chamber. At final harvest, dry weight of each of 15 plants from each of three rows (45 plants total) from each chamber were measured as previously described. Each sampled plant was detached at ground level and dry weights were determined after oven drying at 70°C for 48 h.

Since the response of rice to drought stress depends on crop growth stage (IRRI 1982) the following water management treatments were selected: continuously flooded (CF) controls, drought imposed at panicle initiation (PI), drought imposed at anthesis (ANT) and drought imposed at both panicle initiation and anthesis (PI and ANT). This experiment was conducted with eight chambers, four maintained at each of the two CO_2 treatments. Of the four chambers in each CO_2 treatment, one was kept flooded season-long while the other three chambers experienced drought at either PI, ANT or PI and ANT. In all cases, drought treatments were imposed by shutting off the paddy water supply and unplugging the base of the lysimeter to drain the soil.

Our goal in this experiment was to impose similar levels of physiological stress rather than similar durations of stress. In order to impose similar physiological levels of drought, canopy net photosynthesis (P_n) measured at 300 s intervals (Baker et al., 1997a) was used as a diagnostic to determine the day for terminating the drought by reflooding the chambers. In all cases, the drought was terminated on the day when P_n became negative in high light (e.g., PAR > 1000 μmol photons m^{-2}). At this point them plants were visibly wilted and the leaves tightly rolled.

Statistical Analysis

In the 1994 drought study, where possible, prior to the ANT drought stress treatments, means and standard errors were calculated for the plant parameters measured during the destructive samplings using replicated chambers. This resulted in data points being the mean of four chambers (n = 4) at 19, 32, and 52 DAP and two chambers (n = 2) at 76 and 95 DAP. For the plant sampling data collected at final harvest, means, standard errors and analysis of variance were based on three rows (n = 3) containing 15 plants each for each chamber. As in previous experiments with rice at this location, within chamber estimates of variance were similar to that measured between replicated chambers. To describe trends in the data, linear and polynomial regression equations were fit via least squares linear regression.

RESULTS AND DISCUSSION

Growth and Yield

Shown in Figure 1 are the seasonal trends in aboveground biomass accumulation for the CO_2 and water management treatments. In the continuously flooded controls, CO_2 enrichment resulted in a 29% increase in final aboveground biomass while drought stress reduced biomass accumulation in both CO_2 treatments. Although CO_2 enrichment generally increased biomass production, quite similar patterns in biomass accumulation were observed among the water management treatments in both CO_2 treatments.

In both CO_2 treatments, the ANT drought treatment reduced aboveground biomass far more severely than that of the PI or PI and ANT treatments (Figure 1). These differences in seasonal trends in aboveground biomass among the water management treatments were largely due to differences in leaf area and leaf biomass (Baker et al. 1997a). Visual observations of destructively sampled plants following drought stress showed that the plants responded to the drought largely by shedding lower, older leaves on the culms. This effect was especially pronounced in the ANT drought treatment. These large reductions in biomass and leaf area of the ANT compared with the PI and ANT appear to suggest that the PI drought, in some way, acclimated or hardened the plants to withstand the subsequent ANT drought.

FIGURE 1. Seasonal trends in total above ground biomass for rice grown to maturity in two atmospheric CO_2 concentrations and subjected to four water management treatments. Vertical bars are twice the standard error for each mean. Data points are means of four chambers for 19, 32, and 52 days after planting (DAP), two chambers for 76 and 95 DAP and individual chamber means at final harvest. Final harvest standard error bars were obtain from three rows containing 15 plants each. Horizontal bars labeled PI and ANT indicate average durations for the droughts imposed at panicle initiation and anthesis, respectively. (Adapted from Baker et al. 1997a.)

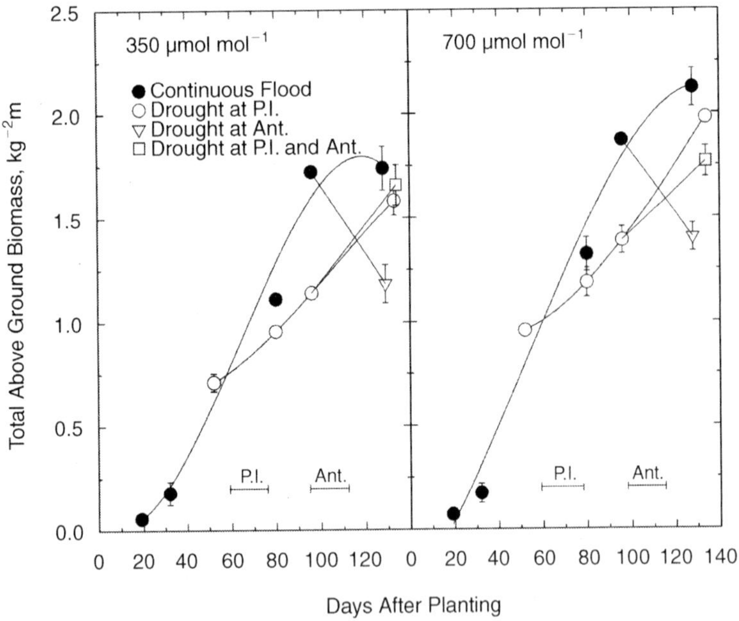

Final grain yield, yield components, final total above ground biomass, and harvest index are shown in Table 1. In the CF treatment, CO_2 enrichment resulted in a 28% increase in grain yield. Drought treatment tended to lower the percentage increases in grain yield due to CO_2 enrichment, especially for the PI and ANT treatment. As in previous experiments at this location with the rice cultivar IR-30 (Baker et al., 1993a), the increases in grain yield with CO_2 enrichment were associated with more panicles plant^{-1}.

In both CO_2 treatments, the effect of drought stress on final seed yield was small except for the ANT drought, which was, reduced by 44 and 53% with respect to the CF treatment, respectively (Table 1). The reductions in seed mass and the lack of significant water management treatment effects on number of

TABLE 1. Rice grain yield, components of yield, total above ground biomass and harvest index for rice (cv. IR-72) grown season-long in two atmospheric CO_2 concentrations and four water management regimes.

CO_2	Water treatment	Grain yield	Panicle plant^{-1}	Filled grain	Grain mass	Biomass	Harvest index
μmol mol^{-1}		Mg ha^{-1}	no. plant^{-1}	no. panicle^{-1}	mg seed^{-1}	kg m^{-2}	kg kg^{-1}
350	CF	6.5	4.3	37.6	19.7	1.7	0.37
	PI	6.6	4.5	36.5	20.0	1.6	0.40
	Ant.	4.5	4.1	32.2	16.3	1.2	0.38
	PI & Ant.	6.7	4.4	40.9	17.5	1.7	0.39
700	CF	8.3	4.9	38.8	20.4	2.2	0.34
	PI	8.3	4.8	44.0	20.1	2.0	0.41
	Ant.	5.4	4.5	36.6	16.1	1.4	0.37
	PI & Ant.	7.2	4.2	41.7	20.5	1.8	0.40
		F-Values					
CO_2 concentration		6.8 *	4.9 *	3.2 NS	4.2 NS	7.3 *	0.1 NS
Water treatment		7.4 **	1.7 NS	3.1 NS	19.5 **	5.9 **	4.5 *
CO_2 * Water		0.6 NS	1.9 NS	0.7 NS	3.4 *	0.6 NS	0.8 NS

**, * Significant at the 0.01 and 0.05 probability levels, respectively. NS = Not significant.

[1] CF = Continuously flooded, PI = drought at panicle initiation, Ant. = drought at anthesis, PI + Ant. = drought at both panicle initiation and anthesis.

filled grain panicle^{-1} (Table 1) indicates that drought in the ANT treatment reduced grain yield mainly by reducing assimilate supply to the grain during seed fill rather than affecting the flowering process. Drought stress during grain fill has also been shown to reduce grain weight in previous drought studies on rice (IRRI, 1982) presumably due to reduced assimilate supply.

Tests for Photosynthetic Acclimation

One method to detect possible canopy-scale photosynthetic acclimation to long-term CO_2 is to grow plants for long periods (weeks or months) at two or more CO_2 and then compare their photosynthetic light responses for brief periods under a common CO_2. The hypothesis being tested here is that previous long-term CO_2 exposure treatment does not affect current short-term photosynthetic rate vs. CO_2 relationships. Additional measurements may then be made to quantify the specific acclimation response. These acclimation responses could conceivably be traced to changes in gross morphology (e.g., changes in leaf thickness or specific leaf weight) or changes in the photosynthetic biochemistry of the plant (e.g., changes in the amount and/or activity of Ruisco).

Shown in Figure 2 are comparisons of canopy P_g vs. PFD for the short-term CO_2 switching experiment conducted during the morning hours from 45 to 49 DAP. The second order polynomial regression models fit each day's data well for both long-term CO_2 treatments with uncorrected (intercept forced to zero) R^2 values exceeding 0.99 in all cases.

In order to develop a photosynthetic response curve to CO_2 at high light, the regression equations in Figure 2 were solved by setting PFD to 1600 μmol (photons) m^{-2} s^{-1}. These estimates of P_g are plotted against the respective short-term CO_2 switch treatment (Figure 3) for both the 350 and 700 μmol mol^{-1} CO_2 long-term treatments. A two stage-segmented model consisting of a second order polynomial for the lower, curved portion of the response and a straight line for the upper, saturated P_g section was iteratively fit simultaneously in order to describe the data. Using this model to estimate P_g for PFD = 1600 μmol (photons) m^{-2} s^{-1} at 350 and 700 μmol mol^{-1} indicates

FIGURE 2. Canopy gross photosynthesis (P_g) vs. photon flux density (PFD) curves for rice grown long-term at 350 and 700 μmol CO_2 mol^{-1} air and subjected to short-term (< 6 h) CO_2 treatments of 160, 250, 350, 700, and 1000 μmol mol^{-1} from 45 to 49 DAP. Each data point is the mean of 4 chambers (n = 4) at a particular PFD level. Error bars are ± SE of the mean. Curves are second-order polynomials fit to each day's data. (Adapted from Baker et al. 1997b.)

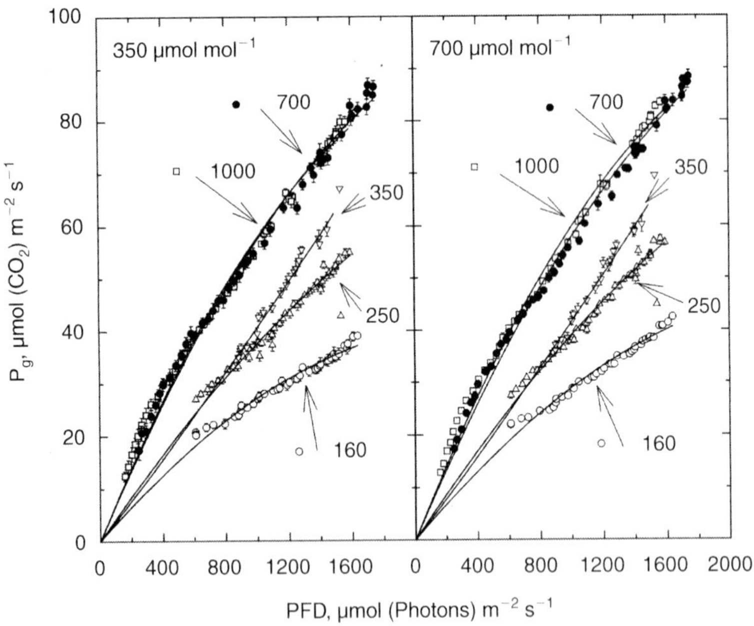

that doubling CO_2 across this range resulted in a 17.4% increase in P_g. These tests for canopy-scale photosynthetic acclimation of rice indicate little difference between long-term CO_2 treatments of 350 and 700 μmol mol^{-1} for the rice cultivar IR-72.

Still, in this same experiment, Vu et al. (1998) found reductions in single-leaf Rubisco content ranging between 6 to 22% for the 700 μmol mol^{-1} compared with the 350 μmol mol^{-1} CO_2 treatments. Thus, while photosynthetic acclimation responses in terms of enzyme down-regulation may be detected at the biochemical level, these effects may or may not result in a detectable loss of canopy-scale photosynthetic capacity when measured using gas exchange techniques. Indeed, Osborne et al. (1998) found that acclimation responses of wheat (*Triticum aestivum* L.) leaf photosynthesis to elevated CO_2 depended on both the vertical position of the leaf within the canopy and the leaf developmental stage.

FIGURE 3. Canopy gross photosynthesis (P_g) for the 350 μmol mol^{-1} (open circle) and 700 (closed circle) long-term CO_2 treatments vs. short-term CO_2 treatments of 160, 250, 350, 700, and 1000 μmol mol^{-1}. Each data point is the solution of polynomial regression models shown in Figure 1 with PFD set to 1600 μmol (photons) m^{-2} s^{-1}. Error bars represent ± 95% confidence limit intervals. The regression curve is a two-stage segmented model consisting of a second order polynomial (lower, curved portion) and a straight line (upper saturated portion). (Adapted from Baker at al. 1997b.)

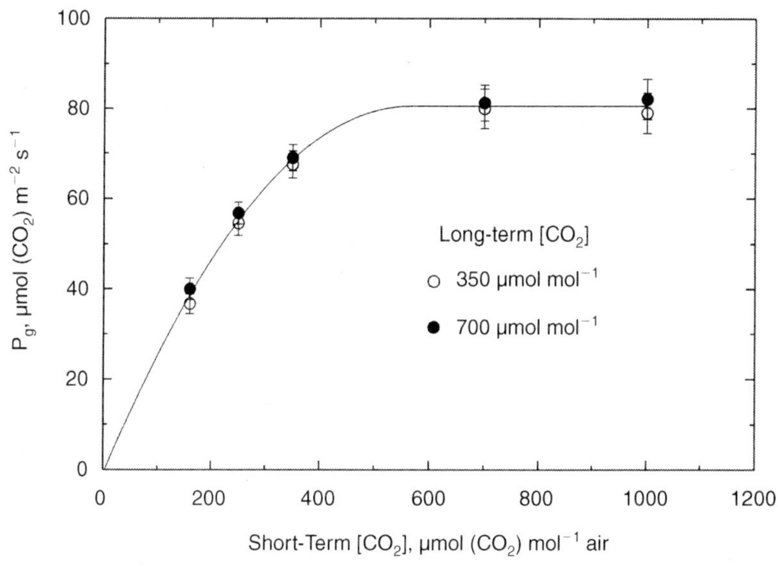

Shown in Figure 4 are similar CO_2 cross-switches to test for photosynthetic acclimation to long-term CO_2 for the rice cultivar IR-30 in 1987 (Baker et al., 1990a). In that experiment, Baker et al. (1990a) grew rice season-long at a wide range of long-term CO_2: subambient (160 and 250 μmol mol^{-1}), ambient (330 μmol mol^{-1}) and superambient (500, 660, and 900 μmol mol^{-1}) (Figure 4a). They compared P_n light responses for all six long-term CO_2 under three common short-term CO_2: 160, 330 and 660 μmol mol^{-1} (Figure 4b). Across this wide range of CO_2, including subambient CO_2, there was clear evidence of a down-regulation of P_n (Figure 4b). In that experiment, total plant tissue nitrogen content (Baker et al., 1992) and both Rubisco content and activity (Rowland-Bamford et al., 1991) decreased with increasing long-term CO_2 treatment.

FIGURE 4. (A). Canopy net photosynthesis (P_n) vs. long-term CO_2 acclimation treatment for rice canopies at 61 days after planting in 1987. (B). Comparison of canopy net photosynthesis (P_n) vs. long-term CO_2 acclimation treatment for short-term CO_2 of 160, 330, and 660 μmol mol^{-1} at 62-67 days after planting. All P_n estimates are from regressions equations of P_n vs. PFD with PFD set to 1500 μmol photons m^{-2} s^{-1}. Vertical bars represent 95% confidence intervals. (Adapted from Baker et al. 1990.)

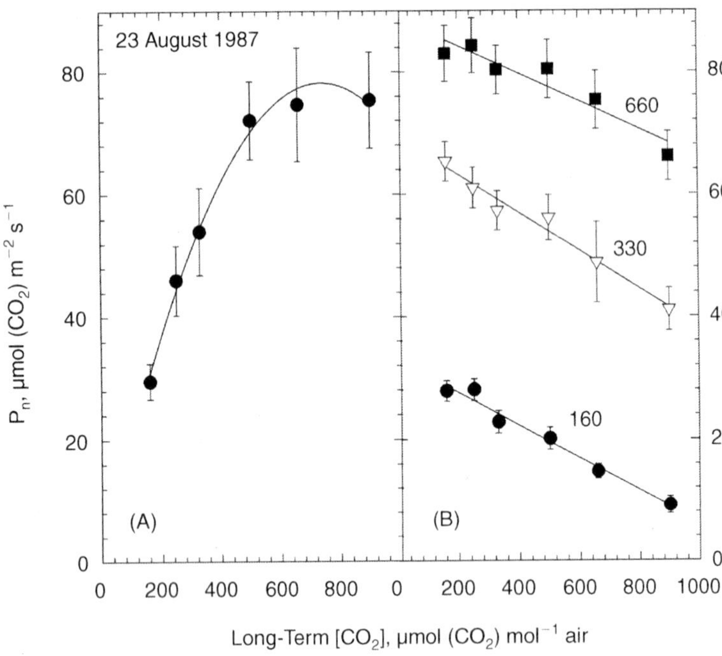

Using the data in Figure 4 and extrapolated to a PFD of 1500 μmol (photons) m^{-2} s^{-1} there was clear evidence of a down-regulation of P_n across the long-term CO_2 from 160 to 900. For example, when measured at a common short-term CO_2 of 160 μmol mol^{-1}, P_n decreased almost three fold from 27.8 to 9.1 μmol m^{-2} s^{-1} across the long-term CO_2 from 160 to 900 μmol mol^{-1}. The majority of this down-regulation response occurred across the subambient long-term CO_2 range from 160 to 330 μmol mol^{-1}. When comparing only the 330 and 660 μmol mol^{-1} long-term CO_2 under the short-term CO_2 of 330 μmol mol^{-1} resulted in P_n estimates were 57.3 and 48.8 μmol m^{-2} s^{-1}, respectively. Under the short-term CO_2 of 660 μmol mol^{-1}, P_n estimates were 80.3 and 75.3 μmol m^{-2} s^{-1}, respectively (Figure 4).

A nearly universal finding in CO_2 enrichment studies, including those on rice, are increases in soluble carbohydrate or total nonstructural carbohydrate (TNC) concentration of leaves and other organs for plants grown under CO_2 enrichment (Rowland-Bamford, 1996; Vu et al., 1997; 1998). The increased TNC is often attributed to enhanced photosynthesis under CO_2 enrichment and carbohydrate concentration is known to modulate the expression of many genes, including photosynthetic genes (Koch, 1996; Jang and Sheen, 1997).

Recently, Gesch et al. (1998) grew rice at ambient (350 μmol mol^{-1}) and elevated (700 μmol mol^{-1}) CO_2. They switched these CO_2 treatments during the late vegetative stage and measured changes in mRNA for the synthesis of the small subunit of Rubisco (*rbcS*) in both growing and fully expanded leaves. They found that within 24 h, plants switched from ambient to elevated CO_2 had a 15 and 23% reduction in *rbcS* mRNA for growing and expanded leaves, respectively. For the plants switched from elevated to ambient CO_2, *rbcS* mRNA was decreased by 27 and 11% for growing and expanded leaves, respectively. They concluded that although soluble carbohydrates may be important in the regulation of *rbcS* expression, changes in total carbohydrate pool size alone could not explain the rapid changes in gene expression that they observed.

These results for the rice cultivar 'IR-30' grown at 330 and 660 μmol mol^{-1} and those for the rice cultivar 'IR-72' grown at 350 and 700 μmol mol^{-1} suggest that across the range from current ambient CO_2 (near 360 μmol mol^{-1}) to the approximate doubling of CO_2 projected for the mid- to late 21st century, canopy-scale photosynthetic acclimation of these rice cultivars to CO_2 may not be a large factor governing rice photosynthesis responses to CO_2. Quantitative differences in these tests for photosynthetic acclimation (e.g., between Figures 2 and 4) may be due to the two different rice cultivars used, differences in crop growth stages when the tests were made and/or the CO_2 ranges used in the two experiments.

The results in Figure 3 also point to a complete CO_2 saturation of P_g near 500 μmol mol^{-1} for the rice cultivar 'IR-72' and support previous findings for the rice cultivar 'IR-30' (Figuire 4a). For 'IR-30' Baker et al. (1990a, b, c; 1992) found that phenological development, growth, tillering, final seed yield, P_n, canopy light utilization efficiency, and canopy dark respiration rates all increased with CO_2 from 160 to 500 μmol mol^{-1} with little change from 500 to 900 μmol mol^{-1}. This very flat leveling off of rice responses to elevated CO_2 contrasts with that of other C_3 crop species such as soybean (Campbell, 1990; Baker et al., 1993a). Due to this leveling off of responses above 500 μmol mol^{-1}, Baker et al. (1990a) suggested selecting or screening rice cultivars for increased response to CO_2 enrichment in order to more fully take advantage of future increases in global atmospheric CO_2 since the photosynthesis and grain yield responses to CO_2 enrichment may not be the same for all rice cultivars. Indeed, Ziska et al. (1996) reported large differences in growth and yield responses of 17 rice cultivars to both temperature and CO_2.

Drought and CO_2 Effects on Canopy Photosynthesis and Evapotranspiration

Carbon dioxide enrichment causes partial stomatal closure, increased stomatal resistance, and can result in reduced canopy or single leaf transpiration (Rogers et al., 1984; Jones et al., 1985; Valle et al., 1985). However, this water savings from CO_2 enrichment is greater on a per unit leaf area basis rather than a total canopy basis. Part of the reason for this difference is that leaf area is also increased so that the surface area available for transpiration increases. This increase in canopy leaf area acts to offset some of the water savings from CO_2 enrichment (Jones et al., 1985; Allen et al., 1985). Furthermore, the stomatal closure caused by CO_2 enrichment results in increased leaf temperature, and thus an increase in the vapor pressure gradient between leaf and air.

Shown in Figure 5 is an example of CO_2 treatment effects on rice canopy evapotranspiration. In this experiment, the water loss represents evaporation from the paddy as well as canopy transpirational water loss. The diurnal trends in ET tracked PFD for both CO_2 treatments, with peak levels of ET for the 350 μmol mol^{-1} CO_2 treatment being greater than that of the 700 μmol mol^{-1} treatment. Comparisons of daily totals of P_n, ET and WUE for this same day (52 DAP) showed that CO_2 enrichment from 350 to 700 μmol mol^{-1} resulted in a significant 20.5% increase in total daily P_n and 10.2% reduction in total ET (Baker et al., 1997b). In general, similar reductions in ET due to CO_2 enrichment, typically ranging from 9 to 11%, were observed throughout most of the growing season prior to the beginning of canopy leaf senescence. These trends in total daily P_n and ET resulted in significantly greater estimates of WUE with CO_2 enrichment (Baker et al., 1997b).

Shown in Figure 6 are diurnal trends in canopy P_n measured near the end of the PI drought at 75 DAP. Throughout most of the day, P_n was higher in the

FIGURE 5. Diurnal trends in photosynthetic photon flux density (PFD, solid line) and evapotranspiration rates for the 350 (open symbols) and 700 (closed symbol) μmol mol^{-1} CO_2 treatments at 52 days after planting, prior to the PI drought stress treatment. Each data point is the mean of 4 chambers (n = 4). (Adapted from Baker et al. 1997b.)

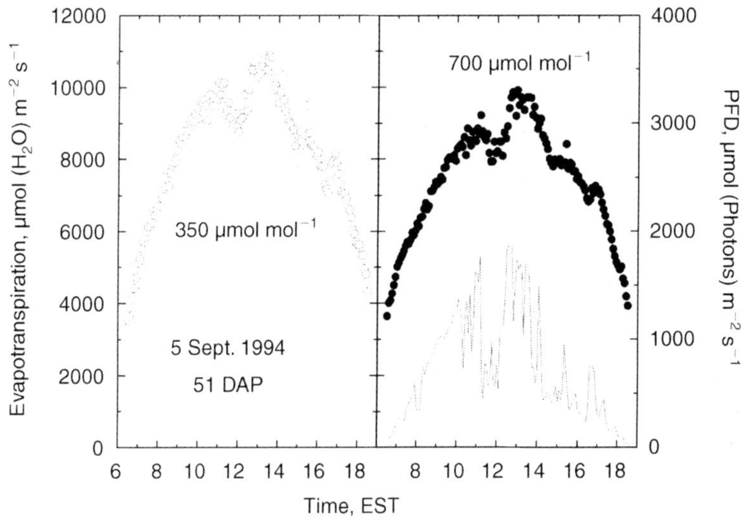

700 than the 350 μmol mol^{-1} CO_2 treatments, for both the CF controls and PI drought treatments. Also, canopy P_n of the PI drought stress treatments were much lower than the flooded controls in the 350 than the 700 μmol mol^{-1} CO_2 treatment (Figure 6). The 350 μmol mol^{-1} PI drought stress treatments were reflooded and drought stress terminated the following day (76 DAP) while the 700 μmol mol^{-1} PI drought stress treatments required an extra 1 to 2 days to exhaust the soil water supply (Baker et al., 1997a) and reduce P_n to similar levels as those shown for the 350 μmol mol^{-1} PI drought treatments. The time courses of soil water potential measurements for the PI drought stress treatments given by Baker et al. (1997a) also indicate that the 700 μmol mol^{-1} treatment lagged that of the 350 μmol mol^{-1} by one to two days in soil water depletion.

Drought and CO_2 Effects on Apparent Canopy Dark Respiration

In discussing CO_2 treatment effects on R_d, a functional distinction is often made between 'direct' and 'indirect' effects of CO_2 on R_d. A short-term or immediate effect of CO_2 on R_d is referred to as a 'direct effect.' An 'indirect' ef-

FIGURE 6. Diurnal trends in photosynthetic photon flux density (PFD, solid line) and canopy net photosynthetic rates (P_n) for the 350 and 700 μmol mol^{-1} CO_2 treatments at 75 DAP, near the end of the PI drought stress treatment. Open and closed symbols are for the continuously flooded control and PI drought stress treatments, respectively. (Adapted from Baker et al. 1997b.)

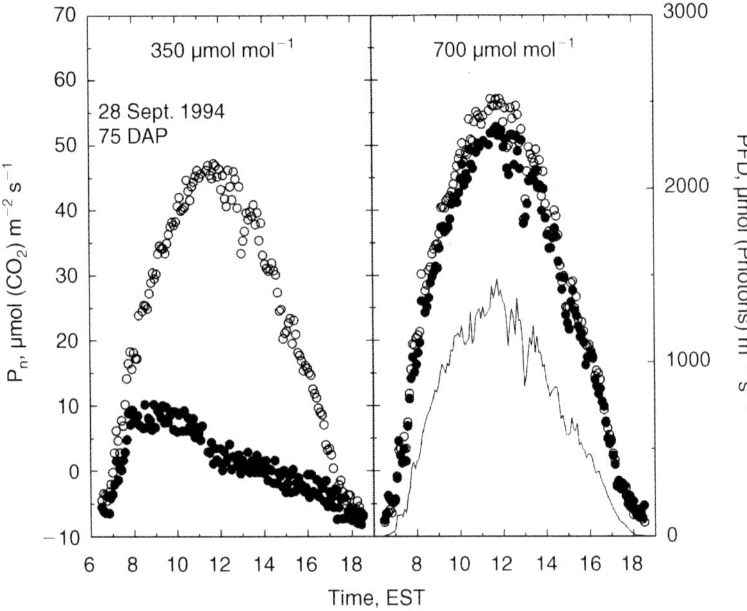

fect of CO_2 on R_d results from growing plants for long periods of time at elevated CO_2. These indirect effects typically manifest themselves in terms of changes in plant biomass, nonstructural carbohydrate concentration, tissue nitrogen concentration or protein content and often result in changes in growth and/or maintenance components of respiration (Amthor, 1991; Baker et al., 1992; Griffin et al., 1993; Amthor, 1997). Work by Bunce and Ziska (1996) and Ziska and Bunce (1998) points to the need for experimental procedures that can distinguish between long-term indirect effects vs. the persistence of a short-term direct effect. To detect a direct effect it is necessary to conduct short-term switching of CO_2 treatments (e.g., ambient treatment measured for a short time at ambient and then elevated CO_2 and elevated treatment measure for a short time at ambient and then elevated CO_2, Ziska and Bunce, 1994).

An example of nighttime chamber CO_2 during the venting procedure used to measure R_d is shown in Figure 7. These estimates of R_d were corrected for chamber leakage rate using the nitrous oxide leak detection system described

FIGURE 7. Example of chamber environmental controls for nighttime dew point air temperature (DPT), dry bulb air temperature (DBT), and CO_2 used to calculated canopy apparent respiration (R_d) on 5 Sept. 1994, 52 days after planting. Open and closed circles indicate rice canopies exposed to 350 and 700 μmol mol^{-1} CO_2 treatments during the day light hours, respectively. (Adapted from Baker et al., 2000a.)

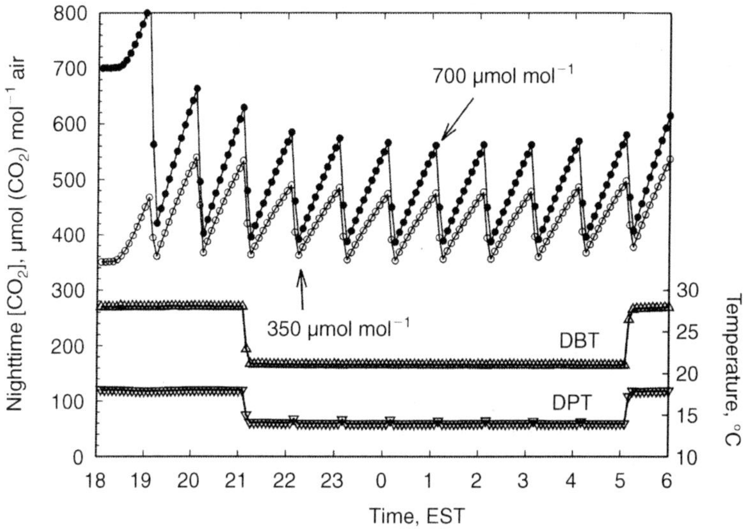

by Baker et al. (2000a). Also, R_d calculated in this way may also include some root respiration since CO_2 exchange through the rice plant aerenchyma system probably occurs as freely as oxygen exchange and CH_4 effluxes (Nouchi et al., 1990; Allen, 1997).

In general, the drought stress treatments reduced R_d due to accelerated leaf senescence and a reduction in above ground biomass (Baker et al., 2000b). Shown in Figure 8 are seasonal trends in R_d at a nighttime dry bulb air temperature of 21°C for both daytime CO_2 treatments in the continuously flooded (CF) control treatments. Maximum seasonal values of R_d occurred at around 30 to 40 DAP for both CO_2 treatments (Figure 8). This time period (30-40 DAP) also coincides with complete canopy closure and seasonal maximum rates of canopy photosynthesis (Baker et al., 1997a,b). For most of the growing season, R_d of the 700 μmol mol^{-1} treatment exceeded that of the 350 μmol mol^{-1} treatment CO_2 by about 1 to 2 μmol m^{-2} s^{-1}. These measurements were made shortly after the vent gates were closed (Figure 7) at a similar and near ambient chamber CO_2. Therefore, these differences in R_d are largely due

FIGURE 8. Seasonal trends in canopy apparent respiration (R_d) for rice exposed to 350 and 700 μmol mol^{-1} CO_2 treatments during the day light hours. Vertical bars are ± S.E. (Adapted from Baker et al. 2000b.)

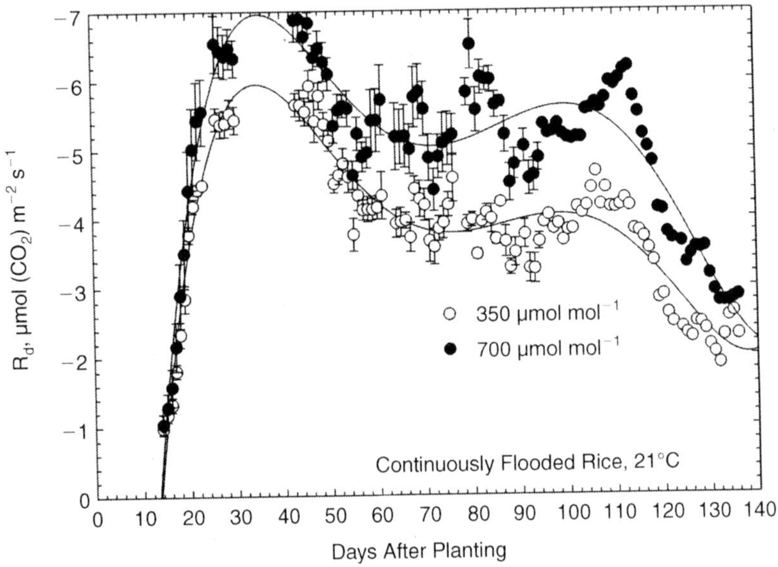

to the secondary effects of CO_2 on plant biomass and plant tissue composition and are thus considered 'indirect effects' (Baker et al., 1992; 2000a, b).

Baker et al. (2000a) tested for 'direct effects' of elevated CO_2 on R_d by utilizing the range of CO_2 produced during the venting procedures (Figure 7). Direct inhibition of respiration by elevated CO_2 is not a universally observed phenomenon (Gifford, 1995; Amthor, 2000; Jahnke, 2001; Bruhn et al., 2002; and reviews by Wullschleger et al., 1994 and Amthor, 1997) and can apparently be altered by air temperature treatment (Ziska and Bunce, 1993) or long-term CO_2 growth concentration (Thomas and Griffin, 1994; Bunce and Ziska, 1996). For example, when whole soybean plants were switched from low to high CO_2, leaf respiration was always reduced but this reduction was 50% less for plants grown in elevated CO_2 compared with near ambient CO_2 (Thomas and Griffin, 1994).

Baker et al. (2000a) measured whole canopy dark respiration (R_d) with rising chamber CO_2 across short-term air temperature switches ranging from 21 to 40°C. They found a rapid and reversible, direct inhibition of canopy dark respiration R_d with rising chamber CO_2 for air temperatures above 21°C. The direct inhibition of R_d by rising chamber CO_2 at air temperatures above 21°C was larger for the 350 compared with the 700 μmol mol^{-1} daytime CO_2 treatment. For example, shown in Figure 9, at 28°C, and across an extrapolated CO_2 range from 350 to 700 μmol mol^{-1} the decline in R_d amounted to a 19 and 11% inhibition of R_d for the 350 and 700 μmol mol^{-1} daytime CO_2 treatments, respectively.

Several mechanisms for this direct effect have been proposed, including dark CO_2 fixation and direct inhibition or deactivation of one or more respiratory enzymes (Amthor, 1991). Work by Gonzalez-Meler et al. (1996) and review by Drake et al. (1999) indicate that direct inhibition of cytochrome *c* oxidase and succinate dehydrogenase by elevated CO_2 may at least be a part of the basis for this response. On the other hand, systematic errors or artifacts in measuring respiration have also been proposed to explain reports of a direct inhibition of R_d with increased CO_2. Two possible artifacts are a dilution of the air surrounding the leaf by transpired water or an unaccounted for leak between the leaf cuvette and the surrounding air (Drake et al., 1999; Jahnke,

FIGURE 9. Canopy apparent respiration (R_d) vs. nighttime CO_2 at a dry bulb air temperature of 28°C for rice canopies maintained during the day at 350 (open circles) and 700 (filled circles) μmol mol^{-1} daytime CO_2 treatments. Each data point is the mean of four replicate chambers averaged at the same point in time. Vertical and horizontal bars are ± SE of the mean. Data are from 46 to 50 days after planting in 1994. (Adapted from Baker et al. 2000a.)

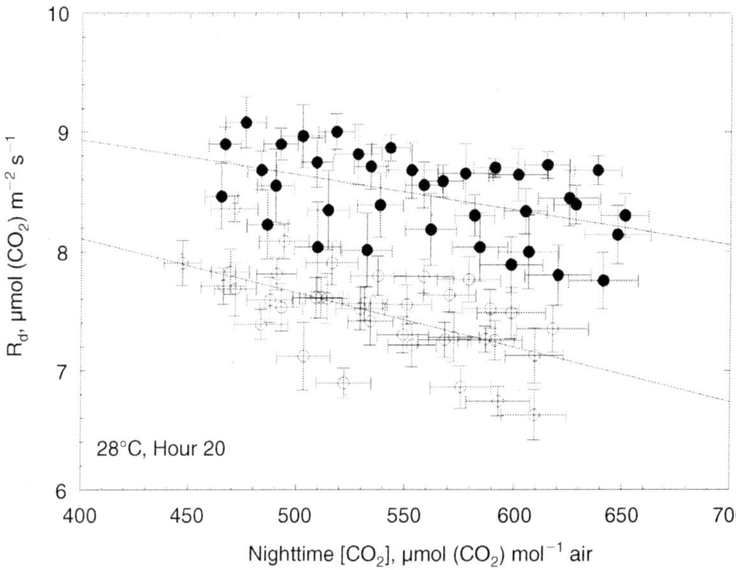

2001; Bruhn et al., 2002). In this experiment (Baker et al., 2000a and Figure 9), moisture was removed from the gas sample by running the sample lines through a refrigerated water trap (4°C) that was automatically drained twice a day while chamber leakage rates were corrected for using a nitrous oxide leak detection system.

CONCLUSIONS

We conclude that in the absence of air temperature increases, future global increases in CO_2 should promote rice growth and yield while providing a modest reduction of near 10% in water use and so increase drought avoidance. Photosynthetic rate saturated with respect to CO_2 near 500 μmol mol^{-1}. Carbon dioxide enrichment significantly increased both canopy net photosynthetic rate (21 to 27%) and water-use efficiency while reducing evapotranspiration by about 10%. This water saving under CO_2 enrichment allowed photosynthesis to continue for about one to two days longer during drought in the enriched compared with the ambient CO_2 control treatments.

Reductions in rice respiration following drought and re-watering appeared to be related to reductions in above-ground biomass due to accelerated leaf senescence. When compared at common, near ambient CO_2, daytime CO_2 enrichment resulted in higher R_d compared with the ambient grown controls. We also detected a rapid and reversible, direct inhibition of R_d with rising chamber CO_2 at an air temperature of 28°C.

REFERENCES

Allen, L.H., Jr., P. Jones and J.W. Jones, J.W. (1985). Rising atmospheric CO_2 and evapotranspiration. *In* Advances in Evapotranspiration. Proceedings of the National Conference on Advances in Evapotranspiration. ASAE Pub. 14-85. Amer. Soc. Agric. Eng., St. Joseph, MI.

Allen, L.H., J.T. Baker and K.J. Boote. (1996). The CO_2 fertilization effect: Higher carbohydrate production and retention as biomass and seed yield. Pp. 64-100 *In*: Global Climatic Change and Agricultural Production. (W.G. Sombroek and F.A. Bazzaz, eds.), (FAO Expert consultation, 7-10 December 1993, Rome, Italy.) John Wiley, New York.

Allen, L.H., Jr. (1997). Mechanisms and rates of O_2 transfer to and through submerged rhizomes and roots via aerenchyma. Soil and Crop Sciences Society of Florida, Proceedings. 56:41-54.

Amthor, J.S. (1991). Respiration in a future, higher-CO_2 world. Plant, Cell and Environment, 14:13-20.

Amthor, J.S. (1997). Plant respiratory responses to elevated CO_2. *In:* Advances in Carbon Dioxide Effects Research (Allen, L.H., Jr., M.B. Kirkham, D.M. Olzyk, and C. Whitman, eds.). American Society of Agronomy, Madison, Wisconsin.

Amthor, J.S. (2000). Direct effect of elevated CO_2 on nocturnal in situ leaf respiration in nine temperate deciduous tree species is small. Tree Physiology. 20:139-144.

Baker, J.T., L.H. Allen, Jr., K.J. Boote, P. Jones and J.W. Jones. (1990a). Rice photosynthesis and evapotranspiration in subambient, ambient, and superambient carbon dioxide concentrations. Agron. J. 82:834-840.

Baker, J.T., L.H. Allen, Jr., K.J. Boote, J.W. Jones and P. H. Jones. (1990b). Developmental responses of rice to photoperiod and carbon dioxide concentration. Agricultural and Forest Meterol. 50:201-210.

Baker, J.T., L.H. Allen, Jr. and K.J. Boote. (1990c). Growth and yield responses of rice to subambient, ambient, and superambient carbon dioxide concentrations. J. Agric. Sci., Camb. 115:313-320.

Baker, J.T., F. Laugel, K.J. Boote and L.H. Allen, Jr. (1992). Effects of daytime carbon dioxide concentration on dark respiration of rice. Plant, Cell and Environment. 15:231-239.

Baker, J.T. and L.H. Allen, Jr. (1993a). Contrasting crop species responses to CO_2 and temperature: rice, soybean and citrus. Vegetatio 104/105:239-260.

Baker, J.T. and L.H. Allen, Jr. (1993b). Effects of CO_2 and temperature on rice: A summary of five growing seasons. J. Agr. Met. 48(5):575-582.

Baker, J.T., L.H. Allen, Jr., K.J. Boote, and N.B. Pickering. (1997a). Rice response to drought under carbon dioxide enrichment: I. Growth and yield. Global Change Biology 3(2):119-128.

Baker, J.T., L.H. Allen, Jr., K.J. Boote and N.B. Pickering. (1997b). Rice response to drought under carbon dioxide enrichment: II. Photosynthesis and evapotranspiration. Global Change Biology 3(2):129-138.

Baker, J.T., L.H. Allen, Jr., K.J. Boote and N.B. Pickering. (2000a). Direct effects of atmospheric carbon dioxide concentration pn whole canopy dark respiration of rice. Global Change Biology 6:275-286.

Baker, J.T., L.H. Allen, Jr., K.J. Boote and N.B. Pickering. (2000b). Whole canopy respiratory responses of rice to daytime carbon dioxide enrichment, temperature and drought. World Resource Review 12(1):171-198.

Boyer J.S. (1982). Plant productivity and environment. Science. 218:443-448.

Besford R.T., L.J. Ludwig, A.C. Withers. (1990). The greenhouse effect: acclimation of tomato plants grown in high CO_2, photosynthesis and ribulose-1, 5-bisphosphate carboxylase protein. Journal of Experimental Botany. 41: 925-931.

Bowes G. (1993). Facing the inevitable: plants and increasing atmospheric CO_2. Annual Review of Plant Physiology and Plant Molecular Biology. 44: 309-332.

Bruhn, D., T.N. Mikkelsen and O.K. Atkin. 2002. Does the direct effect of atmospheric CO_2 concentration on leaf respiration vary with temperature? Responses in two species of *Plantago* that differ in relative growth rate. Physiologgia Plantarum. 114:57-64.

Bryant, J., G. Taylor and M. Frehner. (1998). Photosynthetic acclimation to elevated CO_2 is modified by source:sink balance in three component species of chalk grassland swards grown in free air carbon dioxide enrichment (FACE) experiment. Plant, Cell and Environment. 21: 159-168.

Bunce, J.A. and L.H. Ziska. (1996). Responses of respiration to increased carbon dioxide concentration and temperature in three soybean cultivars. Ann. of Bot. 77: 507-514.

Campbell, W.J., L.H. Allen, Jr. and G. Bowes. (1990). Response of soybean canopy photosynthesis to CO_2 concentration, light, and temperature. J. Exp. Bot. 41:427-433.
Drake, B.G., M.A. Gonzàlez-Meler and S.P. Long. (1997). More efficient plants: a consequence of rising atmospheric CO_2? Annual Review of Plant Physiology and Molecular Biology, 48: 609-639.
Drake, B.G., J.Azcon-Bieto, J. Berry, J. Bunce, P. Dijkstra, J. Farrar, R.M. Gifford, M.A. Gonzalez-Meler, G. Koch, H. Lambers, J. Siedow and S. Wullschleger. (1999). Does elevated atmospheric CO_2 concentration inhibit mitochondrial respiration in green plants? Plant, Cell and Environment. 22: 649-657.
Gesch, R.W., K.J. Boote, J.C.V. Vu, L.H. Allen, Jr. and G. Bowes. (1998). Changes in growth CO_2 result in rapid adjustments of ribulose-1,5-bisphosphate carboxylase/oxygenase small subunit gene expression in expanding and mature leaves of rice. Plant Physiology. 118:521-529.
Gifford, R.M. (1995). Whole plant respiration and photosynthesis of wheat under increased CO_2 concentration and temperature: long-term vs. short-term distinctions for modelling. Global Change Biology. 1:385-396.
Gonzalez-Meler, M.A., M. Ribas-Carbo, J.N. Siedow and B.G. Drake. (1996). Direct inhibition of plant mitochondrial respiration by elevated CO_2. Plant Physiology. 112:1349-1355.
Griffin, K.L., R.B. Thomas and B.R. Strain (1993). Effects of nitrogen supply and elevated carbon dioxide on construction cost in leaves of *Pinus taeda* (L.) seedlings. Oecologia. 95:575-580.
Hansen J., I. Fung, A. Lacis, S. Lebedeff, D. Rind, R. Ruedy, G. Russell and P. Stone. (1988). Global climate changes as forecast by the GISS 3-D model. Journal of Geophysical Research. 98(08):9341-9364.
Houghton, J.T., L.G. Meira Filho, B.A. Callander, N. Harris, A. Kattenburg and K. Maskell. (1996). IPCC Climate change assessment 1995. The Science of Climate Change. Cambridge University Press, Cambridge, UK.
IPCC (2001). Climate Change 2001: The Scientific Basis. *In* J.T. Houghton, Y. Ding, D.J. Griggs, M. Noguer, P.J. van der Linden, X. Da, K. Maskell, C.A. Johnson (Eds.) Cambridge University Press, Cambridge, United Kingdom and New York, NY, USA.
IRRI (International Rice Research Institute) (1975). Major Research in Upland Rice. Los Baños, Philippines.
IRRI (International Rice Research Institute) (1982). Drought Resistance in Crops with Emphasis on Rice. Los Baños, Philippines.
IRRI (International Rice Research Institute). (1984). An Overview of Upland Rice Research. Proceedings of the 1982 Bouaké, Ivory Coast Upland Rice Workshop. Los Baños, Philippines.
Jahnke, S. (2001). Atmospheric CO_2 concentration does not directly affect leaf respiration in bean or poplar. Plant, Cell and Environment. 24:1139-1151.
Jang, J.C. and J. Sheen. (1997). Sugar sensing in higher plants. Trends in Plant Science. 2: 208-214.
Jones, P.D., T.M.L. Wigley and K.R. Briffa. (1994). Global and hemispheric anomalies: land and marine instrumental records. *In* T.A. Boden, D.P. Kaiser, R.J. Sepanski and F.W. Stoss (Eds.) Trends '93: A Compendium of Data on Global

Change. Carbon Dioxide Information Analysis Center, Oak Ridge National Laboratory, Oak Ridge, TN. pp. 603-608.

Jones, P.D., D.E. Parker, T.J. Osborn and K.R. Briffa. (1999). Global and hemispheric anomalies-land and marine instrumental records. *In* Trends: A Compendium of Data on Global Change. Carbon Dioxide Information Analysis Center, Oak Ridge National Laboratory, Oak Ridge, TN. (*http://cdiac.esd.ornl.gov/trends/temp/jonescru/jones.html*).

Jones, P., L.H. Allen, Jr. and J.W. Jones. (1985). Responses of soybean canopy photosynthesis and transpiration to whole-day temperature changes in different CO_2 environments. Agron. J. 77:242-249.

Keeling, C.D., T.P. Whorf, M. Wahlen and J. van der Plicht (1995). Interannual extremes in the rate of rise of atmospheric carbon dioxide since 1980. Nature 375: 660-670.

Koch, K.E. (1996). Carbohydrate-modulated gene expression in plants. Annual Review of Plant Physiology and Plant Molecular Biology. 47:509-540.

Mann, M.E., R.S. Bradley, and M.K. Hughes. (1999). Northern hemisphere temperatures during the past millennium: Inferences, uncertainties, and limitations. Geophys. Res. Lett. 26: 759-762.

Mauney, J.R., G. Guinn, K.E. Fry and J.D. Hesketh. (1979). Correlation of photosynthetic carbon dioxide uptake and carbohydrate accumulation in cotton, soybean, sorghum, and sunflower. Photosynthetica. 13:260-266.

Moore, B.D., S.-H. Cheng, D. Sims and J.R. Seeman. (1999). The biochemical and molecular basis for photosynthetic acclimation to elevated atmospheric CO_2. Plant, Cell and Environment. 22: 567-582.

Nouchi, I., S. Mariko and K. Aoki. (1990). Mechanism of methane transport from the rhizosphere to the atmosphere through rice plants. Plant Physiology.94:59-66.

Osborne, C.P., J. LaRoche, R.L. Garcia, B.A. Kimball, G.W. Wall, P.J. Pinter, Jr., R.L. LaMorte, G.R. Hendrey and S.P. Long. (1998). Does leaf position within a canopy affect acclimation of photosynthesis to Elevated CO_2? Plant Physiol.117: 1037-1045.

O'Toole, J.C. and T.B. Moya. (1978). Genotypic variation in maintenance of leaf water potential in rice. Crop Science. 18:873-876.

Pickering, N.B., L.H. Allen, Jr., S.L. Albrecht, P. Jones, J.W. Jones and J.T. Baker. (1994). Environmental plant chambers: Control and measurement using CR-10T dataloggers. *In* Computers in Agriculture (Eds. Watson D.G., Zazueta F.S., Harrison T.V.), pp. 29-35. Proceedings of the 5th International Conference, Orlando, Florida, Feb. 5-9. American Society of Agricultural Engineers, St. Joseph, Michigan.

Rogers, H.H., N. Sionit, J.D. Cure, J.M. Smith and G.E. Bingham. (1984). Influence of elevated carbon dioxide on water relations of soybeans. Plant Physiol. 74:233-238.

Rowland-Bamford, A.J., J.T. Baker, L.H. Allen, Jr. and K.J. Boote. (1991). Response of carbohydrate status and partitioning in rice to CO_2 concentration. Journal of Experimental Botany. 41: 1601-1608.

Rowland-Bamford, A.J., J.T. Baker, L.H. Allen, Jr. and G. Bowes. (1996). Interactions of CO_2 enrichment and temperature on carbohydrate accumulation and partitioning in rice. Environ. Exp. Bot. 36(1):111-124.

Sage, R.F. (1994). Acclimation of photosynthesis to increasing atmospheric CO_2: the gas exchange perspective. Photosynthesis Research. 39:351-368.

Stitt, M. (1991). Rising CO_2 levels and their potential significance for carbon flow in photosynthetic cells. Plant, Cell and Environment. 14: 741-762.

Thomas, R.B. and K.L. Griffin. (1994). Direct and indirect effects of atmospheric carbon dioxide enrichment on leaf respiration of *Glycine max* (L.) Merr. Plant Physiology. 104:355-361.

Valle, R., J.W. Mishoe, J.W. Jones and L.H. Allen, Jr. (1985). Transpiration rate and water-use efficiency of soybean leaves adapted to different CO_2 environments. Crop Sci. 25:477-482.

Vu, J.C.V., L.H. Allen, Jr., K.J. Boote and G. Bowes. (1997). Effects of elevated CO_2 and temperature on photosynthesis and Rubisco in rice and soybean. Plant, Cell and Environment. 20: 68-76.

Vu, J.C.V., J.T. Baker, A.H. Pennanen, L.H. Allen, Jr., G. Bowes and K.J. Boote. (1998). Elevated CO_2 and water deficit effects on photosynthesis, ribulose bisphosphate carboxylase-oxygenase, and carbohydrate metabolism in rice. Physiol. Plant. 103: 327-339.

Wilson, C.A. and J.F.B. Mitchell. (1987). A doubled CO_2 climate sensitivity experiment with a global climate model, including a simple ocean. Journal of Geophysical Research. 92:13315-13343.

Wullschleger, S.D., L.H. Ziska and J.A. Bunce. (1994). Respiratory responses of higher plants to atmospheric CO_2 enrichment. Physiologia Plantarum. 90:221-229.

Ziska, L.H. and J.A. Bunce. (1993). Inhibition of whole plant respiration by elevated CO_2 as modified by growth temperature. Physiologia Plantarum. 87:459-466.

Ziska, L.H. and J.A. Bunce (1994). Direct and indirect inhibition of single leaf respiration by elevated CO_2 concentrations: Interaction with temperature. Physiologia Plantarum. 90:130-138.

Ziska, L.H., P.A. Manalo and R.A. Ordonez. (1996). Intraspecific variation in the response of rice (*Oryza sativa* L.) to increased CO_2 and temperature: growth and yield response of 17 cultivars. J. Exp. Bot. 47:1353-1359.

Ziska, L.H. and J.A. Bunce. (1998). The influence of increasing growth temperature and CO_2 concentration on the ratio of respiration to photosynthesis in soybean seedlings. Global Change Biology. 4:637-643.

Ziska, L.H., T.B. Moya, R. Wassmann, O.S. Namuco, R.S. Lantin, J.B. Aduna, E. Abao, Jr., K.F. Bronson, H.U. Neue and D. Olzyk. (1998). Long-term growth at elevated carbon dioxide stimulates methane emission in tropical rice. Global Change Biology. 4:657-665.

Response of Rice to $p(CO_2)$ Enrichment: The Relationship Between Photosynthesis and Nitrogen Metabolism

S. Seneweera
A. Makino
T. Mae
A. S. Basra

SUMMARY. Rice (*Oryza sativa* L.) is grown in many different agro-ecological zones and it assimilates CO_2 directly through C_3 photosynthesis. At current CO_2 partial pressure [$p(CO_2)$], rice is not photosynthetically saturated but immediately after exposure to 100 $p(CO_2)$, photosynthetic rates are increased by 48 percent on an average. However, this increase is not sustained during prolonged exposure and is associated with reduced amounts of ribulose-1-5-bisphospate carboxylase/oxygenase (Rubisco) content and its maximum potential activity (V_{cmax}). Rubisco content and V_{cmax} were reduced by 33 and 22%, respectively, at elevated $p(CO_2)$ and this has an additive effect on leaf photosynthesis.

S. Seneweera, A. Makino, and T. Mae are affiliated with the Department of Applied Plant Science, School of Agricultural Science, Tohoku University, 1-1 Tsutsumidori-Amamiyamachi, Sendai 981-8555, Japan (E-mail: seneweera@biochem.tohoku.ac.jp).

A. S. Basra is Senior Scientist, Monsanto Company, 700 Chesterfield Parkway West, Chesterfield, MO 63017.

Address correspondence to: S. Seneweera at the above address.

This work was supported by Grant-in-Aid for JSPS fellows (to S.S.) and for Scientific Research (No. 13001116 to S.S. and A.M., and 14360036 to A.M.) from the Ministry of Education, Culture, Sports, Science and Technology of Japan.

[Haworth co-indexing entry note]: "Response of Rice to $p(CO_2)$ Enrichment: The Relationship Between Photosynthesis and Nitrogen Metabolism." Seneweera, S. et al. Co-published simultaneously in *Journal of Crop Improvement* (Food Products Press, an imprint of The Haworth Press, Inc.) Vol. 13, No. 1/2 (#25/26), 2005, pp. 31-53; and: *Ecological Responses and Adaptations of Crops to Rising Atmospheric Carbon Dioxide* (ed: Zoltán Tuba) Food Products Press, an imprint of The Haworth Press, Inc., 2005, pp. 31-53. Single or multiple copies of this article are available for a fee from The Haworth Document Delivery Service [1-800-HAWORTH, 9:00 a.m. - 5:00 p.m. (EST). E-mail address: docdelivery@haworthpress.com].

http://www.haworthpress.com/web/JCRIP

Digital Object Identifier: 10.1300/J411v13n01_03

Activation state of the Rubisco is also decreased by elevated $p(CO_2)$ but no deactivation of Rubisco occurs in *rbc*S antisense rice with 40% wild type Rubisco suggesting that deactivation is an optimize response. Large amounts of Rubisco (about 80-90%) is synthesized prior to full expansion of leaf when sugar accumulation is minimal. After the leaf has fully expanded, very little Rubisco (about 10-20%) is synthesized and a large amount of sugar accumulates during this period. Therefore, it is unlikely that downregulation of Rubisco at elevated $p(CO_2)$ is caused by decreasing the mRNA transcripts for Rubisco small subunit (*rbc*S) and Rubisco large subunit (rbcL) through soluble sugars. It is more likely that elevated $p(CO_2)$ mediated decline in Rubisco is due to rapid degradation of protein and lower nitrogen (N) allocation to Rubisco due to reduced N afflux. Leaf N content was reduced by 16% in plants grown at elevated $p(CO_2)$ though leaf N to Rubisco ratio remained consistently unchanged regardless of growth $p(CO_2)$ suggesting that N assimilation, distribution and remobilization at the whole plant level is tightly controlled at elevated $p(CO_2)$. Further, changes in N and C metabolisms at elevated $p(CO_2)$ are likely to reduce the critical N concentration by 30% at elevated $p(CO_2)$.

KEYWORDS. Elevated $p(CO_2)$, downregulation, photosynthesis, rice, Rubisco, nitrogen

ABBREVIATIONS. A/Ci, photosynthetic rates against intercellular $p(CO_2)$; A_{max}, photosynthesis at saturating $p(CO_2)$; nitrogen, (N); FACE, free air $p(CO_2)$ enrichment; $p(CO_2)$, CO_2 partial pressure; PCO, photosynthetic carbon oxidation cycle; PCR, photosynthetic carbon reduction cycle; *rbc*S, ribulose-1-5-bisphospate carboxylase/oxygenase small subunit; *rbc*L, Rubisco, ribulose-1-5-bisphospate carboxylase/oxygenase large subunit; V_{cmax}, maximal Rubisco activity estimated from gas exchange

INTRODUCTION

The current atmospheric CO_2 partial pressure [$p(CO_2)$] is expected to nearly double by the end of the 21st century which is likely to have a profound effect on the biochemical makeup, growth and morphology of plants, particularly in C_3 species (Baker et al., 1990; Drake et al., 1997; Moore et al., 1999; Makino and Mae 1999). The altered growth and morphological responses are mainly due to a direct effect of elevated $p(CO_2)$ on leaf photosynthesis (Baker et al., 1990; Seneweera et al., 1994; Drake et al., 1997; Jitla et al., 1997; Seneweera et al., 2003). For short-term $p(CO_2)$ enrichment, photosynthetic rates are accelerated by 56% on an average in C_3 plants species when exposed to double the ambient $p(CO_2)$ (Drake et al., 1997), yet after long-term exposure, depending on species, assimilation declines until it stabilizes (Stitt, 1991; Curtis, 1996; Rey and Jarvis, 1998). Initial stimulation of photosynthesis can be easily explained by gas exchange characteristics and ribulose-1-5-bisphospate carboxylase/oxygenase (Rubisco) kinetic data (Farquhar et al., 1980).

Rubisco is the key enzyme of photosynthesis and about 15-35% total nitrogen (N) is invested in this enzyme in C_3 plants (Evans, 1989; Makino and Osmond 1991). Rubisco content is the rate limiting factor for light saturated photosynthesis at current ambient $p(CO_2)$ (Makino et al., 1995). At elevated $p(CO_2)$, photosynthesis is not limited by Rubisco content whereas electron transport capacity (Farquhar et al., 1980) or availability of Pi in the chloroplast to synthesise ATP limits the photosynthesis (Sharkey, 1985). With long-term exposure to elevated $p(CO_2)$, photosynthetic rates frequently declined and is associated with reduced Rubisco and leaf N content which is more pronounced at low N supply (Stitt, 1991; Bowes, 1991; Webber et al., 1994; Moore et al., 1999; Nakano et al., 1997; Makino and Mae, 1999; Bloom et al., 2002; Seneweera et al., 2002). However, the mechanism of decline in Rubisco and leaf N content by elevated $p(CO_2)$ is still a matter of debate.

The amount of Rubisco in leaves is mainly determined by both synthesis and degradation and it is regulated by internal and external signals. Decline in Rubisco content at elevated $p(CO_2)$ is known to be often associated with accumulation of structural and non structural carbohydrate content (Long and Drake, 1992). The widely accepted hypothesis for the reduced Rubisco content at elevated $p(CO_2)$ is that sugar repression of the *rbc*S and *rbc*L genes expression (Sheen, 1994; Cheng et al., 1998; Moore et al., 1999; Rolland et al., 2002), However, a negative relationship between photosynthesis and soluble sugar content has not always been reported (Ludewig and Sonnewald, 2000; Nakano et al., 2000). Recently with rice, it was clearly demonstrated that there was no significant difference in *rbc*S transcript activity between ambient and elevated $p(CO_2)$ during leaf expansion (Aoki et al., 2003). During this period, large amounts of Rubisco were synthesized but very little soluble sugar accumulated in the leaves (Suzuki et al., 2001; Aoki et al., 2003). Another report indicates that repression of photosynthetic genes only occur during the senescence phase in leaves (Ludewig and Sonnewald, 2000) when Rubisco synthesis is minimal (Suzuki et al., 2001). Therefore, it is unlikely that downregulation of Rubisco at elevated $p(CO_2)$ is caused by sugar accumulation suggesting that other mechanisms may exist to regulate the Rubisco content at elevated $p(CO_2)$, e.g., rapid Rubisco degradation at elevated $p(CO_2)$.

At elevated $p(CO_2)$, leaf N content changes and this affects the rate of photosynthesis by altering N allocation to Rubisco (Wong, 1979; Seneweera et al., 1994; Nakano et al., 1997; Makino and Mae, 1999; Bloom et al., 2001). In most studies, total N uptake (for a given area or on a whole plant basis) increased, but tissue N concentration (on the leaf area or on the dry mass basis) decreased (Makino et al., 1997; Aben et al., 1999; Bloom et al., 2001; Kim et al., 2001; Seneweera et al., 2002). This is partly due to lower N assimilation for a given leaf area or root volume at elevated $p(CO_2)$, the mechanism of which is not clearly understood. Hence, these complex changes in N metabolism at elevated $p(CO_2)$ are likely to optimize photosynthetic response to elevated $p(CO_2)$. In addition, decreases in leaf N content leads to greater N use efficiency at elevated $p(CO_2)$ by which critical N requirements may change at elevated $p(CO_2)$ (Aben et al., 1999). These changes in critical N will have implications for crop production in the 21st century.

In this review, our main focus is on the influence of elevated $p(CO_2)$ on photosynthesis and N metabolism in rice. First, we discuss Rubisco kinetics and gas exchange parameters to short-term $p(CO_2)$ enrichment. Secondly we discuss plant response to long-term $p(CO_2)$ enrichment with major emphasis on the mechanism of photosynthetic downregulation.

SHORT-TERM PHOTOSYNTHETIC RESPONSE TO ELEVATED $p(CO_2)$

The present $p(CO_2)$ concentration of 37 $p(CO_2)$ limits the rate of photosynthesis in C_3 plants and presumably, lower concentrations of $p(CO_2)$ in the recent past were even more limiting (Sage and Coleman, 2001). The currently increasing $p(CO_2)$ will undoubtedly increase the assimilation rates in C_3 plants (Pearcy and Björkman, 1983; Drake et al., 1997; Makino and Mae, 1999). For example, Drake et al. (1997) surveyed 60 experiments on a wide range of plants and reported that photosynthetic rates of C_3 plants were approximately increased by 58% when plants grown at about 37 $p(CO_2)$ were exposed to 70 $p(CO_2)$. However, this initial stimulation of photosynthesis at elevated $p(CO_2)$ was not maintained for longer periods due to morphological, biochemical and molecular adjustments at both the cellular and at the whole plant level (Stitt, 1991; Drake et al., 1997; Cheng et al., 1998; Makino and Mae, 1999; Seneweera et al., 2002). To demonstrate the initial stimulation of photosynthesis at elevated $p(CO_2)$, photosynthetic data was first analyzed using Rubisco kinetics and gas exchange characteristics by Farquhar et al. (1980).

The Kinetics and Regulation of Rubisco at Elevated p(CO_2)

All the biochemical models of leaf photosynthesis are based on the kinetic properties of Rubisco. Rubisco is the key enzyme responsible for both photosynthesis and photorespiration because it has an affinity for CO_2 and O_2 (Lorimer, 1981). Photosynthetic CO_2 assimilation occurs when RuBP is carboxylated by Rubisco to produce two molecules of 3-phosphoglyceric acid (PGA) which is further processed into carbohydrates and used to regenerate RuBP in a reaction sequence requiring ATP and NADPH. When Rubisco takes up O_2, RuBP is oxygenated to form one molecule each of PGA and 2-phosphoglycolate (PG). However, the potential photosynthetic response can be evaluated by measuring the response of photosynthesis in young fully expanded leaves to increased intercellular $p(CO_2)$ (Farquhar et al., 1980). Increased $p(CO_2)$, accelerates the rates of photosynthesis in C_3 plants by raising the carboxylation efficiency of Rubisco. The reason for an increase in carboxylation efficiency is due to an increase in the CO_2/O_2 ratio at the site of the fixation. Further, this leads to reduction in photorespiration of C_3 plants, e.g., at the present atmospheric oxygen concentration of 21 kPa and 37 $p(CO_2)$, a considerable amount of light energy is used up for photorespiration which reduces photosynthetic rates by 20 to 50 percent depending on temperature (Sharkey, 1985). All these physiological adjustments at elevated $p(CO_2)$ have an implication on potential photosynthetic capacity.

Three major limitations for C_3 photosynthesis have been identified (i) the limitation of photosynthesis imposed by Rubisco referred to as the limitation due to supply and utilization of $p(CO_2)$, (ii) the supply and utilization of light which limits the rate of electron transport for regeneration of RuBP, and (iii) the utilization of triose phosphate which limits the availability of Pi in the chloroplast for ATP synthesis to regenerate RuBP (Farquhar and Sharkey, 1982; Sharkey, 1985). At current ambient $p(CO_2)$ and under saturating light conditions Rubisco content limits photosynthesis (Figure 1; Makino et al., 1985). This limitation was clearly illustrated by using rbcS antisense rice (Figure 1; Table 1). The antisense (AS) lines with 40% of wild type Rubisco compared to the wild type (Makino et al., 2000a), showed a remarkable reduction in photosynthetic rates which was further exacerbated at elevated $p(CO_2)$ (Figure 1). Similar results were obtained with rbcS antisense tobacco plants when they were grown at elevated $p(CO_2)$ (Sicher et al., 1994). The second limitation can be caused by low photon flux densities or inability to convert light energy into chemical energy. The third limitation, triose phosphate limitation occurs when there are insufficient sinks for carbon utilization (Sharkey, 1985; Stitt, 1991; Paul and Foyer, 2001). This is very common at elevated $p(CO_2)$ and it has been suggested that photosynthesis at elevated $p(CO_2)$ is limited either by electron transport or triose phosphate utilization (Makino and

FIGURE 1. Photosynthetic response to increasing $p(CO_2)$ for wild type (○ and ●) and *rbc*S antisense (□ and ■) of rice(*Oryza sativa* L. cv Notohikari) grown in ambient (○ and □) or 100 $p(CO_2)$ (● and ■). Arrows indicate the growth $p(CO_2)$ concentration. Gas exchange measurements were made on the 13th leaf after leaf had fully expanded. Measurements were made at 25°C, an irradiance at 1,800 μmol m^{-2} s^{-1} and a leaf to air vapor pressure deficit of 0.9-1.2 kPa. Each A/Ci curves represent three replicates from each $p(CO_2)$ treatment. Plants were grown at 25/21°C for day/night, respectively.

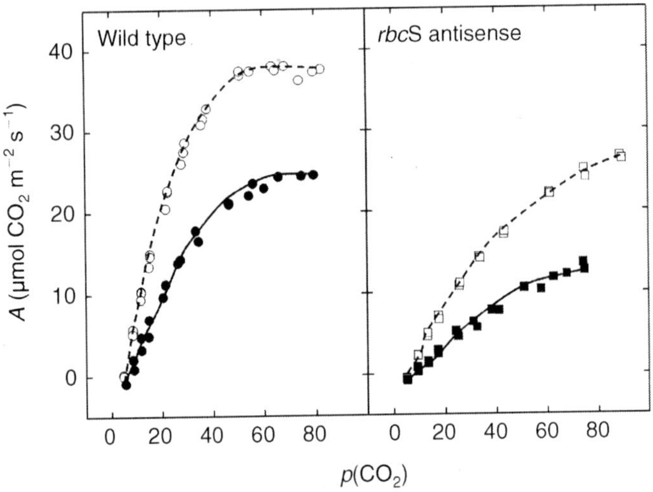

Mae, 1999). These three limitations to leaf photosynthesis were first identified in plants that were grown at ambient $p(CO_2)$ and then transferred to another during photosynthetic measurements (Farquhar and Sharkey, 1982). However, under long-term $p(CO_2)$ enrichment, photosynthesis rates are reduced by altering the growth, morphology and biochemical make up of photosynthetic enzymes (Webber et al., 1994; Drake et al., 1997; Makino and Mae, 1999; Seneweera et al., 2002).

LONG-TERM PHOTOSYNTHETIC RESPONSE TO ELEVATED P(CO_2)

Prolonged exposure to elevated $p(CO_2)$, causes photosynthetic downregulation, although this response is not necessarily reported in all the investigations of plants grown at elevated $p(CO_2)$. Two major factors within the plant have been identified as causes for this downregulation (i) reduced sink capacity, where plants are unable to use the excess carbohydrates produced at ele-

TABLE 1. Gas exchange parameters 13th leaf of rice (*Oryza sativa* L. Notohicari) grown in 39 or 100 $p(CO_2)$. Values are means of three replicates. Asterisks indicate significant differences between CO_2 treatments. *$P \leq 0.05$; $P \leq 0.001$ and ns, not significant. Growth conditions were as described in Figure 1.

Gas exchange parameters	$p(CO_2)$	Wild type	*rbc*S antisense (AS 41)
A_{max}	39Pa	67.9	52.5
(μmol CO_2 m^{-2} s^{-1})	100Pa	48.2***	36.7***
V_{cmax}	39Pa	91.4	54.9
(μmol CO_2 m^{-2} s^{-1})	100Pa	72.1***	38.6***
Conductance (g)	39Pa	0.58	0.65
(μmol CO_2 m^{-2} s^{-1})	100Pa	0.21***	0.19***
J_{max}	39Pa	189	373
(μmol CO_2 m^{-2} s^{-1})	100Pa	187ns	292*

vated $p(CO_2)$, and (ii) reduced Rubisco and N content, the mechanism of which is still a matter of debate. However, these two parameters are inter-dependent. Decline in Rubisco content at elevated $p(CO_2)$ is known to be associated with the accumulation of soluble sugars but other mechanisms may exist. Further, elevated $p(CO_2)$ reduces transpiration and photorespiration by a large magnitude which also has a profound influence on photosynthesis and N metabolism at elevated $p(CO_2)$.

Sink Limitation Causes Accumulation of Carbohydrates in the Leaves at Elevated p(CO₂)

The most significant response to elevated $p(CO_2)$ is an increase in carbohydrate concentration in the source leaves. The extent to which starch and soluble sugars accumulate at elevated $p(CO_2)$ is greatly dependent on species, e.g., some species such as cotton preferentially accumulate starch; fructans and sucrose are mainly accumulated by wheat and rice, respectively. Some species accumulate large amounts of starch in the chloroplast which can cause physical damage to the thylakoid membranes and also hinders CO_2 diffusion into the chloroplast (Makino et al., 1994; Nakano et al., 2000). The damage that occurs to thylakoid membranes can only be seen in starch accumulators. In general, it has been reported that soluble carbohydrate and starch content increased by 52% and 160% on an average, across 32 experiments (Long et al., 1992). In rice, sucrose and starch content increased by 55% and 300%, respectively, when plants were grown at 100 $p(CO_2)$ under high light condition (Nakano et al., 1997). However, this is extremely high when compared to rice grown at 55 $p(CO_2)$ in the FACE experiment (Seneweera et al., 2002). The reason for this different response could have been due to the fact that carbohydrate synthesis at 100 $p(CO_2)$ was in far excess for growth and utilization when compared to the FACE-grown plants (Seneweera et al., 2002). Plants grown at FACE showed increased sucrose content only by 15 and 22%, respectively, at the 8th and 15th leaf when compared to the ambient $p(CO_2)$-grown leaves. In addition, sucrose accumulation in expanding and expanded leaves was minimal when rice plants were grown at 100 $p(CO_2)$ (Table 2). All this data suggests that the small magnitude of sugar accumulation at early stages of plant development at elevated $p(CO_2)$ may not have an influence on photosynthesis or Rubisco content (Seneweera et al., 2002). At elevated $p(CO_2)$, demand for N and C are increased during the reproductive growth stage because more panicles and more grains are produced (Seneweera et al., 1994; Kim et al., 2001). Hence, increased N demand at elevated $p(CO2)$ matched protein degradation and this led to downregulation of Rubisco protein (Seneweera et al., 2002). In

TABLE 2. Concentration of leaf N, Rubisco, soluble protein, starch and sucrose in the 12th (fully expanded) and 13th leaf (expanding) of rice grown in 39 or 100 $p(CO_2)$. Values are means of three replicates. Asterisks indicate significant differences between CO_2 treatments. *$P \leq 0.05$; $P \leq 0.001$ and ns, not significant. Growth conditions were as described in Figure 1. After gas exchange measurements, leaves were sampled for following biochemical measurements.

Gas exchange parameters	$p(CO_2)$	Fully expanded leaf (12th leaf)	Expanding leaf (13th leaf)
Leaf N	39Pa	1.93	2.06
($g\ m^{-2}$)	100Pa	1.677	1.82**
Rubisco	39Pa	2.67	2.76
($g\ m^{-2}$)	100Pa	2.03***	2.23**
Soluble protein	39Pa	3.96	4.15
($g\ m^{-2}$)	100Pa	3.25***	3.90*
Starch	39Pa	5.62	5.92
($g\ m^{-2}$)	100Pa	8.49***	6.44ns
Sucrose	39Pa	5.73	6.18
($g\ m^{-2}$)	100Pa	6.23*	7.62*

addition, Rogers et al. (1998) have shown the importance of C sinks in response to photosynthetic downregulation at elevated $p(CO_2)$ with *Lolium perenne*. They altered the source sink balance by removing 89% of the leaf canopy by applying defoliants, sink to source activity was rapidly increased and no photosynthetic downregulation to elevated $p(CO_2)$ was observed. This data suggests the important role of sink strength on the regulation of photosynthetic activity at elevated $p(CO_2)$. Also, it is widely known that sink limitations can also result from low nutrient supplies (Rogers et al., 1996; Paul and Foyer, 2001) or from low temperatures (Baker et al., 1992; Conroy et al., 1994). However, where environmental factors are optimal for growth, elevated $p(CO_2)$ appears to increase the generation of new sinks and also maintains higher photosynthetic rates during ontogeny.

Rubisco Content and Activity at Elevated p(CO_2)

Rubisco is the most abundant enzyme and constitutes about 56% of all soluble protein and 26% of total leaf N in rice (Mae et al., 1983; Suzuki et al., 2001) and this is somewhat greater in other C_3 plants (Makino et al., 1992). It is the key enzyme in the photosynthesis carbon reduction cycle and a rate-limiting factor for light saturated photosynthesis at current ambient $p(CO_2)$ (Makino et al., 1985). The activity of Rubisco is regulated by CO_2 and O_2 concentrations, the concentration of RuBP, the presence of inhibitors which compete with RuBP binding, e.g., 2-carboxy-D-arabinitol 1-phosphate (CA1P), and other factors such as Rubisco activase (Portis, 1992).

A major feature of photosynthetic downregulation by elevated $p(CO_2)$ is the reduction in Rubisco content (Webber et al., 1994; Nakano et al., 1997; Moore et al., 1999; Seneweera et al., 2002). In a large number of experiments carried out under elevated $p(CO_2)$ it has been shown that Rubisco content was reduced, predominantly in C_3 species, e.g., Rubisco content was decreased by up to 32% in a large fully expanded leaf of rice grown at elevated $p(CO_2)$ (Table 2). Makino et al. (2000b) have also shown that under elevated $p(CO_2)$ and when other conditions are optimum, 30% of Rubisco had been lost before Rubisco limits photosynthesis in rice. Antisense rice with 40% wild type Rubisco, when grown at ambient $p(CO_2)$, the photosynthetic rates were significantly reduced and thereby reduced plant growth (Figure 1) but no differences in growth were observed when these plants were grown at 100 $p(CO_2)$ (Seneweera, Makino and Mae, unpublished). In addition, when rbcS antisense rice plants were exposed to 100 $p(CO_2)$, photosynthesis rates increased by 20 and 85%, respectively, in wild type and antisense plants (Figure 1). Similar data was obtained when rbcS antisense tobacco plants were exposed to elevated $p(CO_2)$ (Sicher et al., 1994). This data suggests that Rubisco content is not necessarily important to maintain high photosynthetic rates at elevated

$p(CO_2)$ and other regulatory mechanisms within the photosynthetic machinery may exist for downregulation of photosynthesis at elevated $p(CO_2)$.

Rubisco has an extremely low catalytic capacity compared to other enzymes. In addition, in vivo carboxylation efficiency is reduced under ambient $p(CO_2)$ because of its poor affinity for $p(CO_2)$ compared to O_2 (Lorimer, 1981). Under any growth conditions, observed photosynthetic rates are relatively lower than that predicted from Rubisco activity, e.g., V_{cmax} was greater than the Amax calculated from A/Ci response curves (Figure 1 and Table 1; Farquhar et al., 1980). With a short-term response of rice to elevated $p(CO_2)$, the maximum Rubisco activity estimated at 100 $p(CO_2)$ from Rubisco kinetics is 1.5~2.0 fold greater than the actual photosynthetic rates observed at the same $p(CO_2)$ suggesting that the efficiency of CO_2-saturated photosynthesis is 50~70% in rice (Makino et al., 2000b). In general, under ambient $p(CO_2)$, Rubisco is a limiting factor for photosynthesis when Rubisco is fully activated (Sage et al., 1988). At elevated $p(CO_2)$, it has been reported that there is a decrease in the activation state of Rubisco in C_3 species $p(CO_2)$ (Von Caemmerer et al., 1986; Sage et al., 1988; Theobald et al., 1998; Rey and Jarvis, 1998). On the contrary, some reports indicate that decrease in Rubisco content is always associated with an increased activation state of the enzyme (Table 1; Cheng and Fuchigami, 2000). However, this is not true under elevated $p(CO_2)$ scenarios because the activation state of Rubisco decreased though the Rubisco content was reduced by 32%, for, e.g., activation state of Rubisco calculated from gas exchange data was 75 and 67% at 39 and 100 $p(CO_2)$, respectively, (Table 1; Table 2). However, antisense rice with 40% wild-type Rubisco showed 100% activation of the enzyme both at 39 and 100 $p(CO_2)$ suggesting that deactivation of Rubisco could be an optimizing response to elevated $p(CO_2)$ (Table 1, Figure 1). On the contrary, decline in Rubisco protein content of leaves at elevated $p(CO_2)$ was not accompanied by a decline in CO_2 assimilation rates because the activation state of Rubisco increased (Rowland-Bamford et al., 1991; Allen et al., 2000). Apart from this, it has been demonstrated that Rubisco content in the leaf is varied during development (Seneweera et al., 2002). At elevated $p(CO_2)$, Rubisco content was significantly reduced at the reproductive stage. These changes in Rubisco content were related to modulation in C and N balance due to changes in source and sink activity during development but the underlying mechanism is not clearly understood (Seneweera et al., 2002). Decrease in Rubisco content and activation state of the enzyme during ontogeny may have accounted for the loss of growth sensitivity to elevated $p(CO_2)$. It has also been suggested that deactivation or activation of Rubisco under elevated $p(CO_2)$ is a secondary response to maintain the balance between Rubisco and other processes that limit photosynthesis (Sage et al., 1988). One possible mechanism at elevated $p(CO_2)$ which causes deactivation of Rubisco is by decreasing ATP/ADP ratio which leads to decreases in

Rubisco activase activity. However, these changes have been documented both in monocotyledons and dicotyledons. In most cases, photosynthetic response to elevated $p(CO_2)$ varies dramatically between species. Hence, these data suggest that downregulation is not controlled by a single factor but by an integration of many factors at the cellular and whole plant level.

Rubisco Turnover at Elevated p(CO_2)

The Rubisco enzyme is composed of a heteromeric structure of eight large and eight small subunits of polypeptides resulting in a native molecular mass of 520 kDa. The small subunit is coded by a multigene family. The genes that code for the large subunit (rbcL) occur in the chloroplast and the genes that code for the small subunit (rbcS) occur in the nucleus (Rodermel, 1999). The catalytic sites are present on the large subunit but catalytic competency requires the presence of the small subunit (Lorimer, 1981). The amount of Rubisco in the leaves is the result of the balance between its synthesis and degradation. The amount of Rubisco in plants varies from leaf emergence to senescence (Mae et al., 1983; Suzuki et al., 2001). Rubisco synthesis is controlled by transcriptional, post transcriptional and translational processes. Rubisco is rapidly synthesized during leaf expansion followed by a gradual degradation as leaf ontogeny progresses. Rubisco degradation mainly starts just prior to the full expansion of leaves and occurs within the chloroplast (Figure 2; Mae et al., 1983). Many environmental factors influence Rubisco synthesis and degradation, e.g., light, nitrogen, atmospheric $p(CO_2)$ and ozone.

At elevated $p(CO_2)$ Rubisco content was reduced by 32% in fully expanded leaves of rice (Table 2). However, it is not known to what extent Rubisco synthesis or degradation was affected by elevated $p(CO_2)$. Nonetheless, it is widely accepted that accumulation of excess soluble carbohydrates, depending on growth $p(CO_2)$, predominantly, when the sinks are limited, leads to repression of rbcS and rbcL genes and thereby reduced the Rubisco content. (Gesch et al., 1998; Cheng et al., 1998; Moore et al., 1999). This repression of rbcS and rbcL genes are mediated through the hexokinase signaling pathway (Jane and Sheen, 1994; Rolland et al., 2002). In that case, sugar metabolism may be central for repression of photosynthetic genes, rather than its total pool size at elevated $p(CO_2)$. When we closely look at the current literature, most of the photosynthesis and Rubisco measurements have been carried out under elevated *p*(CO2), after the leaf had reached full expansion when Rubisco synthesis was minimal (Figure 2; Mae et al. 1983; Suzuki et al., 2001). At this stage, about 80-90% of Rubisco had already been synthesized (Figure 2). This data was further supported by the rbcS and rbcL mRNA, which increased rapidly during the leaf expansion and reached maxima a few days before full expansion (Figure 2; Suzuki et al., 2001). Rubisco content reaches a maximum im-

FIGURE 2. Changes in the amount of Rubisco (○) and relative level of *rbc*S mRNA (●) in the eighth leaf of rice from emergence to senescence. For *rbc*L mRNA, 0.5 μg of total RNA were digested with *Eco*RI and slot-blotted onto a positively charged nylon membrane. Arrow indicates the full expansion of the leaves. Each data point represent is mean of three replicates. Adapted from Suzuki et al., 2001.

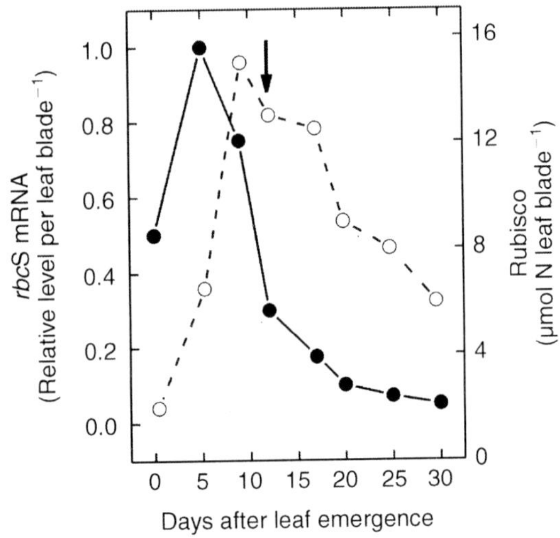

mediately after full expansion when mRNA decreased to a very low level. At the point of full expansion, large amounts of Rubisco is synthesized and soluble carbohydrates start to accumulate. However, there is a big time lag between maximum Rubisco synthesis and soluble carbohydrate accumulation (Figure 2). Therefore, it is unlikely that repression of rbcS and rbcL is caused by soluble sugars at elevated $p(CO_2)$ because soluble sugar accumulation starts after the leaf had fully expanded (Aoki et al., 2003). Recently, it has been reported with tobacco that repression of photosynthetic genes was only apparent in senescing leaves and also found no relationship between gene transcript and soluble sugar (Ludewig and Sonnewald, 2000). These evidences strongly support our argument that downregulation of Rubisco at elevated $p(CO_2)$ are not due to reduced Rubisco synthesis. In addition, an elegant work by Sims et al. (1998) had demonstrated that when a soybean leaflet was exposed to elevated $p(CO_2)$ while the rest of the plant was exposed to ambient $p(CO_2)$, there was a dramatic effect on the soluble carbohydrate accumulation but Rubisco content remained unchanged in the single leaflet. On the other hand, when the rest of the plant was exposed to elevated $p(CO_2)$ and a single leaflet exposed to ambient $p(CO_2)$, there was no sugar accumulation but Rubisco content decreased in the leaf exposed to ambient $p(CO_2)$. These results strongly suggest that there was no suppression of photosynthetic genes by sugar at the single leaf level and suggests that the Rubisco content is controlled by the whole-plant internal signal(s).

Another major factor which controls Rubisco content in leaf is protein degradation. There have been a number of studies on Rubisco degradation in C_3 species under ambient $p(CO_2)$ (Mitsuhashi et al., 1992; Desimore et al., 1996; Ishida et al., 1999). Particularly in rice, Rubisco degradation starts even before the leaf has fully expanded and becomes more pronounced during the senescence phase (Figure 2; Mae et al., 1983; Suzuki et al., 2001). At first, internal and external factors cause reversible or irreversible inactivation of Rubisco (Mae et al., 1983; Eckardt and Pell, 1994). Irreversibly inactivated Rubisco has to be degraded and replaced by de novo synthesis in order to fully re-establish the photosynthetic function. Usually in monocots, Rubisco degradation is always predominant after full expansion of leaf, and will lead to a rapid decline in Rubisco content (Figure 2). This could be an adopted salvage mechanism in terms of nutrient remobilization for sink development.

There has been no report on Rubisco degradation at elevated $p(CO_2)$ but indirect evidence suggests that the Rubisco degradation is rapid at elevated $p(CO_2)$. We strongly support this idea and our hypothesis is based on the following findings. As discussed in the previous section, most of the studies leading to photosynthetic downregulation have been carried out after the leaf had fully expanded. At this stage, the amount of Rubisco is mainly determined by Rubisco degradation rather than its synthesis and this was clearly demonstrated with rice (Figure 2). The magnitude of photosynthetic downregulation is greater in older than in young leaves and also leads to accelerated leaf senescence (Sims et al., 1999; Makino and Mae, 1999; Ludewig and Sonnewald, 2000). In addition, in the expanding leaf of rice, photosynthetic rates or Rubisco content marginally changed between $p(CO_2)$ treatments (Table 2). Similar results have been reported with sunflower (Sims et al., 1999). It is more likely that downregulation of Rubisco in response to elevated $p(CO_2)$ may be related to accelerated Rubisco degradation rather than its reduced synthesis through soluble sugar. To prove this hypothesis, Rubisco synthesis and degradation during leaf initiation to senescence has to be quantitatively analyzed and the mechanism by which elevated $p(CO_2)$ accelerates Rubisco degradation still needs to be found. One possible mechanism of Rubisco degradation at elevated $p(CO_2)$ is by an increase in reactive oxygen species. In different conditions, it has been demonstrated that the large subunit of Rubisco is directly cleaved by reactive oxygen species (Desimone et al., 1996; Ishida et al., 1998). Under elevated $p(CO_2)$, it was reported that superoxide dismutase (SOD) activity reduced (Schwanz et al., 1996; Schwanz et al., 1998; Pritchards et al., 2000). In another experiment with soybean it has been demonstrated that most of the antioxidative enzyme activity like superoxide dismutase, peroxidase, catalase, ascorbate peroxidase, and glutathione peroxidase activity decreased when plants were grown at elevated $p(CO_2)$ (Pritchards et al., 2000). These antioxidative enzymes are known to detoxify highly reactive oxygen species. By decreasing these enzyme activities, reactive oxygen concentration may have been enhanced in the chloroplast in plants grown at elevated $p(CO_2)$ and this may lead to rapid Rubisco degradation. Another possible mechanism of Rubisco degradation may be via increased proteolytic enzyme activity in the chloroplasts (Vierstra, 1993; Kokubun et al., 2002), but its physiological relevance and the regulation of these proteases are still not well understood.

Leaf N Concentrations and Photosynthetic Rates at Elevated p(CO_2)

The relationship between the photosynthetic capacity and unit leaf N content has been extensively investigated whereas light-saturated photosynthetic rates (A_{max}) were linearly correlated with leaf N content under a wide variety of conditions (Figure 2; Sage, 1987; Evans, 1989; Makino et al., 1994; Nakano et al., 1997). However, the leaf N and photosynthetic relationship substantially varies between plant species (Saga, 1987; Evans, 1989). In any species, a large proportion of N is allocated to leaves throughout their life cycle, e.g., in rice, at early panicle initiation stage, about 65-70 % of the total N in the shoot is invested in the leaf blade and 20-30% in the leaf sheath (Seneweera, Makino and Mae unpublished). A major proportion of leaf N (~ 80%) is allocated to chloroplasts and most of the N in the chloroplast is invested in photosynthetic protein, including large amounts in Rubisco and thylakoid protein (Evans, 1989).

The most striking observation at elevated $p(CO_2)$ is the decreased N content and N allocation for the photosynthetic proteins (Nakano et al., 1997; Seneweera et al., 2002). The photosynthesis N relationship remained unchanged at either $p(CO_2)$ (Figure 3, Nakano et al., 1997). The relationship between photosynthesis and leaf N content in leaves during different leaf ages also shows a positive linear relationship with a positive intercept on the N content axis indicating that when leaf photosynthesis is theoretically zero, the leaf would still contain significant amounts of N, probably due to non photosynthetic leaf N (Figure 3). At both the elevated and ambient $p(CO_2)$ theoretical zero photosynthesis was achieved when the leaf remobilized 70% of N compared to the maximum N in the leaf (Seneweera, Makino and Mae, unpublished). Total leaf N content was reduced by 16% in rice grown at 100 $p(CO_2)$ but higher photosynthetic N use efficiency was maintained during the growth (Table 2; Rowland-Bamford et al., 1991; Curtis and Wang, 1998; Roger et al., 1996; Nakano et al., 1997; Aben et al., 1999; Seneweera et al., 2002).

FIGURE 3. Photosynthesis rates of 9th to 13th leaf (Canopy level) against their respective Rubisco content. Plants were grown at either 39 (○) or 100 $p(CO_2)$ (●) at 100 $p(CO_2)$ in nutrient solution for 70 days and measurements were made at 100 $p(CO_2)$. Growth and measurement conditions as described in Figure 1.

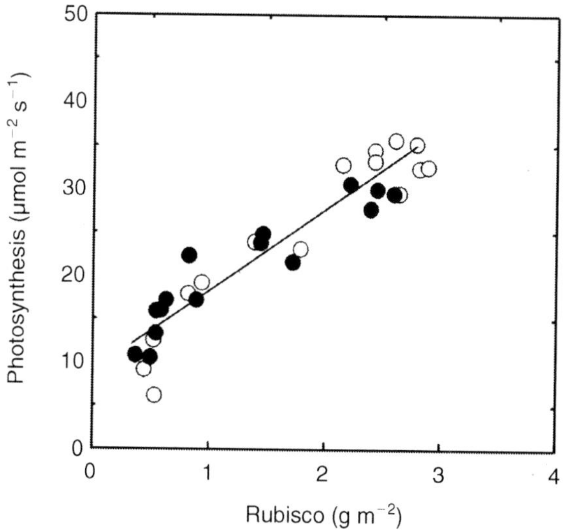

Total N uptake in rice remained the same or increased at elevated $p(CO_2)$, e.g., in the FACE experiment when plants were grown at 20 $p(CO_2)$ above ambient $p(CO_2)$, total N uptake was increased by 15% (per given land area basis) at panicle initiation stage but the N content was reduced by 12% (on leaf area basis) at the same stage (Kim et al., 2001; Seneweera et al., 2002). This decrease was found only after panicle initiation and the reason for these differences could be due to an increase in N demand for larger panicles in plants grown at elevated $p(CO_2)$ (Seneweera et al., 2002). Another possibility is that plants reduced their N allocation for photosynthetic proteins, making nitrogen more freely available for rapid sink development. There are various other explanations for the reduction in leaf N concentration in plants grown at elevated $p(CO_2)$ and one of the common explanations is the dilution of N due to extra carbohydrate accumulation (Conroy, 1992). This hypothesis did not explain low leaf N in wheat leaves because concentrations were reduced by elevated $p(CO_2)$ irrespective of whether N was expressed on a total dry mass or total structural dry mass, or leaf area basis (Rogers et al., 1996). In addition, there was no change in the N concentration of leaf sheaths at elevated $p(CO_2)$ in rice, despite the fact that the sheaths accumulate large amounts of carbohydrate (Seneweera et al., 1994; Aben et al., 1999). Makino et al. (1997) demonstrated that N investment to the leaves is reduced in favor of the sheaths and roots at elevated $p(CO_2)$. Other possibilities are that reduction of N flux through the photorespiratory cycle could lead to a lower N requirement at elevated $p(CO_2)$ (Bowes, 1991; Conroy, 1992). Factors such as reduced transpiration and photorespiration would also have significantly contributed to lower N concentration at elevated $p(CO_2)$ because lower stomatal conductance reduced the transpiration flow thereby lowering the N uptake at elevated $p(CO_2)$ (Table 1). However, this argument is puzzling since other nutrients are not affected by elevated CO_2, if transpiration was the limiting factor for nutrient uptake, then the uptake of other nutrients should also have been lower. A widely known explanation for reduction in leaf N content at elevated $p(CO_2)$ is that accumulation of highly soluble carbohydrates which cause repression of genes involved in photosynthesis and N assimilation (Sheen, 1990; Van Oostern et al., 1994; Stitt and Krapp, 1999; Moore et al., 1999). Because large amounts of leaf N is invested in the photosynthetic machinery a suppression of synthesis of these proteins lead to lower leaf N content. However, this suggestion was not supported by Nakano et al. (1997), because in some circumstances there can be a very large accumulation of carbohydrates without any effect on photosynthesis. In some species, e.g., in pod removed soybean, accumulation of starch led to suppressed photosynthesis but photosynthesis was not associated with the amount of Rubisco or other photosynthetic components (Nakano et al., 2000). Another possibility at elevated $p(CO_2)$ is that N assimilation, distribution and remobilization at the whole plant level is tightly controlled and protein synthesis is matched with N afflux.

Critical N Concentration at Elevated p(CO$_2$)

The critical nutrient concentration is defined as the concentration at which 90% of the maximum growth rate is achieved at a given time. Leaf analysis is used to calculate the plant's critical nutrient requirements and this differs between genotypes due to intrinsic differences in their biochemistry, e.g., C_4 plants generally have lower critical N concentration than C_3 plants. It also varies within the same genotype due to prevailing environmental conditions that invariably affect the growth potential. Deficiency of this nutrient reduces plant growth and alters the allocation of biomass between the shoot and root (Rufty et al., 1984; Makino et al., 1997; Aben et al., 1999).

Critical N concentration and dry matter per unit ground area is the same for a wide range of crops and is independent of many climatic variables. However, it has been shown that elevated $p(CO_2)$ increased the dry mass of rice per unit of leaf N (Figure 4). It was suggested that elevated $p(CO_2)$ changes the physiology of N use, e.g., when rice plants were grown at ambient $p(CO_2)$, the critical values for N were, 4.0% and at double the atmospheric $p(CO_2)$ the critical N requirement had decreased to 2.8% (Figure 4). As a result, critical N requirement for maximum growth was reduced by 30% at mid vegetative phase for plants grown at elevated $p(CO_2)$ when compared to plants grown at 39 $p(CO_2)$. Reduction in critical N at elevated $p(CO_2)$ leads to greater N use efficiencies both at the cellular and whole plant level (Figure 4; Aben et al., 1999). Particularly, N use efficiency was the greatest at elevated $p(CO_2)$ when plants received suboptimal N (Figure 3). At optimum nutrient supplies, the repression of nutrient uptake by elevated $p(CO_2)$ was lower in magnitude (Figure 4; Aben et al., 1999). The importance of high nutrient supplies in regulating the response to high $p(CO_2)$ is by high sink generation. This is particularly true for cereals, which require high N supplies to produce tillers early in ontogeny (Aben et al., 1999; Seneweera et al., 2002). Hence, the response of rice to elevated $p(CO_2)$ is smaller at low N supplies and partly contributed to, by the inhibition of tillering (Aben et al., 1999). As $p(CO_2)$ increases, requirements for N fertilizer may not increase in proportion to dry weight accumulation, but lower concentrations of N in the plant may have implications for rice grain quality.

FIGURE 4. Relationship between leaf N and P concentration and shoot dry mass. Plants were grown at 36 (○) or 70 $p(CO_2)$ (●). Leaf N and P concentration were measured at 30 and 56 DAP, respectively. N data were obtained from Aben et al. (1999). Values are expressed as a percentage of the maximum dry mass. Arrows indicate the foliar N concentration corresponding to 90% of maximum value (Critical concentration).

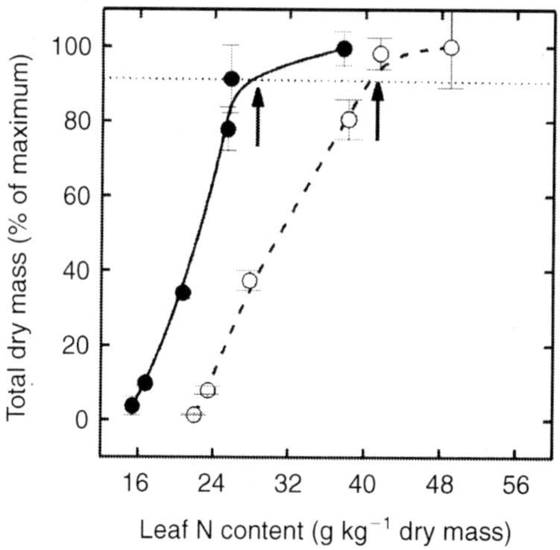

CONCLUDING REMARKS

Photosynthetic downregulation is mainly associated with reduced Rubisco content but a causal relationship between them has not been established. Photosynthesis at elevated $p(CO_2)$ is not limited by Rubisco capacity, which is clearly demonstrated with rbcS antisense rice. It is more likely that downregulation of Rubisco at elevated $p(CO_2)$ is associated with protein degradation and lower N allocation to Rubisco due to reduced N afflux rather than reduced Rubisco synthesis through soluble sugar. Regardless of growth $p(CO_2)$, Rubisco and N ratio does not change which indicates that N assimilation, distribution and remobilization at the whole plant level is strongly affected by elevated $p(CO_2)$ and leads to greater N use efficiency. However, this physiological and biochemical adjustment under elevated $p(CO_2)$ is central to support the accelerated growth, development and yield formation.

REFERENCES

Aben, S.K., S.P., Seneweera, O. Ghannoum, and J.P. Conroy. (1999). Nitrogen requirements for maximum growth and photosynthesis of rice, *Oryza sativa* L. cv. Jarrah grown at 36 and 70 Pa CO_2. *Australian Journal of Plant Physiology* 26: 759-766.

Allen, L.H., A.J. Rowland-Bamford, J.T. Baker, K.J. Boote, and G. Bowes. (2000). Response of rice ribulose-1,5-bisphosphate carboxylase/oxygenase activity to elevated CO_2 concentration and temperature. *Proceeding of Soil & Crop Science Society of Florida*. 59: 46-56.

Baker, J.T., L.H. Allen, and K.J. Boote. (1990). Growth and yield responses of rice to carbon dioxide concentration. *Journal of Agricultural Science*. 115: 313-320.

Baker, J.T., J. Allen, and K.J. Boote. (1992). Response of rice to carbon dioxide and temperature. *Agricultural and Forest Meteorology*. 60: 153-166.

Bloom, A.J., D.R. Smart, D.T. Nguyen, and P.S. Searles. (2002). Nitrogen assimilation and growth of wheat under elevated carbon dioxide. *Proceedings of the National Academy of Sciences of the United States of America*. 99: 1730-1735.

Bowes, G. (1991). Growth at elevated CO_2: photosynthetic responses mediated through Rubisco. *Plant, Cell and Environment*. 14: 795-806.

Cheng, L. and Fuchigami, L.H. (2000) Rubisco activation state decreases with increasing nitrogen content in apple leaves. *Journal of Experimental Botany*.51: 1687-1694.

Cheng, S., B. Moore, J.R. Seemann, and S.H. Cheng. (1998). Effects of short and long-term elevated CO_2 on the expression of ribulose-1,5-bisphosphate carboxylase/oxygenase genes and carbohydrate accumulation in leaves of *Arabidopsis thaliana* (L.) Heynh. *Plant Physiology*. 116: 715-723.

Conroy, J.P. (1992). The influence of elevated atmospheric CO_2 concentrations on plant nutrition. *Australian Journal of Botany*. 40: 445-456.

Curtis, P.S. (1996). A meta-analysis of leaf gas exchange and nitrogen in trees grown under elevated carbon dioxide. *Plant, Cell & Environment*. 19: 127-137.

Curtis, P.S., and X.Z. Wang. (1998). A meta-analysis of elevated CO_2 effects on woody plant mass, form, and physiology. *Oecologia*. 113: 299-313.

Desimone, M., E. Wagner, and U. Johanningmeier. (1988). Degradation of active oxygen-modified Rubisco-1-5-bisphospahe carboxylase/oxygenase by chloroplast proteases requires ATP-hydrolisis. *Planta*. 205: 459-566.

Drake, B.G., M.A. Gonzalezmeler, and S.P. Long. (1997). More efficient plants–a consequence of rising atmospheric CO_2. *Annual Review of Plant Physiology & Plant Molecular Biology*. 48: 609-639.

Eckardt, N.A. and E.J. Pell. (1994). O_3-induced degradation of Rubisco protein and loss of Rubisco mRNA in relation to leaf age in *Solanum tuberosum* L. *New Phytologist*. 127: 741-784.

Evans, J.R. (1989). Photosynthesis and nitrogen relationships in leaves of C_3 plants. *Oecologia*. 78: 9-19.

Farquhar, G.D., S. von Caemmerer, and J.A. Berry. (1980). A biochemical model of photosynthetic CO_2 fixation in C_3 species. *Planta*. 149: 178-190.

Farquhar, G.D. and T.D. Sharkey. (1982). Stomatal conductance and photosynthesis. *Annual Review of Plant Physiology*. 33: 317-345.

Gesch, R.W., K.J. Boote, J.C.V. Vu, L.H. Allen, and G. Bowes. (1998). Changes in growth CO_2 result in rapid adjustments of ribulose-1,5-bisphosphate carboxylase/oxygenase small subunit gene expression in expanding and mature leaves of rice. *Plant Physiology*. 118: 521-529.

Henry, L.T. and D. Raper. (1991). Soluble carbohydrate allocation to root, photosynthetic rates of leaves and nitrogen assimilation as affected by nitrogen stress and irradiance. *Botanical Gazette*. 152: 23-33.

Ishida, H., Y. Nishimori, M. Sugisawa, A. Makino, and T. Mae. (1997). The large subunit of ribulose-1,5-bisphosphate carboxylase/oxygenase is fragmented into 37-kda and 16-kda polypeptides by active oxygen in the lysates of chloroplasts from primary leaves of wheat. *Plant & Cell Physiology*. 38: 471-479.

Ishida, H., A. Makino, and T. Mae. (1999). Fragmentation of the large subunit of ribulose-1,5-bisphosphate carboxylase by reactive oxygen species occurs near Gly-329. *Journal of Biological Chemistry*. 274: 5222-5226.

Jang, J.C. and J. Sheen. (1997). Sugar sensing in higher plants [Review]. *Trends in Plant Science*. 2: 208-214.

Jitla, D.S., G.S. Rogers, S.P. Seneweera, A.S. Basra, R.J. Oldfield, and J.P. Conroy. (1997). Accelerated early growth of rice at elevated CO_2: Is it related to development changes in shoot apex? *Plant Physiology*. 115: 15-22.

Kim, H.Y., M. Lieffering, S. Miura, K. Kobayashi, and M. Okada. (2001). Growth and nitrogen uptake of CO_2 enrich rice under field condition. *New Phytologist*. 150: 223-229.

Koch, K.E. (1996). Carbohydrate modulate gene expression in plants. *Annual Review of Plant Physiology and Plant Molecular Biology*. 46: 115-122.

Kokubun, N., H. Ishida, A. Makino, and T. Mae. (2002). The degradation of large subunit of Rubisco 1-5-bisphosphate carboxylase/oxygenase into the 44-kDa fragment in the lysates of chloroplast incubated in darkness. *Plant Cell Physiology*. 43: 1390-1394.

Long, S.P., G.Y. Nie, B.G. Drake, G. Hendrey, K. Lewin, and N. Murata. (1992). The implications of concurrent increases in temperature and CO_2 concentration for terrestrial C_3 photosynthesis, Proceeding of Research in photosynthesis: Nagoya, Japan, 811-817 September, 1992.

Lorimer, G.H. (1981). The carboxylation and oxygenation of ribulose 1,5-bisphosphate-the primary events in photosynthesis and photo-respiration. *Annual Review of Plant Physiology and Plant Molecular Biology*. 32: 349-383.

Ludewig, F. and U. Sonnewald. (2000). High CO_2-mediated downregulation of photosynthetic gene transcripts is caused by accelerated leaf senescence rather than sugar accumulation. *Federation of European Biochemical Science Letter*. 479: 19-24.

Mae, T., A. Makino, and K. Ohira. (1983). Changes in the rubilose bisphosphate carboxylase synthesis and degradation during life span of rice leaf (*Oryza sativa* L.). *Plant Cell Physiology*. 24: 1079-1086.

Makino, A., T. Mae, and K. Ohira. (1985). Enzymic properties of ribulose-1, 5-bisphosphate carboxylase/oxygenase purified from rice leaves. *Plant Physiology*. 79: 57-61.

Makino, A., H. Nakano, and T. Mae. (1994). Responses of ribulose-1,5-bisphosphate carboxylase, cytochrome f, and sucrose synthesis enzymes in rice leaves to leaf nitrogen and their relationships to photosynthesis. *Plant Physiology*. 105: 173-179.

Makino, A., M. Harada, T. Sato, H. Nakano, and T. Mae. (1997). Growth and N allocation in rice plants under CO_2 enrichment. *Plant Physiology*. 115: 199-203.

Makino, A. and T. Mae. (1999). Photosynthesis and plant growth at elevated levels of CO_2. *Plant & Cell Physiology*. 40: 999-1006.

Makino, A. and B. Osmond. (1991). Effect of nitrogen nutrition on nitrogen partitioning between chloroplast and mitochondria in pea and wheat. *Plant Physiology*. 96: 355-362.

Makino, A., H. Nakano, T. Mae, T. Shimada, and N. Yamamoto. (2000a). Photosynthesis, plant growth and N allocation in transgenic rice plants with decreased Rubisco under CO_2 enrichment. *Journal of Experimental Botany*. 51: 383-389.

Makino, A., M. Harada, K. Kaneko, T. Mae, T. Shimada, and N. Yamamoto. (2000b). Whole-plant growth and N allocation in transgenic rice plants with decreased content of ribulose-1,5-bisphosphate carboxylase under different CO_2 partial pressures. *Australian Journal of Plant Physiology*. 27: 1-12.

Mitsuhashi, W., S.J. Crafts-Brandner, and U. Feller. (1992). Rubisco 1-5-bisphosphate carboxylase/oxygenase degradation in isolated pea chloroplast incubated in light or the dark. *Journal of Plant Physiology*. 139: 2810-2816.

Moore, B.D., S.H. Cheng, D. Sims, and J.R. Seemann. (1999). The biochemical and molecular basis for photosynthetic acclimation to elevated atmospheric CO_2. *Plant, Cell & Environment*. 22: 567-582.

Nakano, H., A. Makino, and T. Mae. (1997). The effect of elevated partial pressures of CO_2 on the relationship between photosynthetic capacity and n content in rice leaves. *Plant Physiology*. 115: 191-198.

Nakano, H., S. Muramatsu, A. Makino and T. Mae. (2000). Relationship between the suppression of photosynthesis and starch accumulation in the pod-removed bean. *Australian Journal of Plant Physiology*. 27: 167-173.

Paul, M.J. and C.H. Foyer. (2001). Sink regulation of photosynthesis [Review]. *Journal of Experimental Botany*. 52: 1383-1400.

Pearcy, R.W. and O. Bjorkman. (1983). Physiological effects. In *CO_2 and plants, The response of plants to rising level of atmospheric carbon dioxide*, ed. E.R. Lemon, Westview Press, Boulder, Colorado, pp. 65-105.

Portis, A.R. (1992) Regulation of ribulose-1,5-bisphosphate carboxyase/oxygenase activity. *Annul Review of Plant Physiology and Plant Molecular Biology*. 43: 415-437.

Pritchard, S.G., Z.L. Ju, E. van Santen, J.S. Qiu, D.B. Weaver, S.A. Prior, and H.H. Rogers. (2000). The influence of elevated CO_2 on the activities of antioxidative enzymes in two soybean genotypes. *Australian Journal of Plant Physiology*. 27: 1061-1068.

Rey, A. and P.G. Jarvis. (1998). Long-term photosynthetic acclimation to increased atmospheric CO_2 concentration in young birch (*Betula pendula*) trees. *Tree Physiology*. 18: 441-450.

Rodermel, S. (1999). Subunit control of Rubisco biosynthesis–a relic of an enndosymbiotic past? *Photosynthesis Research*. 59: 105-123.

Rogers, A., B.U. Fischer, J. Bryant, M. Frehner, H. Blum, C.A. Raines, and S.P. Long (1998). Acclimation of photosynthesis to elevated CO_2 under low-nitrogen nutrition is affected by the capacity for assimilate utilization-perennial ryegrass under free-air CO_2 enrichment. *Plant Physiology*. 118: 683-689.

Rogers, G.S., P.J. Milham, M. Gillings, and J.P. Conroy. (1996). Sink strength may be the key to growth and nitrogen responses in N-deficient wheat at elevated CO_2. *Australian Journal of Plant Physiology*. 23: 253-264.

Rolland, F., B. Moore, and J. Sheen. (2002). Sugar sensing and signaling in plants. *Plant Cell*. 14: 185-205

Rowland-Bamford, A.J., J.T. Baker, J. Allen, and G. Bowes. (1991). Acclimation of rice to changing atmospheric carbon dioxide concentration. *Plant Cell and Environment*. 14: 577-583.

Rufty, T.W., C.D. Raper, and S.C. Huber. (1984). Alteration in internal partitioning of carbon in soybean plants in response to nitrogen stress. *Canadian Journal of Botany*. 62: 501-508.

Sage, R.F. and Peary, R.W. (1987). The nitrogen use efficiency of c-3 and c-4 plants. ii. Leaf nitrogen effects on the gas exchange characteristics of *Chenopodium album* L. and *Amaranthus retroflexus* L. *Plant Physiology*. 84: 959-963.

Sage, R.F., T.D. Sharkey, and J.R. Seemann (1988). The in-vivo response of the ribulose-1,5-bisphosphate carboxylase activation state and the pool sizes of photosynthetic metabolites to elevated CO_2 in *Phaseolus vulgaris*. *Planta*. 174: 407-416.

Sage, R.F. and Coleman, J.R. (2001). Effect of low atmospheric CO_2 on plants: more than a thing of the past. *Trend in Plant Science*. 6: 18-24.

Schwanz, P., P. Catherine, V. Philippe, D. Erwin, and G. Jean. (1996). Responses of antioxidative systems to drought stress in penduncuIate oak and maritime pine as modulated by elevated CO_2. *Plant Physiology*. 110: 393-402.

Schwanz, P. and A. Polle. (1998). Antioxidative systems, pigment and protein contents in leaves of adult mediterranean oak species (*Quercus pubescens* and *Q. ilex*) with lifetime exposure to elevated CO_2. *New Phytologist*. 140: 411-423.

Seneweera, S., P. Milham, and J. Conroy. (1994). Influence of elevated CO_2 and phosphorus nutrition on the growth and yield of a short-duration rice (*Oryza sativa* L. cv Jarrah). *Australian Journal of Plant Physiology*. 21: 281-292.
Seneweera, S., Aben, S., A. Basra, B. Jones, and J. Conroy. (2003). Involvement of ethylene in morphological and developmental responses of rice to elevated atmospheric CO_2 concentration. *Plant Growth Regulation*. 39: 143-153.
Seneweera, S.P., A.S. Basra, E.W. Barlow, and J.P. Conroy. (1995). Diurnal regulation of leaf blade elongation in rice by CO_2–Is it related to sucrose-phosphate synthase activity. *Plant Physiology*. 108: 1471-1477.
Seneweera, S.P., J.P. Conroy, K. Ishimaru, O. Ghannoum, Okada, M. Lieffering, H.Y. Kim, and K. Kobayashi. (2002). Changes in source-sink relations during development influence photosynthetic acclimation of rice to free air CO_2 enrichment (FACE). *Functional Plant Biology*. 29: 945-953.
Sharkey, T.D. (1985). Photosynthesis in intact leaves of C_3 plants: Physics, physiology and rate limitations. *The Botanical Review*. 51: 53-105.
Sheen, J. (1990). Metabolic repression of transcription in high plants. *The Plant Cell*. 2: 1027-1038.
Sheen, J. (1994). Feedback control of gene expression. *Photosynthesis Research*. 39: 427-438.
Sicher, R.C., D.F. Kremer, and S.R. Rodermel. (1994). Photosynthetic acclimation to elevated CO_2 occurs in transformed tobacco with decreased ribulose-1,5-bisphosphate carboxylase/oxygenase content. *Plant Physiology*: 104: 409-415.
Sims, D.A., Y. Luo, and J.R. Seemann. (1998). Importance of leaf versus whole plant CO_2 environment for photosynthetic acclimation. *Plant Cell & Environment*. 21: 1189-1196.
Simes, D.A., W. Cheng, Y. Luo, and J.R. Seemann. (1999). Photosynthetic acclimation to elevated CO_2 in a sunflower canopy. *Journal of Experimental Botany*. 50: 645-653.
Stitt, M. (1991). Rising CO_2 levels and their potential significance for carbon flow in photosynthetic cell. *Plant Cell & Environment*. 14: 741-762.
Stitt, M. and A. Krapp. (1999). The interaction between elevated carbon dioxide and nitrogen nutrition: the physiological and molecular background. *Plant, Cell & Environment*. 22: 583-621.
Suzuki, Y., A. Makino, and T. Mae. (2001). Changes in the turnover of Rubisco and levels of mRNAs of rbcL and rbcS in rice leaves from emergence to senescence. *Plant, Cell & Environment*. 24: 1353-1360.
Theobald, J.C., R.A.C. Mitchell, M.A.J. Parry, and D.W. Lawlor. (1998). Estimating the excess investment in ribulose-1,5-bisphosphate carboxylase/oxygenase in leaves of spring wheat crown under elevated CO_2. *Plant Physiology*. 118: 945-955.
Van Oosten, J.J., D. Wilkins, and R.T. Besford. (1994). Regulation of the expression of photosynthetic nuclear genes by CO_2 is mimicked by regulation by carbohydrates: A mechanism for the acclimation of photosynthesis to high CO_2? *Plant Cell & Environment*. 17: 913-923.
Vierstra, R.D. (1993). Protein degradation in plants. *Annual Review of Plant Physiology and Plant Molecular Biology*. 44: 385-410.

Von Caemmerer, S., and D.L. Edmondson. (1986). Relationship between steady-state gas exchange in-vivo ribulose bisphosphate carboxylase activity and some carbon reduction cycle intermediates in *Raphanus sativus. Australian Journal of Plant Physiology*. 13: 669-688.

Webber, A.N., G.Y. Nie, and S.P. Long. (1994). Acclimation of photosynthetic proteins to rising atmospheric CO_2. *Photosynthesis Research*. 39: 413-425.

Wise, R.R. (1995). Chilling-enhance photooxidation: the production, action and study of reactive oxygen species produced during chilling in the light. *Photosynthesis Research*. 45: 79-97.

Wong, S.C. (1979). Elevated atmospheric partial pressure of CO_2 and plant growth. I. Interaction of nitrogen nutrition and photosynthetic capacity in C_3 and C_4 plants. *Oecologia*. 44: 68-74.

Photosynthetic Capacity and Productivity of CO_2-Enriched Rice (*Oryza sativa* L.) Under Field Conditions: The Role of N Fertilization

Harry Harmens
David J. Marriott
Trevor W. Ashenden

SUMMARY. We studied the impacts of free-air CO_2 enrichment (FACE), and the interaction with nitrogen (N) fertilization, on the photosynthetic

---

Harry Harmens, Plant Ecophysiologist, David J. Marriott, Plant Physiologist, and Trevor W. Ashenden, Professor of Environmental Pollution, are affiliated with the Centre for Ecology and Hydrology, Bangor, UK.

Address correspondence to: Harry Harmens, Centre for Ecology and Hydrology, Bangor, University of Wales, Bangor, Deiniol Road, Bangor, Gwynedd LL57 2UP, UK (E-mail: hh@ceh.ac.uk).

The authors thank the British Council for the travel grant awarded to H.H., D.J.M. and T.W.A. This work was completed whilst D.J.M. was in receipt of postdoctoral fellowship awarded by the Natural Environment Research Council of the UK. They also thank Dr. Kazuhiko Kobayashi (Research Leader of Atmospheric Impacts Unit, National Institute for Agro-Environmental Sciences, Tsukuba, Japan) and Dr. Masumi Okada (Head of the CO_2 and Temperature Research Laboratory, National Agricultural Research Center for Tohoku Region, National Agricultural Research Organization, Morioka, Japan) for supplying the yield data, their support, advice, and hospitality. This work was supported partially by Core Research for Evolutional Science and Technology (CREST) of the Japan Science and Technology Corporation (JST).

[Haworth co-indexing entry note]: "Photosynthetic Capacity and Productivity of CO_2-Enriched Rice (*Oryza sativa* L.) Under Field Conditions: The Role of N Fertilization." Harmens, Harry, David J. Marriott, and Trevor W. Ashenden. Co-published simultaneously in *Journal of Crop Improvement* (Food Products Press, an imprint of The Haworth Press, Inc.) Vol. 13, No. 1/2 (#25/26), 2005, pp. 55-72; and: *Ecological Responses and Adaptations of Crops to Rising Atmospheric Carbon Dioxide* (ed: Zoltán Tuba) Food Products Press, an imprint of The Haworth Press, Inc., 2005, pp. 55-72. Single or multiple copies of this article are available for a fee from The Haworth Document Delivery Service [1-800-HAWORTH, 9:00 a.m. - 5:00 p.m. (EST). E-mail address: docdelivery@haworthpress.com].

http://www.haworthpress.com/web/JCRIP

Digital Object Identifier: 10.1300/J411v13n01_04

capacity of the flag leaf, above-ground dry matter production and grain yield of rice (*Oryza sativa* L.). Plant were exposed to ambient or elevated [CO_2] (ambient + 200 μmol mol^{-1}) and 4, 9 or 15 g N m^{-2}. We found no evidence of down-regulation of the photosynthetic capacity of the flag leaf in FACE. Although a reduction in N fertilization reduced the content of nitrogen and photosynthetic pigments, this hardly affected the photosynthetic capacity of the flag leaf. Both FACE and N enrichment stimulated the above-ground dry matter production and grain yield, however, the stimulation of grain yield by FACE was not significant at low N supply.

KEYWORDS. Chlorophyll, elevated [CO_2], free-air CO_2 enrichment (FACE), nitrogen, photosynthesis, photosynthetic capacity, rice, yield

INTRODUCTION

Rice (*Oryza sativa* L.) is the most important crop in Asia, in terms of both area and tonnage harvested, providing a significant portion of the dietary needs for this region (Alexandratos, 1995). The predicted increase in the global atmospheric carbon dioxide concentration ([CO_2]) (IPCC, 2001) is expected to increase the dry matter production and yield of agricultural crops (Kimball, 1983; Lawlor and Mitchell, 1991; Drake, Gonzàlez-Meler and Long, 1997). However, the response of plants to CO_2 enrichment is species specific and varies with environmental conditions. Studies on rice in controlled environments have found that CO_2 enrichment enhances dry matter production (Baker et al., 1996; Makino et al., 1997; Moya et al., 1998), and increases grain yield ranging from 5 to 60% (Horie, 1993; Baker et al., 1996; Jitla et al., 1997; Ziska et al., 1997; Moya et al., 1998). The considerable variation in the magnitude of the response of grain yield to atmospheric CO_2 enrichment can be attributed to differences in experimental conditions, such as air temperature (Baker et al., 1996; Kim et al., 1996; Lal et al., 1998) and fertilization (Seneweera and Conroy, 1997), developmental stage at the start of the exposure (Jitla et al., 1997) and cultivars (Ziska et al., 1996; Moya et al., 1998).

Most previous studies investigating the effects of elevated [CO_2] on rice used enclosures to contain the added CO_2. Enclosures have been shown to affect microclimatological factors, which may influence the response of plants to elevated [CO_2] (McLeod and Long, 1999). There is little information on the interactions between atmospheric CO_2 enrichment and nitrogen (N) availability in rice. Ziska et al. (1996) grew plants in open top chambers and found that the photosynthetic and dry matter responses of rice to elevated [CO_2] were greater with higher levels of N supply. Using free-air CO_2 enrichment (FACE), Kim et al. (2001) showed that to maximise rice grain yield under elevated [CO_2], it is important to supply sufficient N over the whole growing season, in order to maintain enhanced dry matter production. In addition, N fertilization must be co-ordinated with the developmental stage of the crop. FACE stimulated the N use efficiency of rice, which is consistent with effects on other species (Drake, Gonzàlez-Meler and Long, 1997), and FACE crops were more responsive to N in terms of grain yield (Kim et al., 2001).

Crop dry matter production is usually positively related to canopy photosynthesis. In controlled environments an increase in biomass of rice by elevated [CO_2] resulted from an increase in carbon gain at early rather than later stages of growth (Makino et al., 1997; Sakai et al., 2001). The enhancement of daily canopy gross photosynthesis by CO_2 enrichment was 33% for the first 3 weeks, but it declined gradually and disappeared by the heading stage. Although the enhancement of daily canopy net photosynthesis by CO_2 enrichment decreased with time as well, it showed a peak near heading stage, after which it declined to negative values (Sakai et al., 2001).

Little is known about the responses of the photosynthetic capacity of rice leaves to CO_2 enrichment, and the interaction with N fertilization. Many longer term studies in controlled environments have reported a decline in photosynthetic capacity of leaves by CO_2 enrichment (Sage, 1994; Drake, Gonzàlez-Meler and Long, 1997), resulting in a smaller increase in the rate of net photosynthesis than expected. The magnitude of the reduction in photosynthetic capacity due to CO_2 enrichment is variable and depends on species, the crop developmental stage and environmental conditions. A decrease in photosynthetic capacity might be an artefact of restricted root growth in pots (Arp, 1991), or could be due to nutrient limitation (Sage, 1994; Harmens et al., 2000). Little or no decrease in photosynthetic capacity has been observed in the field with ample nutrient supply (Sage, 1994). Mechanisms for down-regulation of photosynthetic capacity by elevated [CO_2] are not fully understood. There may be a decline in the amount and/or activity of Rubisco at elevated [CO_2] (Bowes, 1991; Webber, Nie and Long, 1994; Harmens et al., 2000). Plants may not be able to utilize all the extra assimilates produced at elevated [CO_2] due to sink limitations, resulting in an accumulation of non-structural carbohydrates and a negative feedback to photosynthetic enzymes (Stitt, 1991; Krapp et al., 1993; Sheen, 1994; Moore et al., 1999; Stitt and Krapp, 1999). However, Nakano et al. (1997) showed that in rice accumulation of non-structural carbohydrates did not directly affect photosynthetic capacity at a given leaf nitrogen content. Observed decreases in the photosynthetic capacity at elevated [CO_2] could simply be accounted for by a decrease in leaf nitrogen content, independent of atmospheric [CO_2], in agreement with results reported for some other species (Theobald et al., 1998; Harmens et al., 2000).

In this case study we examined the impacts of elevated atmospheric [CO_2], and the interaction with N fertilization, on the photosynthetic capacity of the flag leaf of rice, grown in the field in FACE (Kim et al., 2001; Okada et al., 2001). Rice crops were grown under either ambient or ambient + 200 μmol mol^{-1} CO_2 (target concentration), with three different N applications (4, 9, and 15 g N m^{-2}). Impacts of treatments on the photosynthetic capacity of the flag leaf, however, were studied at high and low N supply only. As components of the photosynthetic system may adjust independently to growth in elevated [CO_2] (Bunce and Ziska, 1999), photosynthetic acclimation of the following parameters was determined: (1) the apparent quantum yield for CO_2 assimilation; (2) the light-saturated rates of net photosynthesis (A_{sat}) at ambient and elevated [CO_2]; (3) the maximum apparent carboxylation velocity ($V_{c,max}$), and (4) the light- and CO_2-saturated rate of net photosynthesis (A_{max}). We hypothesized that growth under FACE will result in a reduction in the photosynthetic capacity of the flag leaf at low N availability only.

MATERIALS AND METHODS

Experimental Site

A full description of the rice FACE facility is provided by Okada et al. (2001). Briefly, the facility is located in northern Japan in an area representative of the agro-climatic region that grows a large proportion of Japan's rice crop. Plants were exposed in four paddies to elevated atmospheric [CO_2] by growing them within 12 m diameter 'rings' which sprayed pure CO_2 towards the centre from peripheral emission tubes; these are referred to as FACE plots. In another four paddies, plants were grown under ambient atmospheric [CO_2] with no ring structures in place; these are referred to as ambient plots. There were four blocks, each consisting of an ambient and FACE plot in paddies having similar soils and agronomic histories. The target [CO_2] at the centre of the FACE plots was 200 μmol mol^{-1} above ambient [CO_2]. In 1999 the actual season-long average [CO_2] in the FACE subplots ranged from 230-300 μmol mol^{-1} above ambient [CO_2]; season-long average in the ambient plots was 392 ± 45 μmol mol^{-1} CO_2. In 1999 the average temperatures during the growing seasons was 20.8°C and the total solar radiation was 1877 MJ m^{-2} (Kim et al., 2001).

Crop History

A detailed description of the crop history is provided by Kim et al. (2001). Briefly, seeds of rice (*Oryza sativa*) cv. Akitakomachi, which were obtained locally, were germinated under running water and just after radicle emergence, the seedlings were sown into trays containing fertilizer-impregnated rockwool mats. Trays were placed into plastic chambers fumigated with either ambient air or ambient air + 200 μmol mol^{-1} CO_2. On 20 May 1999, i.e., 23 d after emergence, seedlings were hand-transplanted into groups of three (referred to as a hill) into either ambient or FACE plots at a density of 19 hills m^{-2}. Nitrogen was supplied as ammonium sulphate at a rate of 4 (low), 9 (medium) or 15 g m^{-2} (high). N was supplied as a basal dressing (63% of the total), at mid-tillering (20%) and at panicle initiation (17%). To minimize mixing of the paddy water between N treatments, areas receiving low and high N were separated from the rest of the plots (which received medium N) by a 30-cm PVC barrier pushed 10 cm into the ground. At all N levels, 48 g P_2O_5 m^{-2} and 15 g K_2O m^{-2} was applied. The experimental plants were surrounded by machine-transplanted border plants, which were treated similarly to the experimental plants using local agronomic techniques.

Gas Exchange

Net photosynthesis was determined *in situ* for 3 cm-long mid-sections of fully-expanded flag leaves using two portable open gas exchange systems (LI-6400; LI-COR, Inc., Lincoln, USA). Measurements were made over a nine-day period (9th-17th August) on plants receiving the low and high N supply in two rings for each CO_2 treatment (rings C, D, Y and Z; Okada et al., 2001), with 4-6 replicates per N treatment per ring. Measurement dates coincided with the late heading-early ripening stage of the crop. Measurements were made first in the FACE plots and then two days later in the ambient plots to account for the accelerated development of plants in the FACE plots. Measurements were made with a relative humidity within the cuvette of 65 ± 5%, a leaf temperature of 28.6 ± 2.4°C, and with the assumption that stomata were evenly distributed on both surfaces of the leaves.

The response of net photosynthesis to varying photosynthetic photon flux density (PPFD) was determined at the targeted growth [CO_2] (ambient: assumed to be 360 μmol mol^{-1}, FACE: assumed to be 560 μmol mol^{-1}), as well as at the opposing [CO_2] (i.e., grown under ambient [CO_2] and measured under FACE and vice-versa). Measurements were made over ten step-changes in PPFD between 100 and 2000 μmol m^{-2} s^{-1} supplied by a LI-COR red-blue LED (LI-COR, Inc., Lincoln, USA). The light response curves were fitted by a quadratic equation, after Long and Hällgren (1995), and the apparent quantum yield was determined from the initial slope and the light-saturated rate of net photosynthesis (A_{sat}) was determined from the upper asymptote of the light response curve of net photosynthesis.

In addition, the light-saturated rate of net photosynthesis (A) was measured at 1700 μmol m^{-2} s^{-1} PPFD over a range of intercellular [CO_2] (C_i) between 50 and 1000 μmol mol^{-1} for the construction of A/C_i-curves. The A/C_i-curves were analysed as described by Stirling et al. (1997); CO_2- and RuBP-saturated rates of CO_2 uptake *in vivo* ($V_{c,max}$, the maximum apparent carboxylation velocity) were estimated from the initial slope and light- and CO_2-saturated rates of photosynthesis (A_{max}) were estimated from the upper asymptote of the A/C_i-curve. $V_{c,max}$ provided an estimate of the activity of Rubisco *in vivo*.

Tissue Sampling and Analyses

Flag-leaves were removed from the plant immediately after gas exchange measurements, and a 4 cm-long mid-section, which included the 3 cm-long section on which gas exchange measurements were made, was harvested. A sub-section was frozen in liquid N_2 for determination of photosynthetic pigments (chlorophyll and carotenoids), the remainder was dried at 65°C for 48 h for total carbon and nitrogen analysis. Dried sections of known length and width were used to determine the specific leaf area (SLA = leaf area per unit dry weight).

Two 1 cm-long segments of frozen leaf were rapidly homogenized in ice-cold 80% (v/v) acetone under dim lighting (MacKinney, 1941). Extracts were made up to 10 ml with 80% (v/v) acetone and centrifuged for 4 min at 5000 rpm at 4°C. Immediately after centrifugation, absorbence of the supernatant was determined against acetone blanks at 470, 646, 663, and 750 nm (Beckman DU7500 Spectrophotometer). Following the subtraction of absorbence by contaminants at 750 nm, this procedure allowed quantification of chlorophyll *a*, chlorophyll *b* and total carotenoids (xanthophylls and carotenes) after Lichtentaler and Wellburn (1983). One 2 cm-long segment of dried leaf was homogenized and analyzed for total carbon and nitrogen content (CHN-2000 combustion analyser).

Statistical Analyses

The experiment was a completely randomised block design with two atmospheric [CO_2] replicated twice. ANOVA (split-plot) was used to determine differences between treatment means with [CO_2] as the whole plot and N fertilization as the sub-plots (Genstat 5.4.1 for Windows, Lawes Agricultural Trust, Rothamsted, UK). For the determination and analyses of dry matter production and grain yield we refer to Kim et al. (2001).

RESULTS

Gas Exchange

The rate of net photosynthesis of the flag leaf was stimulated by elevated long-term growth [CO_2] (i.e., FACE) at all light intensities, due to both a significantly higher apparent quantum yield ($P = 0.012$) as well as a higher light-saturated rate of photosynthesis ($P < 0.001$; Figure 1). However, when measured at the same [CO_2] inside the leaf cuvette (i.e., either 360 or 560 μmol mol^{-1}), FACE did not significantly affect the apparent quantum yield and A_{sat} (Table 1, Figure 1). In general, FACE had no significant effect on the photosynthetic capacity of the flag leaf, as both $V_{c,max}$ and A_{max} were not significantly affected by FACE (Tables 1, 2). However, for $V_{c,max}$ there was a

FIGURE 1. (A) Light-saturated rate of net photosynthesis (A_{sat}) and (B) quantum yield of the flag leaf of rice, grown in ambient [CO_2] (AMB) or FACE at 4 (−) or 15 g N m^{-2} (+). A_{sat} and quantum yield were measured at 360 (black bars) and 560 μmol mol^{-1} CO_2 (gray bars). Values are the mean of two rings (± one standard error).

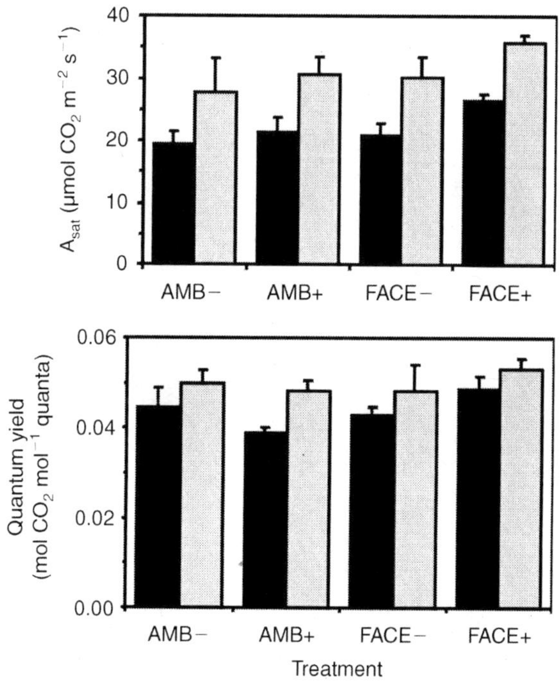

TABLE 1. Statistical significance of the effects of atmospheric $[CO_2]$ and N supply on several parameters of the flag leaf of rice. The light-saturated rate of net photosynthesis (A_{sat}) and apparent quantum yield (Q_{yield}) were determined at 360 and 560 μmol mol^{-1} CO_2 (see text for further details); ns = not significant, # P < 0.1, * P < 0.05, ** P < 0.01.

Parameter	Source of variation		
	CO_2	N	$CO_2 \times N$
$V_{c,max}$	ns	*	#
A_{max}	ns	ns	ns
A_{sat}-360	ns	ns	ns
A_{sat}-560	ns	ns	ns
Q_{yield}-360	ns	ns	#
Q_{yield}-560	ns	ns	ns
Chlorophyll *a*	ns	*	#
b	ns	#	ns
total	ns	*	#
Carotenoids	ns	*	#
N content - %	**	#	ns
- $g\ m^{-2}$	ns	#	ns
C:N	**	#	ns
SLA	ns	ns	#

TABLE 2. The maximum carboxylation velocity ($V_{c,max}$) *in vivo* and the light- and CO_2-saturated rate of net photosynthesis (A_{max}) of the flag leaf of rice, grown in ambient $[CO_2]$ or FACE at 4 or 15 g N m^{-2}. Values are the mean of two rings (± one standard error).

CO_2	N ($g\ m^{-2}$)	$V_{c,max}$	A_{max}
Ambient	4	91.5 (2.1)	27.6 (0.8)
	15	107.5 (6.0)	31.9 (3.0)
FACE	4	113.3 (7.3)	34.8 (3.0)
	15	116.2 (5.3)	35.5 (0.3)

significant (P = 0.09) interaction between atmospheric $[CO_2]$ and N fertilization. Whereas FACE did not affect $V_{c,max}$ at high N supply, FACE stimulated $V_{c,max}$ at low N supply. A similar, though not significant, trend was found for A_{max} (Tables 1, 2). Whereas high N stimulated $V_{c,max}$ in ambient $[CO_2]$, high N did not affect $V_{c,max}$ in FACE; a similar trend was found for A_{max}.

Leaf Chemistry

Although FACE reduced the total nitrogen content on a dry weight basis, FACE had no affect on the total nitrogen content of the flag leaf on a leaf area basis (Tables 1, 3). This was due to a significant reduction of the specific leaf area (SLA) of the flag leaf in FACE at low N supply (Tables 1, 3). High N supply enhanced the total nitrogen content (per unit leaf area as well as per unit leaf dry weight) of the flag leaf, independent of atmospheric [CO_2] (Tables 1, 3). Evidently a reduction in total nitrogen content of the flag leaf had hardly any effect on its photosynthetic capacity (Figure 2), indicating an increase in the photosynthetic nitrogen use efficiency of the flag leaf at low compared with high N fertilization, in particular in FACE. Whereas FACE significantly enhanced the C:N ratio, N enrichment reduced the C:N ratio of the flag leaf (Tables 1, 3). Atmospheric [CO_2] and N fertilization did not affect the total carbon content of the flag leaf (data not shown).

In general, FACE had no significant effect on the content of photosynthetic pigments in the flag leaf (Figure 3, Table 1). However, there was a tendency of FACE to enhance the content of photosynthetic pigments at low N supply, whereas the opposite was found a high N supply (Figure 3), hence the $CO_2 \times N$ interactions (Table 1). The up-regulation of A_{max} by FACE at low N is in accordance with the increase in the content of photosynthetic pigments. High N fertilization enhanced the content of photosynthetic pigments in the flag leaf.

TABLE 3. The N content per unit leaf dry weight (%) or leaf area (g m^{-2}), C:N ratio and specific leaf area (SLA) of the flag leaf of rice, grown in ambient [CO_2] or FACE at 4 or 15 g N m^{-2}. Values are the mean of two rings (± one standard error).

		N (g m^{-2})	
Parameter	CO_2	4	15
N (%)	Ambient	2.61 (0.14)	3.15 (0.15)
	FACE	2.25 (0.04)	2.94 (0.01)
N (g m^{-2})	Ambient	1.38 (0.10)	1.72 (0.08)
	FACE	1.33 (0.01)	1.63 (0.04)
C:N	Ambient	15.9 (0.87)	13.5 (0.85)
	FACE	18.0 (0.15)	14.1 (0.02)
SLA (cm^2 g^{-1})	Ambient	190.5 (4.7)	183.4 (0.0)
	FACE	169.8 (4.3)	181.1 (5.0)

FIGURE 2. (A) Maximum carboxylation velocity ($V_{c,max}$) *in vivo* and (B) the light- and CO_2-saturated rate of net photosynthesis (A_{max}) plotted against the N content of the flag leaf of rice. Plants were grown in ambient [CO_2] (AMB) or FACE at 4 (−) or 15 g N m^{-2} (+). Values are the mean of two rings (± one standard error). Equations of the fitted linear regression lines for $V_{c,max}$ and A_{max} are, respectively: $y = 72.2 + 22.6x$ ($r^2 = 0.19$) and $y = 25.6 + 4.5x$ ($r^2 = 0.07$).

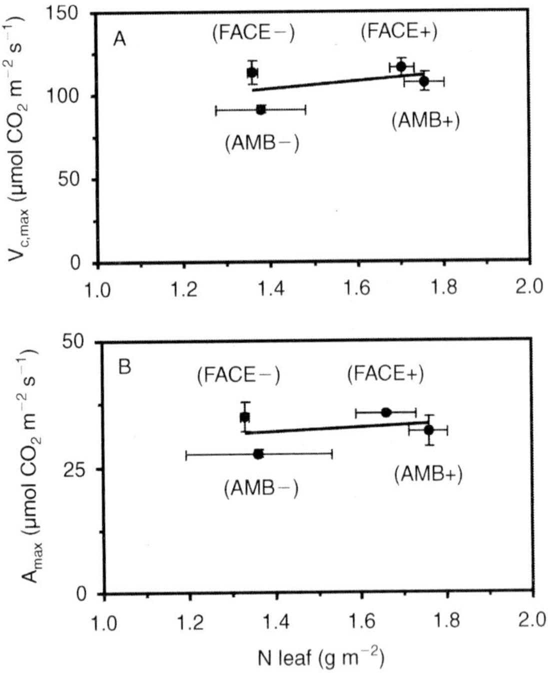

Crop Production

FACE stimulated the total dry matter production at panicle initiation and grain maturity (Kim et al., 2001), which was due to a stimulation of both below (Kim et al., 2001) and above-ground dry matter production (Figure 4). The stimulation of the total dry matter production by enhanced N fertilization was primarily due to a stimulation of above-ground dry matter production (Kim et al., 2001; Figure 4). Whilst both FACE and enhanced N fertilization stimulated grain yield, the stimulation by FACE at low N supply was not significant, resulting in a significant $CO_2 \times N$ interaction (Kim et al., 2001; Figure 4).

FIGURE 3. Chlorophyll (total, *a* and *b*) and carotenoids content of the flag leaf of rice. Plants were grown in ambient [CO_2] (AMB) or FACE at 4 (−) or 15 g N m^{-2} (+). Values are the mean of two rings (± one standard error).

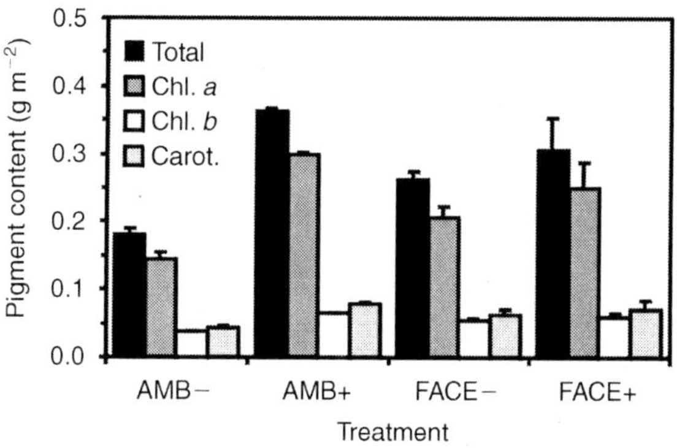

DISCUSSION

We found no evidence of down-regulation of the photosynthetic capacity of the flag leaf of rice at the stage of late heading-early ripening in FACE. In contrast, FACE tended to stimulate $V_{c,max}$ and A_{max} at low N fertilization. Despite a significant lower nitrogen content of the flag leaf at low compared with high N supply, this hardly affected its photosynthetic capacity (Figure 1). The nitrogen content of the flag leaf of rice was in a range similar to the nitrogen content of the youngest fully expanded leaf of *Dactylis glomerata*, where Harmens et al. (2000) found hardly any changes in the photosynthetic capacity of the leaf, despite an increase in its nitrogen content. In contrast, Nakano et al. (1997) found a strong linear relationship between the nitrogen content of rice leaves and its net rate of photosynthesis at light saturation at ambient and elevated [CO_2], but this was observed over a much wider range of nitrogen content of the leaves (0.8-2.7 g m^{-2}). Similarly, Theobald et al. (1998) found a strong linear relationship between the nitrogen content and $V_{c,max}$ of spring wheat leaves when the nitrogen content of leaves was in the range of 0.3-2.2 g m^{-2}. Our results are in agreement with those reported for spring wheat grown in FACE: there was no acclimatory loss of photosynthetic capacity in FACE, in contrast to some controlled-environment and field-enclosure studies (Garcia et al., 1998).

Discrepancies between our results and those of Nakano et al. (1997) might be explained to some extend by differences in experimental set-up. Nakano et al. (1997) grew rice plants in nutrient solutions in controlled environments and

FIGURE 4. (A) Shoot dry weight at panicle initiation (circles) and maturity (squares) and (B) grain yield of rice. Plants were grown in ambient [CO_2] (solid lines) or FACE (dashed lines) at 4, 9 or 15 g N m^{-2}. Values are the mean of four rings, LSD values are indicated by the vertical lines on the right (adapted from Kim et al., 2001).

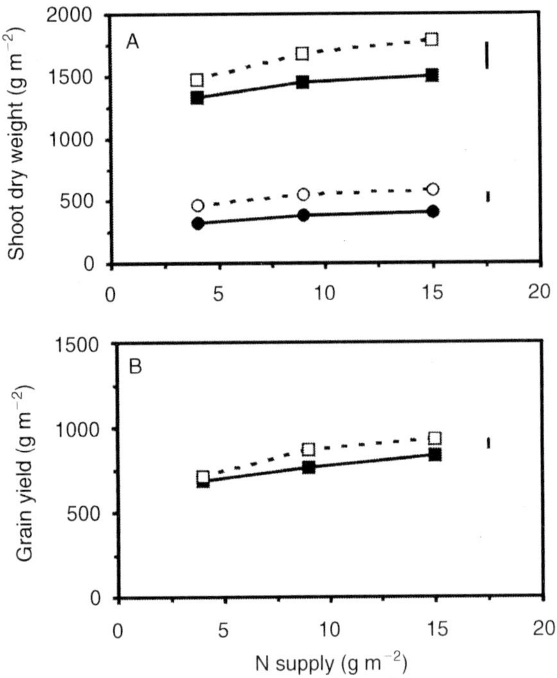

did their measurements on young, fully expanded leaves of plants germinated and grown until leaf eight at ambient [CO_2]. Measurements were conducted about 3 weeks after transfer to elevated [CO_2]. Despite a transient increase in the relative growth rate in the first week after transfer, elevated [CO_2] did not affect tiller numbers, as it did in plants exposed to elevated [CO_2] during (Kim et al., 2001) or immediately after germination (Jitla et al., 1997). Accelerated development in the shoot early on is crucial for obtaining maximum increase in grain yield at elevated [CO_2], caused by an increase in tillering, and therefore panicle numbers (Jitla et al., 1997; Kim et al., 2001). A delay of exposure of rice plants to elevated [CO_2] will inhibit a positive tillering response to elevated [CO_2] (Jitla et al., 1997). Therefore, the phenology of rice plants is likely to be different when exposed to elevated [CO_2] from germination or later on during plant development, which might affect the carbon source to sink relationship and the photosynthetic responses to CO_2 enrichment.

Although we did not find any evidence of down-regulation of the photosynthetic capacity of the flag leaf at the stage of late heading-early ripening, down-regulation might occur later on in its developmental stage. In wheat down-regulation of photosynthetic capacity of the flag leaf in FACE was not observed until the stage of senescence, which occurred earlier in FACE compared with ambient [CO_2] (Garcia et al., 1998). Similarly, enhanced N fertilization might have the greatest impact on the photosynthetic capacity of the flag leaf when it senesces (Delgado et al., 1994). In rice the rate of net photosynthesis decreases with development from heading to grain maturity (Sasaki, Samejima, and Ishii, 1996). In addition, cultivar differences in the rate of net photosynthesis of the flag leaf of rice at the heading stage have been observed and are primarily caused by differences in the resistance for CO_2 fixation in mesophyll cells, which are closely correlated with their Rubisco contents (Sasaki, Samejima, and Ishii, 1996).

In general, growth at elevated [CO_2] leads to an accumulation of non-structural carbohydrates in the source leaves, which might result in a negative feedback to the synthesis of photosynthetic enzymes (Stitt, 1991; Krapp et al., 1993; Sheen, 1994; Moore et al., 1999; Stitt and Krapp, 1999). Although starch accumulation is generally thought to physically distort the chloroplast and possibly hinder CO_2 diffusion, there is no evidence of a causal relationship between starch accumulation and down-regulation of the photosynthetic capacity of leaves (Stitt, 1991). In spring wheat grown in FACE there was only a weak relationship between increased soluble carbohydrate contents and decreased levels for nuclear gene transcripts of photosynthetic enzymes. It was suggested that, whereas repression of photosynthetic genes by increased soluble carbohydrate content may explain acclimation of photosynthesis to elevated CO_2 in model laboratory systems, many other factors might modify its significance in the field (Nie et al., 1995). Nakano et al. (1997) showed that, in spite of accumulation of starch and sucrose in rice leaves, this did not affect the net rate of photosynthesis for a given leaf nitrogen content. They concluded that the decrease in photosynthetic rate per unit of leaf area by CO_2 enrichment was the result of a decrease in the absolute amount of nitrogen in the leaves and not directly affected by the accumulation of carbohydrates. A similar result was reported by Harmens et al. (2000) for *Dactylis glomerata*. However, it cannot be excluded that accumulation of carbohydrates directly or indirectly leads to a decrease in leaf nitrogen content, resulting in reduced photosynthetic capacity.

The absolute amount of carbohydrates accumulating in rice leaves is small compared with other species, and rice plants can accumulate a great deal of their carbohydrates in their leaf sheaths (Nakano, Makino and Mae, 1997). Furthermore, in rice the partitioning of total non-structural carbohydrates between plant tissues changes with developmental stage and atmospheric [CO_2]

(Rowland-Bamford et al., 1990). Early in the growing season most total non-structural carbohydrates in rice are present in leaf tissue, but prior to panicle initiation 40 to 60%, dependent on atmospheric [CO_2], is present in culms and leaf sheaths. From panicle initiation onwards the majority of the total non-structural carbohydrates can be found in the panicle. By 71 days after planting leaf starch and sucrose levels were similar across [CO_2] and much lower than at 19 days after planting (Rowland-Bamford et al., 1990).

In rice leaves, the amount of Rubisco appears to be determined by the amount of nitrogen, regardless of irradiance, temperature and atmospheric [CO_2], and rice plants do not optimize nitrogen allocation for Rubisco under variations in those environmental conditions (Mae, 1997). In rice (Nakano, Makino and Mae, 1997) and spring wheat (Theobald et al., 1998), a linear relationship was found between the nitrogen and Rubisco content of leaves, which was independent of growth [CO_2]. In *Dactylis glomerata*, this relationship was best described by a second order polynomial (Harmens et al., 2000). Therefore, a decrease in the nitrogen content of the flag leaf of rice at low compared with high N supply was most likely accompanied by a reduction in the Rubisco content. This, however, had no effect on the Rubisco activity *in vivo* in FACE, as indicated by the absence of a negative effect of low N supply on $V_{c,max}$. This suggests that in FACE a decrease in the Rubisco content was compensated for by an increase in the amount of active Rubisco. Previously, Nakano et al. (1997) found that the high activation state of Rubisco at elevated [CO_2] declined significantly with increasing N fertilization. Apparently, at high N fertilization the Rubisco content of the flag leaf was in excess of that required to support the observed photosynthetic rates (Stitt, 1991; Theobald et al., 1998; Harmens et al., 2000). This might explain why a reduction in nitrogen content of the flag leaf had hardly any impact on its photosynthetic capacity. In rice both the absolute amount and the proportion of Rubisco-N to total nitrogen in fully expanded leaves increased from 22 to 32% with increased N fertilization (Mae, 1997), suggesting excessive investment in Rubisco at high N supply. In spring wheat excess investment in Rubisco was greater in leaves grown with high than low N application, and declined as leaves senesced. The fraction of excess Rubisco was strongly dependent on leaf nitrogen content: 5% at 1 g N m^{-2} to 40% at 2 g N m^{-2} (Theobald et al., 1998).

The up-regulation of A_{max} in FACE at low N supply was in agreement with an increase in photosynthetic pigments. As reported previously by Nakano et al. (1997), there was a significant linear relationship between the content of nitrogen and photosynthetic pigments of the leaves (data not shown). From the apparent quantum yield and A_{sat}, it is evident that FACE stimulated the rate of net photosynthesis of the flag leaf more at high than low N supply. FACE stimulated the apparent quantum yield and A_{sat} by 9 and 54%, respectively, at low N supply, and by 36 and 68%, respectively, at high N supply. Furthermore, these data show clearly that FACE stimulated the net rate of photosynthesis more at high than low light intensity, which was to be expected as at low light intensity the net rate of photosynthesis is more limited by the light intensity than by the atmospheric [CO_2]. The response of the apparent quantum yield and A_{sat} to FACE supports the suggestion by Kim et al. (2001), that FACE not only increased radiation interception, but also radiation use efficiency. Although Baker et al. (1996) reported down-regulation of canopy photosynthesis in rice at elevated atmospheric [CO_2], this was mainly observed at a growth [CO_2] of 900 μmol mol^{-1}. When the canopy photosynthesis of rice, grown at 330 μmol mol^{-1} CO_2 and measured at 330 and 660 μmol mol^{-1} CO_2, was compared with the canopy photosynthesis of rice, grown at 660 μmol mol^{-1} CO_2 and measured at 330 and 660 μmol mol^{-1} CO_2, there was no evidence of down-regulation of canopy photosynthesis.

As shown before for other species (Harmens et al., 2001, and references therein), FACE reduced the nitrogen content of the flag leaf of rice when expressed on a leaf dry weight basis, but not when expressed on a leaf area basis. This was primarily due to a decrease (although not significant) in the SLA caused by FACE, which might be due to an accumulation of non-structural or structural carbohydrates, or both (Harmens et al., 2000; Marriott, Stirling and Farrar, 2001). On the other hand, CO_2 enrichment might reduce investment of nitrogen in leaf blades and increase allocation into leaf sheaths and roots (Makino et al., 1997). Thus the decrease in leaf nitrogen might be due to a change in nitrogen allocation at the whole plant level. Growth responses of rice to CO_2 enrichment might be controlled primarily by leaf area expansion and nitrogen allocation into leaf blades at the whole plant level. The higher photosynthetic rates at the leaf level in FACE, in combination with no impact of FACE on the nitrogen content of leaves per unit leaf area, most likely contributed positively to the greater nitrogen use efficiency of the crop in FACE compared with ambient [CO_2] (Kim et al., 2001), a phenomenon reported for other species as well (Drake, Gonzàlez-Meler and Long, 1997; Stitt and Krapp, 1999; Harmens et al., 2001; Marriott, Stirling and Farrar, 2001).

In conclusion, in this case study we found no evidence of down-regulation of any of the studied components of the photosynthetic system of rice in FACE, not even at low N fertilization. This, in addition to an accelerated development of rice during early stages of growth (Jitla et al., 1997), resulted in a stimulation of total dry matter production at both high and low N fertilization. However, grain yield was not stimulated significantly by FACE at low N fertilization, which was due the absence of a stimulation of panicle number by FACE (Kim et al., 2001).

REFERENCES

Alexandratos, N. (1995). *World agriculture. Towards 2010.* Chichester, UK: FAO-Wiley.

Arp, W.J. (1991). Effects of source-sink relations on photosynthetic acclimation to elevated CO_2. *Plant, Cell and Environment* 14: 1003-1006.

Baker, J.T., L.H. Allen, Jr., K.J. Boote, and N.B. Pickering. (1996). Assessment of rice response to global climate change: CO_2 and temperature. In *Carbon dioxide and terrestrial ecosystems*, eds. G.W. Koch and H.A. Mooney. San Diego, USA: Academic Press, pp. 265-282.

Bowes, G. (1991). Growth at elevated CO_2: photosynthetic response mediated through Rubisco. *Plant, Cell and Environment* 14: 795-806.

Bunce, J.A., L.H. Ziska (1999). Impact of measurement irradiance on acclimation of photosynthesis to elevated CO_2 concentration in several plant species. *Photosynthetica* 37: 509-517.

Delgado, E., R.A.C. Mitchell, M.A.J. Parry, S.P. Driscoll, V.J. Mitchell, and D.W. Lawlor. (1994). Interacting effects of CO_2 concentration, temperature and nitrogen supply on the photosynthesis and composition of winter wheat leaves. *Plant, Cell and Environment* 17: 1205-1213.

Drake, B.G., M.A. Gonzàlez-Meler, and S.P. Long. (1997). More efficient plants: a consequence of rising atmospheric CO_2? *Annual Review of Plant Physiology and Plant Molecular Biology* 48: 609-639.

Garcia, R.L., S.P. Long, G.W. Wall, C.P. Osborne, B.A. Kimball, G.Y. Nie, P.J. Pinter, R.L. Lamorte, and F. Wechsung. (1998). Photosynthesis and conductance of spring-wheat leaves: field response to continuous free-air atmospheric CO_2 enrichment. *Plant, Cell and Environment* 21: 659-669.

Harmens, H., C. Marshall, C.M. Stirling, and J.F. Farrar. (2001). Partitioning and efficiency of use of N in *Dactylis glomerata* as affected by elevated CO_2: interactions with N supply? *International Journal of Plant Sciences* 162: 1267-1274.

Harmens, H., C.M. Stirling, C. Marshall, and J.F. Farrar. (2000). Does down-regulation of photosynthetic capacity by elevated CO_2 depend on N supply in *Dactylis glomerata*? *Physiologia Plantarum* 108: 43-50.

Horie, T. (1993). Predicting the effects of climate variation and elevated CO_2 on rice yield in Japan. *Journal of Agricultural Meteorology* 48: 567-574.

IPCC. (2001). *Climate change 2001: the scientific basis. Contribution of working group I to the third assessment report of the Intergovernmental Panel on Climate Change*, eds. J.T. Houghton, Y. Ding, D.J. Griggs, M. Noguer, P.J. van der Linden, X. Dai, K. Maskell, C.A. Johnson. Cambridge, UK: Cambridge University Press, 881 pp.

Jitla, D.S., G.G. Rogers, S.P. Seneweera, A.S. Basra, R.J. Oldfield, and J.P. Conroy. (1997). Accelerated early growth of rice at elevated CO_2. *Plant Physiology* 115: 15-22.

Kim, H.Y., T. Horie, H. Nakagawa, and K. Wada. (1996). Effects of elevated CO_2 concentration and high temperature on growth and yield of rice. II. The effect on yield and its components. *Japanese Journal of Crop Science* 65: 644-651.

Kim, H.Y., M. Lieffering, S. Miura, K. Kobayashi, and M. Okada. (2001). Growth and nitrogen uptake of CO_2-enriched rice under field conditions. *New Phytologist* 150: 223-229.

Kimball, B.A. (1983). Carbon dioxide and agricultural yield: an assemblage and analysis of 430 prior observations. *Agronomy Journal* 75: 779-788.

Krapp, A., B. Hofmann, C. Schaefer, and M. Stitt. (1993). Regulation of the expression of rbcS and other photosynthetic genes by carbohydrates: a mechanism for the 'sink regulation' of photosynthesis? *Plant Journal* 3: 817-828.

Lal, M., K.K. Singh, L.S. Rathore, G. Srinivasan, and S.A. Saseendran. (1998). Vulnerability of rice and wheat yields in NW India to future changes in climate. *Agricultural and Forest Meteorology* 89: 101-114.

Lawlor, D.W and R.A.C. Mitchell. (1991). The effects of increasing CO_2 on crop photosynthesis and productivity: a review of field studies. *Plant, Cell and Environment* 14: 807-818.

Lichtenthaler, H.K. and A.R. Wellburn. (1983). Determinations of total carotenoids and chlorophyll a and b of leaf extracts in different solvents. *Biochemical Society Transactions* 11: 591-592.

Long, S.P. and J.-E. Hällgren. (1995). Measurement of CO_2 assimilation by plants in the field and the laboratory. In *Photosynthesis and production in a changing environment: a field and laboratory manual*, eds. D.O. Hall, J.M.O. Scurlock, H.R. Bolhàr-Nordenkampf, R.C. Leegood, and S.P. Long. London, UK: Chapman and Hall, pp. 129-167.

MacKinney, G. (1941). Absorption of light by chlorophyll solutions. *Journal of Biological Chemistry* 140: 315-322.

Mae, T. (1997). Physiological nitrogen efficiency in rice: nitrogen utilization, photosynthesis, and yield potential. *Plant and Soil* 196: 201-210.

Makino, A., M. Harada, T. Sato, H. Nakano, and T. Mae. (1997). Growth and N allocation in rice plants under CO_2 enrichment. *Plant Physiology* 115: 199-203.

Marriott, D.J., C.M. Stirling, and J. Farrar (2001). Constraints to growth of annual nettle (*Urtica urens*) in an elevated CO_2 atmosphere: decreased leaf area ratio and tissue N cannot be explained by ontogenetic drift or mineral N supply. *Physiologia Plantarum* 111:23-32.

McLeod A.R. and S.P. Long. (1999). Free-air carbon dioxide enrichment (FACE) in global change research; a review. *Advances in Ecological Research* 28: 1-56.

Moore, B.D., S.-H. Cheng, D. Sims, and J.R. Seemann. (1999). The biochemical and molecular basis for photosynthetic acclimation to elevated atmospheric CO_2. *Plant, Cell and Environment* 22: 567-582.

Moya, T.B., L.H. Ziska, O.S. Namuco, and D. Olszyk. (1998). Growth dynamics and genotypic variation in tropical, field-grown paddy rice (*Oryza sativa* L.) in response to increasing carbon dioxide and temperature. *Global Change Biology* 4: 645-656.

Nakano, H., A. Makino, and T. Mae. (1997). The effect of elevated partial pressures of CO_2 on the relationship between photosynthetic capacity and N content in rice leaves. *Plant Physiology* 115: 191-198.

Nie, G.Y., D.L. Hendrix, A.N. Webber, B.A. Kimball, and S.P. Long. (1995). Increased accumulation of carbohydrates and decreased photosynthetic gene transcript levels in wheat grown at an elevated CO_2 concentration in the field. *Plant Physiology* 108: 975-983.

Okada, M., M. Lieffering, H. Nakamura, M. Yoshimoto, H.Y. Kim, and K. Kobayashi. (2001). Free-air CO_2 enrichment (FACE) with pure CO_2 injection: rice FACE system design and performance. *New Phytologist* 150: 251-260.

Rowland-Bamford, A.J., L. H. Allen, Jr., J.T. Baker, and K.J. Boote. (1990). Carbon dioxide effects on carbohydrate status and partitioning in rice. *Journal of Experimental Botany* 41: 1601-1608.

Sage, R.F. (1994). Acclimation of photosynthesis to increasing atmospheric CO_2: the gas exchange perspective. *Photosythesis Research* 39: 351-368.

Sakai, H., K. Yagi, K. Kobayashi, S. Kawashima. (2001). Rice carbon balance under elevated CO_2. *New Phytologist* 150: 241-249.

Sasaki, H., M. Samejima, and R. Ishii. (1996). Analysis of $\delta^{13}C$ measurement on mechanism of cultivar difference in leaf photosynthesis of rice (*Oryza sativa* L.). *Plant and Cell Physiology* 37: 1161-1166.

Seneweera, S.P. and J.P. Conroy. (1997). Growth, grain yield and quality of rice (*Oryza sativa* L.) in response to elevated CO_2 and phosphorus nutrition. *Soil science and plant nutrition* 43: 1131-1136.

Sheen, J. (1994). Feedback control of gene expression. *Photosynthesis Research* 39: 427-438.

Stirling, C.M., P.A. Davey, T.G. Williams, and S.P. Long. (1997). Acclimation of photosynthesis to elevated CO_2 and temperature in five British native species of contrasting functional type. *Global Change Biology* 3: 237-246.

Stitt, M. (1991). Rising CO_2 levels and their potential significance for carbon flow in photosynthetic cells. *Plant, Cell and Environment* 14: 741-762.

Stitt, M. and A. Krapp. (1999). The interaction between elevated carbon dioxide and nitrogen nutrition: the physiological and molecular background. *Plant, Cell and Environment* 22: 583-621.

Theobald, J.C., R.A.C. Mitchell, M.A.J. Parry, and D.W. Lawlor. (1998). Estimating the excess investment in ribulose-1,5-bisphosphate carboxylase/oxygenase in leaves of spring wheat grown under elevated CO_2. *Plant Physiology* 118: 945-955.

Webber, A.N., G.Y. Nie, and S.P. Long. (1994). Acclimation of photosynthetic proteins to rising atmospheric CO_2. *Photosynthesis Research* 39: 413-425.

Ziska, L.H., O. Namuco, T. Moya, and J. Quilang. (1997). Growth and yield responses of field-grown tropical rice to increasing carbon dioxide and air temperature. *Agronomy Journal* 89: 45-53.

Ziska, L.H., W. Weerakoon, O.S. Namuco, and R. Pamplona. (1996). The influence of nitrogen on the elevated CO_2 response on field-grown rice. *Australian Journal of Plant Physiology* 23: 45-52.

CO_2 Enrichment Effects on Forage and Grain Nitrogen Content of Pasture and Cereal Plants

H. J. Weigel
R. Manderscheid

SUMMARY. Increasing atmospheric CO_2 concentrations [CO_2] have the potential to enhance growth and yield of agricultural plants. Concomitantly plants grown under high [CO_2] show significant changes of the chemical composition of their foliage and of other plant parts. Particularly, high [CO_2] result in a decrease of plant nitrogen (N) concentration, which may have serious consequences for crop quality. This presentation summarizes the results of a variety of CO_2 enrichment studies with pasture plants (*Lolium* spp., *Trifolium repens*) and cereal species (*Triticum aestivum, Hordeum vulgare*) which were conducted at our laboratory under different growth and CO_2 exposure conditions ranging from controlled environment studies to investigations under free air carbon dioxide enrichment (FACE). With the exception of clover in all experiments a CO_2-induced decline of forage and grain N concentration was observed. The magnitude of this reduction differed between species, cultivars, management conditions (N fertilization) and CO_2 exposure conditions. No unambiguous evidence was obtained whether N fertilization can contribute to meet the quality requirements for cereals

---

H. J. Weigel and R. Manderscheid are Senior Research Scientists, Institute of Agroecology of the Federal Agricultural Research Centre (FAL), Bundesallee 50, 38116 Braunschweig, Germany (E-mail: hans.weigel@fal.de).

[Haworth co-indexing entry note]: "CO_2 Enrichment Effects on Forage and Grain Nitrogen Content of Pasture and Cereal Plants." Weigel, H. J., and R. Manderscheid. Co-published simultaneously in *Journal of Crop Improvement* (Food Products Press, an imprint of The Haworth Press, Inc.) Vol. 13, No. 1/2 (#25/26), 2005, pp. 73-89; and: *Ecological Responses and Adaptations of Crops to Rising Atmospheric Carbon Dioxide* (ed: Zoltán Tuba) Food Products Press, an imprint of The Haworth Press, Inc., 2005, pp. 73-89. Single or multiple copies of this article are available for a fee from The Haworth Document Delivery Service [1-800-HAWORTH, 9:00 a.m. - 5:00 p.m. (EST). E-mail address: docdelivery@haworthpress.com].

http://www.haworthpress.com/web/JCRIP

Digital Object Identifier: 10.1300/J411v13n01_05

and grass monocultures with respect to tissue N concentrations in a future high-CO_2 world. As shown in the FACE experiments current application rates of N fertilizers are inadequate to achieve quality standards.

KEYWORDS. Elevated atmospheric CO_2, nitrogen content, wheat, barley, ryegrass, clover, grain and forage quality

INTRODUCTION

There is no doubt that the atmospheric carbon dioxide (CO_2) concentration will continue to rise rapidly during the next decades and is expected to reach ca. 550 ppm by the middle of the present century (Houghton et al., 2001). Evidence based on numerous experimental studies clearly indicates that along with the enhancement of net photosynthesis, elevated CO_2 concentrations have the potential to increase growth and biomass of plants. This is particularly true for agricultural plants under conditions of favorable supply of other growth resources (Rosenzweig and Hillel, 1998; Reddy and Hodges, 2000).

Concomitantly to an increase in biomass production, a frequently observed phenomenon is that plants grown in CO_2-enriched air exhibit significant changes of the chemical composition of their foliage and of other plant parts (Idso and Idso, 2001), which may have serious ecological and economic consequences. One of the most often observed consequence of a CO_2 enrichment is a decrease of the foliar nitrogen (N) concentration and also of the N content of seeds and grains, respectively (Conroy, 1992; Owensby et al., 1996; Cotrufo et al., 1998). Among the explanations of possible causes of such a decline of N concentrations are a "biomass dilution effect" due to an increase in total nonstructural carbohydrate status of the plants, reductions in Calvin cycle enzymes, particularly of ribulose-bisphosphate carboxylase/oxygenase and/or reduced uptake of nitrate due to CO_2-induced lower transpiration rates (Conroy, 1992; Idso and Idso, 2001). Moreover, plant N concentration depends on the developmental stage and the total biomass (Greenwood et al., 1990). Thus, CO_2-induced reductions in plant N concentration can also result from a stimulation in plant development and biomass (Colemann et al., 1993).

With regard to the nutritive value of fodder crops CO_2-induced alterations in leaf N concentration may be critical. When plant foliage is directly consumed by animals the forage nutritive value is highly dependent on the leaf protein and N content, respectively. This is of particular relevance for forage quality of plant species of productive temporary and permanent grasslands as wells as of rangeland ecosystems which provide the majority of food resources, e.g., for ruminants (Owensby et al., 1996; Nösberger et al., 2000). Beside fiber content, non-structural carbohydrates, minerals and secondary compound concentration of crude protein is one of the key quality parameters for grassland forage. A variety of experiments have shown changes in mineral concentrations and particularly negative effects on forage quality as plant N usually declines at elevated CO_2 over a wide range of N availability (e.g., Overdieck, 1993; Cotrufo et al., 1998; Polley et al., 2000; Lilley et al., 2001). While individual species may be affected positively or negatively by elevated [CO_2], the total quantity of nutrients on offer to a grazing animal is determined by the relative abundance of a particular plant species in a vegetation type. Shifts in species composition of grassland under CO_2 enrichment have repeatedly been observed (e.g., Clark et al., 1995; Lüscher et al., 1996) and this may also have consequences for animal nutrition. Consequently, one might speculate that the productivity of domestic and native animals in a future high CO_2 world may be affected by changes in leaf composition of individual species as well as changes in species distribution.

With respect to cereals and particularly wheat, quality and related bread-making quality might be changed as a consequence of a CO_2 induced reduction of grain N and grain protein. However, reports on the effects of elevated CO_2 concentrations on wheat grain quality are contradictory (Idso and Idso, 2001). While Havelka et al. (1984), Rogers et al. (1998), and Dijkstra et al. (1999) found little or no changes of wheat grain nutritional or bread-making quality, significant decreases in the protein concentration along with reduced nutritional and bread-making quality of wheat grown under elevated CO_2 concentrations were found, e.g., by Conroy et al. (1994), Blumenthal et al. (1996), Thompson and Woodward (1994), Fangmeier et al. (1999), and Pleijel et al. (1999). While all these studies were conducted in some type of chamber environments, recent results of free air CO_2 enrichment (FACE) experiments from Arizona/USA revealed only a slight decrease of wheat grain protein concentration at ample supply of soil N (Kimball et al., 2001). However, in these experiments, which may be regarded as most relevant to real agronomic conditions, high CO_2 concentrations exacerbated deleterious effects of low soil N supply on grain protein. Thus, whether or not future atmospheric CO_2 concentration will affect grain quality remains a matter of debate.

CO_2-induced changes in grain N and consequently grain protein content have to be related to the general observation that there is a close relationship between decreasing grain protein concentration and increasing grain yield (Evans, 1993) irrespective of what are the causes of the biomass increase. Moreover, this relationship is greatly affected by soil N availability. Hence one might expect a decline of grain N concentrations along with any increase in grain biomass due to a CO_2 enrichment of ambient air (Pleijel et al., 1999). Overall, responses of the element concentration in forage and grain tissues to a CO_2 enrichment vary with species, cultivars, growth stage and limitations by other resources, e.g., nutrients, water, light as well as with inter-and intraspecific competition and with the overall growth conditions, i.e., the CO_2 exposure system used in a particular experiment (Cotrufo et al., 1998).

It is, therefore, important to investigate the effects of elevated CO_2 concentrations on food and fodder crops in order to develop scenarios of how future agricultural management considerations have to be adjusted in order to meet quality requirements. The objective of the present paper is to provide a compilation of results of various experiments carried out in our laboratory during recent years on the effects of elevated atmospheric CO_2 concentrations on foliage N content of pasture species (*Lolium* ssp., *Trifolium repens*) and on grain N content of cereal species (wheat and barley). The experiments described cover a range of conditions which are known to influence CO_2 effects on tissue N concentrations, including N supply via the growth medium, cultivar selection and growing season. In addition, it is shown if CO_2 effects are consistent across various CO_2 exposure systems and growth conditions (e.g., pot studies in controlled environments to field studies in open-top chambers and under free air CO_2 enrichment).

MATERIALS AND METHODS

CO_2 enrichment experiments described in the present paper (Table 1) were carried out either in controlled environment chambers (continuous stirred tank reactors, CSTR; Heck et al., 1978), open-top field chambers (OTC; 3.2 m diameter; 3.0 m height) equipped with a CO_2 dispensing and monitoring system as described by Weigel et al. (1992) or using a large free air carbon dioxide enrichment (FACE; Hendrey, 1993) facility (20 diameter rings) installed on a 20 ha field plot in a local crop rotation of winter barley, sugar beet and winter wheat (Weigel and Dämmgen, 2000).

While in the CSTR studies plants were grown in 2.8 l pots, in the OTC studies plants were grown either in pots (cereals 4.2 l volume; grass and clover 12 l volume) or in natural soil at a wheat field site. Pot studies were done using either a standard prefertilized mix of peat and soil (ED73 containing 0.3 g l^{-1} P_2O_5, 0.4 g l^{-1} K_2O, and 0.4 g l^{-1} plant available N), with additional N supply (NPK fertilizer) if necessary or a field soil (cambisol, pH 6.5; organic matter content 1.5%) with appropriate fertilization. Fertilization in the field experiments was carried out according to local agricultural practice including organic and mineral fertilizers. CO_2 treatment concentrations indicated in Table 1 are round values based on 24-h (chamber experiments) or daylight-hour

TABLE 1. Summary of the different experiments contributing to the present study.

Species	Cultivar	Growth conditions	Year of study	Soil type	CO_2 (ppm)	Remarks	Ref.
Cereals							
Triticum aestivum	Turbo	OTC/pot study	1992	ED 73	365/700	> 500 kg ha^{-1} N	1
	Turbo	OTC/pot study	1994	ED 73	365/700		2
	Nandu	OTC/pot study	1994	ED 73	365/700		3
	Minaret	OTC/pot study	1994	ED 73	365/700		3
	Janetzkis Früher 1914	OTC/pot study	1994	ED 73	365/700		2
	Heines Kolben 1890	OTC/pot study	1994	ED 73	365/700		2
	Nandu	OTC/pot study	1994	Cambisol	365/700	200 kg ha^{-1} N	4
	Minaret	OTC/pot study	1994	Cambisol	365/700	200 kg ha^{-1} N	4
	Minaret	OTC/field sown	1999	Cambisol	365/660	200 kg ha^{-1} N	5
Hordeum vulgare	Alexis	OTC/pot study	1992	ED 73	365/700	> 500 kg ha^{-1} N	1
	Alexis	OTC/pot study	1995	ED 73	365/700		6
	Arena	OTC/pot study	1992	ED 73	365/700		1
	Krona 1990	OTC/pot study	1995	ED 73	365/700		6
	Baronesse 1989	OTC/pot study	1995	ED 73	365/700		6
	Franken 1930	OTC/pot study	1995	ED 73	365/700		6
	Intensiv 1921	OTC/pot study	1995	ED 73	365/700		6
	Heines Hanna 1899	OTC/pot study	1995	ED 73	365/700		6
	Theresa	FACE/field sown	99/00	Cambisol	370/550	260/130 kg ha^{-1} N	7
	Eckendorfer Mammut (1934)						
Pasture species							
Trifolium repens	Karina	CSTR/pot study	1993	Cambisol	365/700		9
	Karina	OTC/pot study	1993/1994	Cambisol	380/700	Seeding density	8
Lolium perenne	Parcour	CSTR/pot study	1993	Cambisol	365/700	Seeding density	9
	Parcour	OTC/pot study	1993/1994	Cambisol	380/700	110-700 kg ha^{-1} N	10
L.p./T.r. (mini swards)	Parcour/Karina 75:25	OTC/pot study	1992 +1993	Cambisol	365/700	Species mixture	11
Lolium multi-florum	Lippstädter Futtertrio	FACE/field sown	2000	Cambisol	370/550	195/100 kg ha^{-1} N	12

1. Manderscheid et al. (1995); 2. Manderscheid and Weigel (1997); 3., 4. Unpublished; 5. Manderscheid et al. (2000); 6. Unpublished; 7. Manderscheid et al. (2001); 8. Manderscheid et al. (1997b); 9. Schenk et al. (1995); 10. Manderscheid et al. (1997a); 11. Schenk et al. (1997); 12. Aulrich et al. (2001)

(FACE experiments) seasonal mean CO_2 concentration values. Details describing materials and methods of the different experiments can be found in the papers listed in Table 1.

RESULTS AND DISCUSSION

Response of Foliage N Content of Pasture Species

Cultivation of clover in controlled environments (CSTR) at different plant densities did not result in a significant decrease in shoot N concentration due to the CO_2 enrichment (Table 2). In open-top chambers experiments over two growing seasons a clear decrease in shoot N concentration was observed only at the beginning but not at the end of the growing season. This may be attributed to the fact that N supply of the plants via N_2 fixation was not yet fully developed at the begin of the growing season. Averaged over all experiments and additional modifying factors like intraspecific competition, growing season and harvest date, tissue N concentration of clover monocultures decreased by ca. −5%.

A decline in percent (%) N with increasing foliage biomass similar to the decline found by Greenwood et al. (1990) for well-fertilized C_3-plants was also observed in the present study with clover (Figure 1). However, this decrease in percent (%) N was smaller under CO_2 enrichment than under ambient CO_2 and resulted in a greater tissue N concentration of the high CO_2 plants at a similar plant dry weight. Based on a linear regression analysis between percent (%) N and plant dry weight, the N concentration of the high CO_2 plants increased by about 4% and 26% at a plant dry weight of 1 and 4 t/ha, respectively. Such an effect might be explained by a CO_2-induced increase in leaf area ratio which was observed in our (data not shown) as well as in other studies (Ryle et al., 1992). The effect might also contribute to increase the competitive ability of clover in mixed grass/clover swards.

A large growth enhancing effect of ca. + 27% ± 49% of the high CO_2 treatments was found for clover. This is in line with other studies where a variation in the response of clover to elevated CO_2 concentrations has been found to range between 9% and 75% (e.g., Nijs et al., 1988; Nösberger et al., 2000). These results again support the assumption that legumes will profit more from the rise of the atmospheric CO_2 concentration than non-legume plant species (see below) because of the large sink-strength of the N_2 fixing species (Lüscher et al., 2001). White clover is the most important pasture legume in temperate zones and contributes significantly to the economy of pastures by fixation of atmospheric N.

TABLE 2. Summary of effects of elevated CO_2 concentrations on shoot N concentration (%), yield (g pot^{-1}; FACE experiment g m^{-2}) and N yield (mg pot^{-1}; FACE experiment g m^{-2}) of clover (*Trifolium repens* L. "Karina") and grass (*Lolium perenne* L. "Parcour"; FACE experiment = *L. multiflorum*, 2 harvests). Plants were exposed to ambient (AC) and elevated (EC) CO_2 concentrations under different growth conditions as described in Table 1. R = ratio EC/AC. N treatments = kg ha^{-1} N; Plant density treatment = plants pot^{-1}. OTC = open top chamber. P = pot. * Is indicating a significant effect of the elevated CO_2 treatment. N yield was calculated from mean values without statistical analysis.

		% N			Yield			N yield		
		AC	EC	R	AC	EC	R	AC	EC	R
					Clover					
Effects of plant density	4	3.94	3.60	0.91	4.47	6.58	1.47*	176	236	1.36
	36	3.60	3.63	1.00	10.4	13.1	1.27*	374	475	1.28
Effects of season and harvest	1/93	3.28	3.06	0.93*	14.8	21.1	1.43*	485	646	1.33
	9/93	4.64	4.37	0.95*	17.1	25.4	1.49*	793	1110	1.40
	1/94	4.19	3.75	0.89*	7.29	10.6	1.45*	305	398	1.30
	6/94	4.86	4.80	0.98	5.46	7.99	1.46*	265	384	1.45
Mean of all clover studies				***0.95***			***1.42***			***1.35***
					Grass					
Effects of plant density	4	3.26	2.77	0.84*	4.20	4.89	1.16	136	135	0.99
	16	2.01	1.58	0.78*	6.77	8.28	1.22*	136	130	0.96
	36	1.63	1.42	0.87*	8.14	9.06	1.11*	132	128	0.97
Effects of N supply	110 OTC; pot	2.29	2.24	0.98	2.62	2.46	0.94	60	55	0.93
	310	4.15	3.46	0.83*	3.01	3.86	1.28*	124	133	1.07
	700	4.56	4.41	0.97	2.58	3.33	1.29*	118	147	1.25
	195[a] FACE	3.41	2.75	0.80*	373	460	1.23	12.7	12.6	0.94
	100[a]	2.57	2.27	0.88*	350	379	1.08	8.99	8.60	0.95
	195[b]	2.65	2.20	0.83*	484	543	1.12	12.9	11.9	0.92
	100[b]	1.97	1.63	0.83*	484	531	1.10	9.53	8.61	0.90
Mean of all grass studie				***0.87***			***1.16***			***1.03***
				Grass/clover mixture						
1992	grass	1.76	1.59	0.90*	57.8	57.3	0.99	1018	912	0.89
	clover	3.74	3.70	0.98	59.5	83.4	1.40*	2227	3107	1.39
	75:25 mixture	2.75	2.80	1.01	118	143	1.21*	3245	4019	1.23
1993	grass	1.82	1.59	0.87*	46.5	49.2	1.05	848	783	0.92
	clover	3.70	3.60	0.97	45.0	64.2	1.42*	1716	2314	1.34
	75:25 mixture	2.75	2.71	0.98	93.2	114	1.22*	2564	3098	1.20

[a] Harvested 48 days after emergence, [b] Harvested 71 days after emergence

FIGURE 1. Relationship between above ground biomass and N concentration of white clover grown at ambient (open circles) and elevated CO_2 concentrations (675 ppm; filled circles). Data were taken from several harvests during two consecutive growing seasons. Linear regression analysis yielded for ambient CO_2 (solid line): $y = 5.5-0.691*x$, $r^2 = 0.73$; for high CO_2 (dotted line): $y = 5.53-0.522*x$, $r^2 = 0.81$. The dashed line shows the general relationship for C_3 plants under ambient CO_2 ($y = 5.7*x^{-0.5}$) as described by Greenwood et al. (1990). Adapted from Manderscheid et al. (1997b).

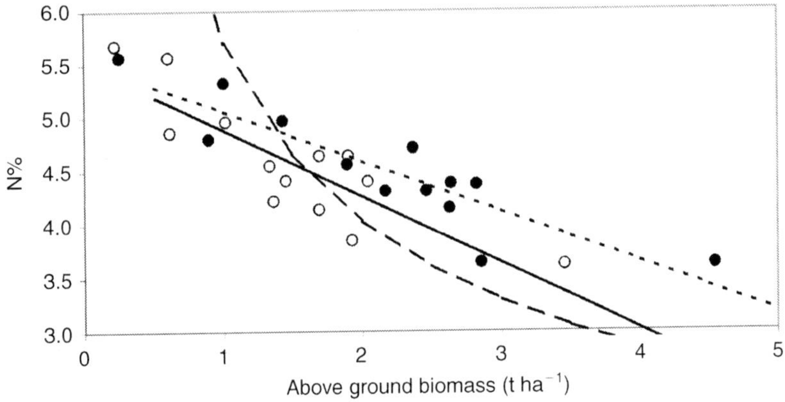

As a consequence of the small effect of CO_2 on leaf N concentration and the strong growth enhancement, total N yield of the clover swards increased by more than 30% which has also been shown in other studies (Murphy, 1986; Ryle and Powell, 1992; Lilley et al., 2001). Consequently, under future atmospheric CO_2 concentrations clover monocultures could contribute to additional sequestration of N into grassland ecosystems.

Percent tissue N concentration of the grass *Lolium perenne* tested in our studies was more negatively affected in monocultures by a CO_2 enrichment than clover (Table 2). Irrespective of the growth conditions in the individual experiments leaf N concentration decreased by ca. −13%. This again supports the common feature of declining tissue N concentrations (Conroy and Hocking, 1993; Cotrufo et al., 1998) with increasing CO_2 concentrations. While variations in N availability by altering either plant density in controlled environment studies (CSTR) or nutrient addition in OTC experiments clearly affected percent (%) shoot N, the relative CO_2 response in these experiments was not related to the N availability.

In comparison to clover the growth enhancement by elevated CO_2 concentrations of grass monocultures was considerably smaller (average ca. +16%). The stimulation of above-ground biomass production was most evident at ample soil N supply, which supports the overall assumption that the magnitude of a CO_2 induced growth stimulation of ryegrass is determined by the N supply of the plants (Schenk et al., 1996; Hebeisen et al., 1997; Daepp et al., 2001). As shown in other studies (Soussana et al., 1996; Zanetti et al., 1997), average N yields remained unaffected or were hardly affected as a result of the CO_2 enrichment (average ca. +3%). A strong positive response (+25%) on N yield only occurred under ample N supply as used in the OTC study.

The CO_2 response observed in the FACE study in the field with a mixture of three cultivars of *Lolium multiflorum*, was quite similar to the above mentioned findings with *Lolium perenne*. High CO_2 concentrations decreased foliage N by more than 10% and this effect was independent of the N supply. Due to insignificant effects of elevated CO_2 concentrations on biomass yield there was hardly any effect on N yield of the grass mixture.

It may thus be speculated that pure ryegrass stands may only benefit from the increase of the atmospheric CO_2 if there is sufficient N supply from the soil. On the other hand, especially under low N input scenarios via mineral fertilization which is one option of future sustainable agricultural management, yield losses of ryegrass may be partly compensated by increasing CO_2 levels, however, with severe problems for the nutritive quality of this grass species.

As mixed perennial ryegrass (*L. perenne*)/white clover (*T. repens*) swards form one of the most important associations in cool temperate grasslands with a high productivity (Nösberger et al., 2000), a model study over two growing seasons with a grass dominated (75%) mixture of grass and clover was carried out. As shown in Table 2, N concentrations and biomass yields of the monocultures responded in a similar manner as in the above mentioned studies with these species. However, in the mixtures N concentration of the forage total yield was not decreased under CO_2 enrichment and there was a significant increase of biomass and N yield.

Response of Cereal Grain N Content

Wheat grain protein concentration (% N $\times$ 5.7) in control plants of the present studies met the values (12%-14%) required for high bread-baking quality (Engel and Zubrinski, 1982). It has been argued that elevated CO_2 concentrations will decrease the grain N and hence protein concentrations irrespective of the N availability (Conroy, 1992), i.e., the reduction in grain quality may not easily be overcome by additional N fertilization (Fangmeier et al., 1999). In our pot studies the negative CO_2 effect on wheat grain N was greater at low N supply levels (Table 3). Increasing the amount of N fertilization to levels, which are no longer common in current agricultural practice (> 500 kg ha^{-1} N), ameliorated the CO_2-induced reduction of grain N of the wheat cultivars tested (Table 3). On the other hand, no such an effect could be observed in the

TABLE 3. Summary of effects of elevated CO_2 concentrations on grain N concentration (%), yield (g pot^{-1}; g m^{-2} for field grown wheat) and N yield (mg pot^{-1}; g m^{-2} for field grown wheat) of different wheat (*Triticum aestivum* L.) cultivars. Plants were exposed to ambient (AC) and elevated (EC) CO_2 concentrations under different growth conditions as described in Table 1. R = ratio EC/AC. HN, LN = high, low N supply. * Is indicating a statistical significant effect ($P < 0.05$) of the elevated CO_2 treatment. N yield was calculated from mean values without statistical analysis.

	% N			Yield			N yield		
	AC	EC	R	AC	EC	R	AC	EC	R
Effects of N supply									
Nandu LN	3.01	2.29	**0.74***	10.4	15.6	**1.51***	313	357	**1.14**
Nandu HN	3.23	3.02	**0.93***	14.4	19.9	**1.38***	465	601	**1.29**
Minaret LN	2.42	1.92	**0.79***	5.24	7.66	**1.46***	127	147	**1.16**
Minaret HN	2.71	2.70	**0.99**	6.91	8.80	**1.27***	187	238	**1.27**
Comparison of old and modern cultivars									
Nandu (1988)	3.23	3.02	**0.93***	14.4	19.9	**1.38***	465	601	**1.29**
Turbo (1979)	3.20	2.97	**0.92**	15.7	18.8	**1.20***	504	558	**1.11**
Janetzkis (1914)	3.42	3.30	**0.96**	15.2	22.3	**1.47***	520	736	**1.42**
Früher Heines Kolben (1890)	3.42	3.33	**0.97***	14.8	20.6	**1.39***	506	686	**1.36**
Mean old cvs.	3.42	3.31	**0.97**	15.0	21.5	**1.43**	513	711	**1.39**
Mean modern cvs.	3.22	3.00	**0.93**	15.1	19.3	**1.28**	485	580	**1.20**
Comparison of growth condition and growing season									
Minaret (pot)	2.42	1.92	**0.79***	5.24	7.66	**1.46***	127	147	**1.16**
Minaret (field)	2.13	1.60	**0.75***	826	968	**1.17**	17.6	15.5	**0.88**
Turbo (1992)	2.42	1.75	**0.72***	24.3	30.8	**1.27**	740	666	**0.90**
Turbo (1994)	3.20	2.97	**0.92**	15.7	18.8	**1.20***	504	558	**1.11**
Mean of all wheat studies			**0.87**			**1.35**			**1.17**

FACE experiment (Table 4) with a current and an old winter barley cultivar. In this experiment N supply of the control treatment was related to local agronomic practices (265 kg ha^{-1} N) and resulted in average final grain yields. However, a significant reduction of grain N (ca. −15%) under the high CO_2 treatments was observed at both N supply levels. This result is in contrast to the findings of the Arizona wheat FACE experiments where high N fertilization rates (ca. 350 kg ha^{-1}) nearly completely ameliorated negative CO_2 effects on wheat grain N concentrations (−11% reduction) under low N supply rates (Kimball et al., 2001).

TABLE 4. Summary of effects of elevated CO_2 concentrations on grain N concentration (%), yield (g pot^{-1}; g m^{-2} = field grown) and N yield (mg pot^{-1}; g m^{-2} = field grown) of different barley (*Hordeum vulgare* L.) cultivars. Plants were exposed to ambient (AC) and elevated (EC) CO_2 concentrations under different growth conditions as described in Table 1. R = ratio EC/AC. HN, LN = high, low N supply. E.M.II = Eckendorfer Mammuth II, cultivar released 1934. * Is indicating a statistical significant effect ($P < 0.05$) of the elevated CO_2 treatment, n.d. = not determined. N yield was calculated from mean values without statistical analysis.

		% N			Yield			N yield		
		AC	EC	R	AC	EC	R	AC	EC	R
Comparison of old and modern cultivars										
Krona	(1990)	2.47	2.31	**0.93**						
Baronesse	(1989)	2.28	1.94	**0.85***						
Alexis	(1986)	2.21	1.95	**0.88***						
Franken	(1930)	2.64	2.25	**0.85***			n.d.			
Intensiv	(1914)	2.49	2.21	**0.88***						
Heines Hanna	(1890)	2.95	2.39	**0.81***						
Cultivar comparison										
Arena 1992		2.73	2.25	**0.82**	11.2	21.2	**1.89**	0.45	0.85	**1.78**
Alexis 1992		2.17	1.65	**0.76***	18.0	27.4	**1.52**	0.58	0.68	**1.16**
Effects of N supply (FACE)										
Theresa	HN	1.93	1.64	**0.85***	952	1023	**1.08***	18.8	16.8	**0.91**
	LN	1.65	1.42	**0.86***	784	850	**1.09**	12.9	12.1	**0.94**
E. M. II	HN	1. 96	1.76	**0.90***	530	564	**1.06**	10.4	9.92	**0.95**
	LN	1. 94	1.60	**0.82***	359	461	**1.28**	6.96	7.37	**1.06**
Mean of all barley studies				**0.85**						

For cereals and particularly wheat there are indications that sink strength of current wheat cultivars has declined during the recent decades, e.g., as plant breeding has improved grain yield primarily by increasing harvest index and leaf area duration after ear emergence. Concomitantly, a decline of the grain N (i.e., protein) concentration has occurred. To test whether there are relationships between the relative effects of elevated CO_2 concentrations and the year of introduction of wheat and barley cultivars and how this is related to yield and grain quality, a comparison was made of the CO_2 response of old and modern wheat and barley cultivars, i.e., cultivars introduced between 1890-1990 (Tables 3 and 4).

The mean reduction of grain N content (ca. 3%-8%) due to the CO_2 enrichment was small for the wheat cultivars tested (Table 3), with a slightly higher reduction in the modern compared to the old cultivars, which resulted in a significant CO_2 × cultivar interaction (data not shown). Under comparable growth conditions as for wheat (i.e., pot study with similar N supply) grain N concentration in old and modern barley cultivars (Table 4) was more affected by elevated CO_2 concentrations (ca. 7%-19%). However, it has to be taken into account that the experiments with the wheat and the barley cultivars were performed in different years. Although there was a tendency that old cultivars were more affected, there was no significant CO_2 × cultivar interaction. In the FACE experiment with barley there were also no differences between the old and modern cultivar with respect to the reaction of grain N content to high CO_2 concentrations. As with the old wheat cultivars, the old barley cultivars had higher grain N concentrations compared to the modern ones, but there was no difference in the reaction to elevated CO_2 concentrations.

Irrespective of whether the plants were grown in pots or under common agricultural management conditions in the field, a similar magnitude of a CO_2 induced reduction of grain N concentration (≥ 20%) was observed for the wheat cultivar 'Minaret'. 'Minaret' was used as a common test plant in the European, multiple-site experiment ESPACE-Wheat (Bender et al., 1999; Fangmeier et al., 1999) and has not been in use in Germany as a common cultivar. As demonstrated for the wheat cultivar 'Turbo' significant differences in the magnitude of the reaction of grain N to elevated CO_2 between two growing seasons could be observed although the plants were grown under near identical conditions.

Wheat grain yield was significantly (up to +51%) stimulated by elevated CO_2 under the different N supply levels. Consequently N yield, i.e., the unit area production of protein of wheat was also enhanced. A similar large CO_2 response was observed for barley cultivars grown in OTC studies with pots. On the other hand, in the FACE experiment with the barley cultivars only a small increase of grain yield (< +10%) was observed. Consequently, N yields of barley remained unchanged by the CO_2 treatment under free air enrichment conditions or were even reduced (−9%) for the modern cultivars under high N supply. There seemed to be a tendency that N yield of the old barley cultivar was less negatively affected by the FACE treatment than that of the modern one. At elevated CO_2 concentration the mean increase in wheat grain yield was higher (ca. 43%) for the old (1890-1914) than for the modern cultivars (ca. 28%).

Averaged over all experiments it can be concluded that in almost all cases season-long CO_2 fertilization resulted in a considerable decrease of grain N concentration of wheat and barley (Tables 3 and 4), albeit the magnitude of this reduction differed due to differences, e.g., in nutrient supply and between cultivars investigated. The mean reduction of grain N concentration as a result of the CO_2 enrichment amounted −13% for the wheat cultivars and −15% for the barley cultivars investigated in the present studies. For wheat, these changes in grain N content are in a similar order of magnitude as those found in the studies of Conroy et al. (1994) (−1% to −14% depending on CO_2 and N level), Blumenthal et al. (1996) (−14%), Monje and Bugbee (1998) (−9%) and Hakala (1998) (−7%) and Fangmeier et al. (1999) (−15%). However, as outlined in the introduction, there is a considerable number of high CO_2 experiments that failed to detect a reduction of wheat grain N. There are no comparable results for barley.

In the pot studies, the reduction in grain N almost always was accompanied by a considerable increase of grain yield (+34% for wheat, > +50% for barley), while the yield enhancing effect was much smaller when plants were grown in the field soil (OTC, FACE). However, when the ratio % N of elevated/ambient CO_2 of all wheat results was plotted against the corresponding ratio for wheat yields, no consistent relationship was found (data not shown). Hence the results do not clearly support the overall assumption that increased yields brought about by elevated CO_2 concentrations might reduce grain quality (Rogers et al., 1998; Pleijel et al., 1999).

CONCLUSION

The present compilation of results of a number of CO_2 enrichment studies conducted over several years with different cultivars of pasture and cereals species under different growth conditions indicates that future atmospheric CO_2 concentrations may affect the quality of fooder and food crops. Except for clover, in all experiments a CO_2-induced decline of forage and grain N concentration, which is one of the key quality parameters for these plants, could be observed, albeit the magnitude of this reduction differed between species, cultivars, management conditions and CO_2 exposure conditions. No unambiguous evidence was obtained whether N fertilization can contribute to meet the quality requirements for cereals and grass monocultures with respect to tissue N concentrations in a future high-CO_2 world. However, as shown in the FACE experiments under real agricultural management conditions, current application rates of N fertilizers are inadequate to achieve this quality standard. Due to a lack of relevant field studies, it remains open whether this is a general phenomenon for other cereal species and whether the production of N and hence protein per unit area will increase under future CO_2 concentrations.

REFERENCES

Aulrich, K., R. Manderscheid und H.J. Weigel. (2001). Einfluss erhöhter CO_2-Konzentrationen auf Nährstoffgehalte von Wintergerste und Weidelgras. VDLUFA-Kongress, Berlin, in press.

Bazzaz, F.A. and K.D.M. McConnaughay. (1992). Plant-plant interactions in elevated CO_2 environment. Australian Journal of Botany 40: 547-563.

Bender, J., U. Hertstein and C.R. Black. (1999). Growth and yield of spring wheat to increasing carbon dioxide, ozone and physiological stress: a statistical analysis of "ESPACE-wheat" results. European Journal of Agronomy 10: 185-195.

Blumenthal, C., H.M. Rawson, E. McKenzie, P.W. Gras, E.W.R. Barlow and C.W. Wrigley. (1996). Changes in wheat grain quality due to doubling the level of atmospheric CO_2. Cereal Chemistry 73: 762-766.

Clark, H., P.C.D. Newton, C.C. Bell and E.M. Glasgow. (1995). The influence of elevated CO_2 and simulated seasonal changes in temperature on tissue turnover in pasture turves dominated by perennial ryegrass (*Lolium perenne*) and white clover (*Trifolium repens*). Journal of Applied Ecology 32: 128-136.

Coleman, J.S., K.D.M. McConnaughay and F.A. Bazzaz. (1993). Elevated CO_2 and plant nitrogen use–is reduced tissue nitrogen concentration size-dependent ? Oecologia 93: 195-200.

Conroy, J.P. (1992). Influence of elevated atmospheric CO_2 concentration on plant nutrition. Australian Journal of Botany 40: 445-456.

Conroy, J.P. and P. Hocking. (1993). Nitrogen nutrition of C_3 plants at elevated atmospheric CO_2 concentrations. Physiologia Plantarum 89: 570-576.

Conroy, J.P., S. Seneweera, A.S. Basra, G. Rogers and B. Nissen-Wooller. (1994). Influence of rising atmospheric CO_2 concentrations and temperature on growth, yield and grain quality of cereal crops. Australian Journal of Plant Physiology 21: 741-758.

Cotrufo, M.F., P. Ineson and A. Scott. (1998). Elevated CO_2 reduces the nitrogen concentration of plant tissue. Global Change Biology 4: 43-54.

Daepp, M., J. Nösberger and A. Lüscher. (2001). Nitrogen fertilization and developmental stage alter the response of *Lolium perenne* to elevated CO_2. New Phytologist 150: 347-358.

Dijkstra, S., A.H.C.M. Schapendonk, K.O. Groenwold, M. Jansen and S.C. van de Geijn. (1999). Seasonal changes in the response of winter wheat to elevated atmospheric CO_2 concentrations grown in open-top chambers and field tracking chambers. Global Change Biology 5: 563-576.

Engel, R.E., and J.C. Zubrinski. (1982). Nitrogen concentrations in spring wheat at several growth stages. Communications in Soil Science and Plant Analysis 13: 531-544.

Evans, L.T., ed. (1993). Crop evolution, adaptation and yield., Cambridge, UK: Cambridge University Press.

Fangmeier, A., L. DeTemmermann, L. Mortensen, K. Kemp, J. Burke, R. Mitchell, M. VanOijen and H.J. Weigel. (1999). Effects on nutrients and on grain quality in spring wheat crops grown under elevated CO_2 concentrations and stress conditions in the European, multiple-site experiment "ESPACE-wheat." Journal of Agronomy 10: 215-229.

Greenwood, D.J., G. Lemaire, G. Gosse, P. Cruz, A. Draycott and J.J. Neeteson. (1990). Decline in percentage N of C_3 and C_4 crops with increasing plant mass. Annales of Botany 66: 425-436.

Hakala, K. (1988). Growth and yield potential of spring wheat in a simulated changed climate with increased CO_2 and higher temperature. European Journal of Agronomy 9: 41-52.

Havelka, U.D., V.A. Wittenbach and M.G. Boyle. (1984). CO_2 enrichment effects on wheat yield and physiology. Crop Science 24: 1163-1168.
Hebeisen, T.H., A. Lüschers, S. Zanetti, B.U. Fischer, U.A. Hartwig, M. Frehner, G.R. Hendrey, H. Blum and J. Nösberger. (1997). Growth response of *Trifolium repens* L. and *Lolium perenne* L. as monocultures and bi-species mixture to free air CO_2 enrichment and management. Global Change Biology 3: 149-160.
Heck, W.W., R.B. Philbeck and J.A. Dunning. (1978). A continuous stirred tank reactor (CSTR) system for exposing plants to gaseous air contaminants. Principles, construction and operation. New Orleans-Agricultural Research Service: US Department of Agriculture.
Hendrey, G.R. (ed.) (1993). Free-Air Carbon Dioxide Enrichment for Plant Research in the Field. Boca Raton, FL, USA: C.K. Smoley.
Houghton, J.T., Y. Ding, D.J. Griggs, M. Noguer, P.J. van der Linden, X. Dai, K. Maskell and C.A. Johnson. (2001). Climate Change 2001. The Scientific Basis. Cambridge, UK: Cambridge University Press.
Idso, S.B. and K.E. Idso. (2001). Effects of atmospheric CO_2 enrichment on plant constituents related to animal and human health. Environmental and Experimental Botany 45: 179-199.
Kimball, B.A., C.E. Morris, P.J. Pinter, G.W. Wall, D.J. Hunsacker, F.J. Adamsen, R.L. LaMorte, S.W. Leavitt, T.L. Thompson, A.D. Matthias and T.J. Brooks. (2001). Elevated CO_2, drought and soil nitrogen effects on wheat grain quality. New Phytologist 150: 295-303.
Lilley, J.M., T.P. Bolger, M.B. Peoples and R.M. Gifford. (2001). Nutritive value and the nitrogen dynamics of *Trifolium subterraneum* and *Phalaris aquatica* under warmer, high CO_2 conditions. New Phytologist 150: 385-395.
Lüscher, A., T. Hebeisen, S. Zanetti, U.A. Hartwig, H. Blum, G.R. Hendrey and J. Nösberger. (1996). Difference between legumes and nonlegumes of permanent grassland in their response to free-air carbon dioxide enrichment. In Carbon Dioxide, Populations, and Communities, eds. C. Körner and F.A. Bazzaz, New York: Academic Press, Inc., pp. 287-300.
Lüscher, A., U.A. Hartwig, D. Suter and J. Nösberger. (2001). Direct evidence that symbiontic N_2 fixation is an important trait for a strong response of the plant to elevated atmospheric CO_2. Global Change Biology 6: 655-662.
Manderscheid, R., J. Bender, H.J. Jäger and H.J. Weigel. (1995). Effects of season long CO_2 enrichment on cereals: II. Nutrient concentration and grain quality. Agriculture, Ecosystems & Environment 54: 175-178.
Manderscheid, R. and H.J. Weigel. (1997). Photosynthetic and growth responses of old and modern spring wheat cultivars to atmospheric CO_2 enrichment. Agriculture, Ecosystems & Environment 64: 65-73.
Manderscheid, R., U. Schenk, S. Burkart, S. Obenauf und H.J. Weigel. (1997a). Photosyntheserate und pflanzeninterne Biomasseverlagerung bei Weidelgras (Lolium perenne) in Abhängigkeit von der Stickstoff-und CO_2-Versorgung. Landschaftsentwicklung & Umweltforschung 107: 23-31.
Manderscheid, R., J. Bender, U. Schenk and H.J. Weigel. (1997b). Response of biomass and nitrogen yield of white clover to radiation and atmospheric CO_2 concentration. Environmental and Experimental Botany 38: 131-143.

Manderscheid, R., S. Burkart, R.A.C. Mitchell and M. Schütz. (2000). Experiments in controlled environments and in open-top chambers. IMPETUS-Project: Final-Report, IACR Rothamsted (UK), December 2000: 6-41.

Manderscheid, R., C. Frühauf, K. Aulrich und H.J. Weigel. (2001). Wechselwirkungen von CO_2-Anreicherung unter Feldbedingungen (FACE) und Stickstoffversorgung auf oberirdische Trockenmasseakkumulation, Strahlungsausnutzungseffizienz und Ertrag bei zwei Wintergerstesorten. Mitteilungen der Gesellschaft für Pflanzenbauwissenschaften 13: 127-128.

Monje, O. and B. Bugbee. (1998). Adaptation to high CO_2 concentration in an optimal environment: radiation capture, canopy quantum yield and carbon use efficieny. Plant Cell and Environment 21: 315-324.

Murphy, P.M. (1986). Effect of light and atmospheric carbon dioxide on nitrogen fixation by herbage legumes. Plant and Soil 95: 399-409.

Nijs, I., I. Impens and T. Behaeghe. (1989). Effects of different CO_2 environments on the photosynthesis-yield relationship and the carbon and water balance of white clover (*Trifolium repens* L. cv. Blanca) sward. Journal of Experimental Botany 40: 353-359.

Nösberger, J., H. Blum and J. Fuhrer. (2000). Crop ecosystem response to climatic change: productive grassland. In Climate Change and Global Crop Productivity, eds. K.R. Reddy and H.F. Hodges, Wallingford Oxon, UK: CAB International, pp. 271-291.

Overdieck, D. (1993). Elevated CO_2 and the mineral content of herbaceous and woody plants. Vegetatio 104/105: 403-411.

Owensby, C.E., R.C. Cochran and L.M. Auen. (1996). Effects of elevated carbon dioxide on forage quality for ruminants. In Carbon Dioxide, Populations, and Communities, eds. C. Körner and F.A. Bazzaz, San Diego: Academic Press, Inc., pp. 363-371.

Pleijel, H., L. Mortensen, J. Fuhrer, K. Ojanpera and H. Danielson. (1999). Grain protein accumulation in relation to grain yield of spring wheat (*Triticum aestivum* L.) grown in open-top chambers with different concentrations of ozone, carbon dioxide and water availability. Agriculture, Ecosystems & Environment 72: 265-270.

Polley, H.W., J.A. Morgan, B.C. Campbell and M.S. Smith. (2000). Crop ecosystem reponses to climatic change: rangelands. In Climate Change and Global Crop Productivity, eds. K.R. Reddy and H.F. Hodges, Wallingford Oxon, UK: CAB International, pp. 293-331.

Reddy, K.R. and H.F. Hodges. (2000). Climate Change and Global Crop Productivity. Wallingford Oxon, UK: CAB International.

Rogers, G.S., P.W. Gras, I.L. Batey, P.J. Milham, L. Payne and J.P. Conroy. (1998). The influence of atmospheric CO_2 concentration on the protein, starch and mixing properties of wheat flour. Australian Journal of Plant Physiology 25: 387-393.

Rosenzweig, C. and D. Hillel. (1998). Climate Change and the Global Harvest, New York, Oxford: Oxford University Press.

Ryle, G.J.A. and C.E. Powell. (1992). The influence of elevated CO_2 and temperature on biomass production of continuously defoliated white clover. Plant Cell and Environment 15: 593-599.

Schenk, U., R. Manderscheid, J. Hugen and H.J. Weigel. (1995). Effects of CO_2-enrichment and intraspecific competition on biomass partitioning, nitrogen content and microbial carbon in soil of perennial ryegrass and white clover. Journal of Experimental Botany 46: 987-993.

Schenk, U., H.J. Jäger and H.J. Weigel. (1996). Nitrogen supply determines responses of yield and biomass partitioning of perennial ryegrass to elevated atmospheric CO_2 concentrations. Journal of Plant Nutrition 19: 1423-1440.

Schenk, U., H.J. Jäger and H.J. Weigel. (1997). The response of perennial ryegrass/ white clover swards to elevated atmospheric CO_2 concentrations. II. Effects on yield, fodder quality and water use. Grass and Forage Science 52: 232-241.

Soussana, J.F., E. Casella and P. Loiseau. (1996). Long-term effects of CO_2 enrichment and temperature increase on a temperate grass sward. II. Plant nitrogen budgets and root fraction. Plant and Soil 182: 101-114.

Thompson, G.B. and F.I. Woodward. (1994). Some influences of CO_2 enrichment, nitrogen nutrition and competition on grain yield and quality in spring wheat and barley. Journal of Experimental Botany 45: 937-942.

Weigel, H.J., G.J. Mejer und H.J.Jäger. (1992). Auswirkungen von Klimaänderungen auf die Landwirtschaft. I. Open-top Kammern zur Untersuchung von Langzeitwirkungen erhöhter CO_2-Konzentrationen auf landwirtschaftliche Pflanzen. Angewandte Botanik 66: 135-142.

Weigel, H.J. and U. Dämmgen. (2000). The Braunschweig Carbon Project: Atmospheric Flux Monitoring and Free Air Carbon Dioxide Enrichment (FACE). Journal of Applied Botany 74: 55-60.

Zanetti, S., U.A. Hartwig, C. van Kessel, A. Lüscher, T. Hebeisen, B.U. Fischer, G.R. Hendrey, H. Blum and J. Nösberger. (1997). Does nitrogen nutrition restrict the CO_2 response of fertile grassland lacking legumes ? Oecologia 112: 17-25.

The Effect of Elevated Levels of Carbon Dioxide on Potato Crops: A Review

J. M. Finnan
A. Donnelly
M. B. Jones
J. I. Burke

SUMMARY. Experiments, which have investigated the effect of elevated concentrations of atmospheric CO_2 on the physiology, growth and yield of potato (*Solanum tuberosum* L.), are reviewed. These studies were conducted in controlled environment chambers, in glasshouses, in open top field chambers as well as using free air carbon dioxide (FACE) exposure systems. In general, photosynthesis is stimulated by elevated CO_2 initially although long-term exposure leads to acclimation. The stomata of potato leaves partially close in response to elevated CO_2 and starch granules build up in the chloroplasts. Although above- and below-ground biomass production is stimulated, accelerated senescence limits

---

J. M. Finnan is affiliated with the Environmental Protection Agency, Richview, Clonskeagh, Dublin 14, Ireland (E-mail: j.finnan@epa.ie).

A. Donnelly and M. B. Jones are affiliated with the School of Botany, Trinity College, Dublin 2.

J. I. Burke is affiliated with the Teagasc, Oak Park Research Centre, Carlow.

Address correspondence to: J. M. Finnan at the above address.

The authors gratefully acknowledge the assistance of L. Temmerman, J. Craigon, K. Vandermeiren, V. Vorne, M. Bindi and J. Wolf.

[Haworth co-indexing entry note]: "The Effect of Elevated Levels of Carbon Dioxide on Potato Crops: A Review." Finnan, J. M. et al. Co-published simultaneously in *Journal of Crop Improvement* (Food Products Press, an imprint of The Haworth Press, Inc.) Vol. 13, No. 1/2 (#25/26), 2005, pp. 91-111; and: *Ecological Responses and Adaptations of Crops to Rising Atmospheric Carbon Dioxide* (ed: Zoltán Tuba) Food Products Press, an imprint of The Haworth Press, Inc., 2005, pp. 91-111. Single or multiple copies of this article are available for a fee from The Haworth Document Delivery Service [1-800-HAWORTH, 9:00 a.m. - 5:00 p.m. (EST). E-mail address: docdelivery@haworthpress.com].

http://www.haworthpress.com/web/JCRIP

Digital Object Identifier: 10.1300/J411v13n01_06

growth towards the end of the growing season exposure to elevated CO_2 stimulates tuber yield, the magnitude of which depends on agronomic practise, cultivar choice and growing conditions. The beneficial effects of elevated CO_2 may be reduced by interaction with other components of climate change, such as drought stress. Modelling of the effects of climate change on potato yield has predicted an increase in yields in northern Europe with little change in central and southern Europe. It is suggested that further research is needed to understand the reasons for photosynthetic acclimation, field trials are also needed to understand and quantify the interaction between elevated CO_2 and drought stress.

KEYWORDS. Potato (*Solanum tuberosum* L.), elevated CO_2, photosynthesis, tuber yield

INTRODUCTION

Atmospheric concentrations of carbon dioxide (CO_2) are higher now (360 μmol mol^{-1}) than at any time in recent geological history (at least 420,000 and probably 20,000,000 years) and the pre-industrial revolution levels of 280 μmol mol^{-1} are very likely to more than double by the end of this century (IPCC 2001). Carbon dioxide affects plants indirectly through climate change, by acting as a greenhouse gas, and directly as a substrate in photosynthesis and by causing partial closure of stomata. Net photosynthesis rates for many C3 crop plants are limited by present day concentrations of CO_2 so that the future increases in CO_2 should lead to marked stimulations in plant growth. In fact, a projected doubling of present-day CO_2 could boost biomass production and marketable yields of agricultural crops by one-third or even more if other constraints to productivity are not limiting (Idso & Idso, 1994). However, plant species vary in their response to elevated CO_2 and, although still an area of contention, there is a general consensus that inherently fast-growing species are stimulated more in biomass production at elevated CO_2 than slow-growing species and plants growing at high nutrient supply respond more strongly than nutrient-stressed plants (Poorter,1993 and 1998). Also indeterminate crops such as potato are expected to show a larger response than determinate crops such as cereals (Lawlor & Mitchell, 1991).

Experiments to determine the effects of elevated CO_2 concentrations on the growth of potato have been carried out in controlled environments (Wheeler et al., 1991; Cao et al., 1994; Wheeler et al., 1994; Mackowiak & Wheeler, 1996; Wheeler & Tibbitts, 1997) and in glasshouse facilities (Goudrian & de Ruiter, 1983). Most studies have been conducted in field-based open top chambers (Schapendonk et al., 2000; Sicher & Bunce, 1999; Hacour et al., 1998, 2000; Fangmeier et al., 1998, 2000; Pleijel et al., 1998; Lawson et al., 2001a; Donnelly et al., 2001a; Finnan et al., 2002). Potato plants have also been grown in free air carbon dioxide enrichment (FACE) systems in Italy (Miglietta et al., 1998; Bindi et al., 1998, 2000) and in Germany (Fangmeier et al., 1998, 2000). Recently, there has been a concerted effort to understand the effects of elevated CO_2 on potato within Europe in a European Commission funded project known as CHIP (Changing Climate and Potential Impacts on Potato Yield and Quality). This was a research program, run over two years, in which a series of field experiments at several sites in different climatic regions of Europe, along with modelling, were used to investigate the combined and interactive effects of elevated CO_2 and elevated O_3 on potato (*Solanum tuberosum*, cv. Bintje) (Temmerman et al., 2002a).

The purpose of this review is to survey the findings of the research carried out under a wide range of experimental conditions and to summarise the current state of knowledge on the effects of elevated CO_2 on potato growth and development, as well as to identify future research needs. The approach is to consider first the effects of elevated CO_2 at the leaf level and then to review the implications of these responses for whole plant growth and development. Finally, the consequences for crop quality and yield are considered.

LEAF PROCESSES

It has been frequently demonstrated that increases in the concentration of atmospheric CO_2 lead to a short-term increase in the light saturated rate of photosynthesis in plants possessing the C_3 pathway of CO_2 fixation (Long et al., 1993). This effect results from the fact that the photosynthetic apparatus of C_3 plants is not fully saturated with CO_2 at present day concentrations (Webber et al., 1994). According to Bunce (1992) the saturation CO_2 concentration for C_3 species is approximately 650 μmol mol^{-1}. Consequently, any increase in atmospheric CO_2 up to that concentration should result in an increase in the photosynthetic rate. However, long-term exposure (continuous exposure over at least several days) to elevated CO_2 may result in acclimation causing a decline in photosynthetic performance (Long & Drake, 1992). Another common observation is a reduction in stomatal conductance in response to elevated CO_2 due to partial stomatal closure. This reduces the conductance for CO_2 diffusion into the leaf and therefore reduces the photosynthetic rate (Morison & Gifford, 1984).

Photosynthetic Rate

Short-term exposure of potato to elevated CO_2 has frequently been shown to increase leaf photosynthetic rates (Sage et al., 1989; Sicher & Bunce, 1999; Donnelly et al., 2001b; Vandermeiren et al., 2002). Leaves grown at ambient CO_2 concentrations showed an 80-100% stimulation in photosynthetic rates when exposed for a short period to elevated CO_2 concentrations (Sicher & Bunce, 1999, 2001; Donnelly et al., 2001b). However, there is considerable evidence of down-regulation (acclimation) when plants are grown continuously in elevated CO_2. Most studies have reported acclimation after an initial stimulation of photosynthesis (Ludewig et al., 1998; Schapendonk et al., 2000, Sicher & Bunce, 1999, Lawson et al., 2001b and Vandermeiren et al., 2002). Donnelly et al. (2001b) found that light saturated rates of photosynthesis for leaves grown for the entire growing season under elevated CO_2 were consistently lower than in equivalent leaves grown under ambient CO_2. Schapendonk et al. (2000) found that this acclamatory response was greater in growing seasons with high irradiance and temperature. Sicher and Bunce (2001) investigated the reversibility of photosynthetic acclimation by transferring plants grown in ambient CO_2 concentrations to an elevated CO_2 environment and *vice versa*. The inhibition of the photosynthetic rate of leaves grown at elevated CO_2 was partially or completely reversed 3 days after being transferred to ambient CO_2 conditions. However, little or no evidence of acclimation was observed three days after plants were transferred from ambient to elevated CO_2 concentrations. Sicher and Bunce (1999) suggested that the acclamatory response to elevated CO_2 resulted from a decrease in the activity of Rubisco rather than any decrease in the leaf content of this protein. In contrast, Schapendonk et al. (2000) found acclimation to be accompanied by a reduction in Rubisco and they concluded that acclimation is a complex mechanism resulting from negative feedback of sink-source balance induced by high irradiance and temperature. A model, which simulated the effect of acclimation on yield, showed that photosynthetic acclimation reduced the positive effect of elevated CO_2 on tuber yield by 50%. Ludewig et al. (1998) suggested that acclimation was caused by a reduction in the end product rate of synthesis. Sage et al. (1989) concluded that Rubisco was not involved in the acclimation of potato plants to elevated CO_2 but that an increase in Pi-regeneration capacity was responsible. It is clear from these results that there is little doubt that acclimation of photosynthesis occurs in potato plants grown (long-term) under elevated CO_2. However, the underlying mechanisms have yet to be elucidated.

A/c_i Response Curves

Analysis of the A/c_i response curves (the response of assimilation (A) to changes in intracellular CO_2 concentration (c_i) of potato leaves at elevated

CO_2 has shown inconsistent results. The maximum rate of photosynthesis (A_{max}) was either unaffected by elevated CO_2 concentrations (Donnelly et al., 2001b) or showed a faster rate of decline than leaves at ambient CO_2 during the growing season (Lawson et al., 2001b). Several studies (Sicher & Bunce, 1999 and 2001; Lawson et al., 2001b; Vandermeiren et al., 2002) reported a reduction in the maximum rate of carboxylation (Vc_{max}) in response to elevated CO_2 whereas Ludewig et al. (1998) and Donnelly et al. (2001b) found no response. Donnelly et al. (2001b) and Lawson et al. (2001b) reported no response in the maximum rate of electron transport (J_{max}) to elevated CO_2 but Vandermeiren et al. (2002) reported a reduction in this parameter. As many of the contradictory reports emanated from experiments which used the same cultivar it is possible that differences in experimental conditions contributed to the diversity in response. Respiration rates (R_d) were found to increase in the latter stages of the growing season in response to elevated CO_2 (Donnelly et al., 2001b) and Vandermeiren et al. (2002), it was concluded that this reflected an increased demand for maintenance support as a result of accelerated senescence.

Stomatal Conductance

Stomatal conductance of potato leaves has been found to decrease at elevated (up to concentrations of 700 $\mu mol\ mol^{-1}$) and super-elevated (1000-10,000 $\mu mol\ mol^{-1}$) CO_2 (Mackowiak & Wheeler, 1996; Sicher & Bunce, 1999; Donnelly et al., 2001b; Finnan et al., 2002; Lawson et al., 2001b; Vandermeiren et al., 2002). Short-term exposure (periods of exposure of a number of hours) of plants grown at ambient CO_2 to elevated CO_2 did not affect stomatal conductance (Sicher & Bunce 1999; Donnelly et al., 2001b). However, higher stomatal conductances have been recorded in potato leaves grown and measured at elevated CO_2 compared to those grown at ambient CO_2 and measured at elevated CO_2 (Sicher and Bunce 1999; Bunce, 2001; Donnelly et al., 2001b) suggesting stomatal acclimation to elevated CO_2. The magnitude of the response of stomatal acclimation to enhanced CO_2 increased in association with the decrease in photosynthetic rate and as the leaf-to-air water vapour pressure difference increased (Bunce, 2001; Donnelly et al., 2001b). Bunce (2001) concluded that two types of stomatal acclimation occurred in potato. The first resulted from very little response of stomatal conductance of ambient grown plants to short-term exposure to elevated CO_2. This effect was also found by Donnelly et al. (2001b). The second related to photosynthetic acclimation, the magnitude of stomatal acclimation being correlated with the magnitude of photosynthetic reduction resulting from growth at elevated CO_2 (Bunce, 2001). Nearly all the field studies mentioned above report the use of irrigation to maintain the soil at field capacity. Decreases in stomatal conductance on long-term exposure to elevated CO_2 could be expected to improve the water use efficiency (WUE) of potato crops. However, this important parameter has not been calculated in studies conducted to date.

Leaf Protein/Chlorophyll Content

Sage et al. (1989) found no change in leaf Rubisco content in response to elevated CO_2 although a reduction in the activation state was evident. In contrast, Ludwig et al. (1998) reported no change in the activation of Rubisco but a reduction in total protein content in leaves grown in elevated CO_2. Sicher and Bunce (1999) reported that neither leaf soluble protein or Rubisco protein content differed between ambient and elevated treatments. However, Schapendonk et al. (2000) found a consistent reduction in Rubisco content of leaves throughout the growing season in response to elevated CO_2. Contradictory results have been reported both among field studies and among controlled environment studies, it is possible that the different results may be indicative of different cultivar responses.

Leaf chlorophyll content in potato has been shown to decrease (Sicher & Bunce, 2001; Lawson et al., 2001b), increase (Bindi et al., 2002) or remain unchanged (Sicher & Bunce, 1999; Donnelly et al., 2001b; Finnan et al., 2002; Sage et al., 1989) in response to elevated CO_2. The reduction in chlorophyll content reported by Lawson et al. (2001b) was for older leaves only and occurred as the crop approached maturity, a response which suggests an acceleration of senescence. However, Sicher and Bunce (1999) concluded that there was no evidence of accelerated senescence in response to elevated CO_2 as none of the major leaf proteins (chlorophyll and Rubisco) were affected by this treatment. Bindi et al. (2002) analysed the chlorophyll content of potato leaves from seven sites across Europe for two growing seasons in the CHIP project. They divided the growing season into three phases; emergence to tuber initiation, tuber initiation to maximum leaf area and maximum leaf area to above ground biomass harvest. In the first phase, the overall effect was for a slight increase in chlorophyll content in response to elevated CO_2 whereas there was a reduction in chlorophyll content in the second and third phases. The authors suggested that the increased rate of senescence followed the pattern of rates of photosynthesis, which increased at the beginning of the season before declining. These results suggest that the pattern of potato leaf chlorophyll content in response to elevated CO_2 varied according to the development stage of the crop.

Starch/Carbohydrate Content

Many plants show an increase in leaf starch content following long-term exposure to elevated CO_2 (Farrar & Williams, 1991). Goudriaan and de Ruiter

(1983) first reported the presence of high concentrations of starch in potato leaves exposed to elevated CO_2 and postulated that the growing starch grains might cause physical damage to chloroplasts. Sicher and Bunce (1999) also found an increase in starch content, although they observed that the concentration of soluble carbohydrates (sucrose, glucose, fructose) was unaffected. Miglietta et al. (1998) reported an increase in total non-structural carbohydrate concentration in the leaves of potato plants exposed to elevated CO_2 in FACE rings. Ludewig et al. (1998) and Schapendonk et al. (2000) also found an increase in leaf starch content together with a concurrent increase in sucrose content in response to elevated CO_2. Schapendonk et al. (2000) observed that the accumulation of leaf carbohydrate coincided with photosynthetic acclimation.

Section Summary

The assimilation rate of potato leaves is initially stimulated by elevated CO_2 but this increase is short lived and is followed by a decline, a process known as acclimation. Elevated CO_2 induces a decrease in stomatal conductance in potato leaves and an increase in starch accumulation in chloroplasts. Decreases in stomatal conductance could be expected to improve the water use efficiency of potato crops.

CHANGES IN PLANT GROWTH AND DEVELOPMENT

There are numerous reports in the literature of increased growth and yield in C3 crops exposed to elevated CO_2 (Cure & Acock, 1986; Poorter, 1993; Bender et al., 1999). Increased leaf area index (LAI) and accelerated development have also been reported. Miglietta et al. (1998) found that elevated CO_2 advanced the flowering date and accelerated canopy sensecence in potato crops. Bindi et al. (1998, 2000) in a subsequent experiment conducted at the same site, also found that canopy senesence was accelerated at elevated CO_2 although crop phenology was not affected. Canopy senescence of a late sown cultivar was found to be accelerated by elevated CO_2 by Schapendonk et al. (2000) although senescence was unaffected in an earlier sown cultivar. Crop phenology was found to be advanced by elevated CO_2 in two consecutive years of experiments carried out in Belgium as part of the CHIP project (Hacour et al., 1998 and 2000).

Most studies have reported either no effect on plant height (Miglietta et al., 1998; Lawson et al., 2001a; Finnan et al., 2002) or a reduction in the height of potato plants exposed to elevated CO_2 (Donnelly et al., 2001a; Hacour et al., 1998; Fangmeier et al., 1998, 2000; Pleijel et al., 1998). One experiment reported an increase in plant height during one year (Bindi et al., 1998) and a decrease in plant height after exposure to elevated CO_2 the following year (Bindi et al., 2000). Hacour et al. (2002) found that both levels of elevated CO_2 (550 μmol mol^{-1} and 680 μmol mol^{-1}) used in the CHIP experiment reduced plant height and attributed this reduction to a premature ending of active plant growth.

Cao et al. (1994) reported increases in leaf area per plant when the CO_2 concentration was increased from 500 μmol mol^{-1} to 1000 μmol mol^{-1}. Increases in LAI were reported at two sites in the CHIP experiment (Hacour et al., 1998, 2000; Lawson et al., 2001a; Donnelly et al., 2001a). These increases in LAI were recorded during the earlier part of the growing season up until the time of maximum leaf area. Most studies have not recorded any change in LAI during the early part of the growing season (Schapendonk et al., 2000; Miglietta et al., 1998, Finnan et al., 2002), an analysis of the combined results from the CHIP experiment showed no change in LAI at full canopy development (Hacour et al., 2002). Miglietta et al. (1998) reported a faster decline in LAI towards the end of the growing season under elevated CO_2. This effect was also noted in subsequent Italian experiments (Bindi et al., 1998 and 2000) as well as in experiments conducted in Belgium (Hacour et al., 1998 and 2000). LAI was found to be lower under elevated CO_2 during crop senescence in an analysis of the overall results of the CHIP project (Hacour et al., 2002) although the effect was not statistically significant. There is one report of an increase in the number of green leaves under elevated CO_2; Lawson et al. (2001a) reported a statistically significant increase in the number of green leaves at 550 μmol mol^{-1} but not at 680 μmol mol^{-1}. In contrast, Miglietta et al. (1998) reported no effect of elevated CO_2 on green leaf number while an analysis of results of the CHIP experiment showed that elevated CO_2 had no effect on green leaf number during the early part of the growing season, but significantly decreased green leaf number at full canopy and during crop senescence (Hacour et al., 2002). Many studies have shown a decrease in specific leaf area (SLA) in potato leaves exposed to elevated CO_2 (Miglietta et al., 1998; Schapendonk et al., 2000; Donnelly et al., 2001a; Bindi et al., 1998; 2000; Hacour et al., 1998 and 2000). There is one report of an increase in SLA under elevated CO_2 (Ojanpera & Vorne, 2000) whilst other studies have reported no change (Fangmeier et al., 1998; 2000). Decreases in SLA are most likely a consequence of accumulation of starch in chloroplasts (see above).

Fangmeier et al. (2002) reported on the effect of elevated CO_2 on a wide range of both macro- and micro-nutrients in above and below ground tissue of potato in the CHIP project. They found a decrease in both phosphorous and potassium concentrations in above-ground biomass in response to elevated CO_2 at an intermediate (maximum leaf area) harvest at five sites across Europe. A reduction in nitrogen, potassium and calcium concentration in tubers was also reported at this harvest. Nitrogen, manganese and iron concentrations in above-ground biomass at final harvest were all significantly reduced in response to elevated CO_2 whereas tuber nitrogen, potassium and magnesium concentrations were also reduced (Fangmeier et al., 2002).

Growth stimulations in potato plants exposed to elevated CO_2 were first reported by Collins (1976). Lawson et al. (2001a) reported substantial stimulation in above-ground biomass and tuber dry weight at harvests at tuber initiation and at 500 degree days. Significant effects on tuber yield and above-ground biomass, however, were no longer apparent at final harvest. Donnelly et al. (2001a) working at the same site in the following year also reported significant stimulation of above-ground biomass and tuber yield at the 500 degree day harvest but these effects were still apparent at final harvest. An overall assessment of biomass effects in the CHIP experiment showed significant stimulation of above-ground biomass, tuber dry weight and tuber number at an intermediate harvest at maximum leaf area. However, above-ground biomass was found to be reduced at final harvest. It was also noted that the CO_2 effect on tuber yield swas less at final harvest than at intermediate harvest suggesting that growth was limited towards the end of the growing season (Craigon et al., 2002).

Section Summary

Elevated levels of CO_2 can be expected to stimulate both above- and below-ground biomass of potato crops, particularly during the early part of the growing season. The period of active plant growth, however, appears to end prematurely and senescence begins earlier, limiting growth rates towards the end of the growing season.

EFFECTS ON CROP YIELD

Studies carried out in controlled environment chambers as well as in the field have shown consistently that elevated levels of carbon dioxide increase tuber yield.

A wide range of positive yield responses (Table 1) have been reported from all methods of plant exposure to elevated CO_2 (controlled environment chamber, open top chamber, free air carbon dioxide enrichment). Tuber and root crops have previously been identified as being particularly responsive to elevated concentrations of CO_2 (Kimball, 1983; Miglietta et al., 2000). It has been suggested that crops such as potato with large below-ground sinks for carbon (Korner et al., 1995; Farrer, 1996) and with apoplastic mechanisms of phloem loading (Komor et al., 1996) are likely to be the best candidates for a large response to rising atmospheric CO_2. Only two studies have reported a

TABLE 1. Responses of final harvest tuber yield, tuber number and above-ground biomass to elevated CO_2. Changes in yield, tuber number and above-ground biomass are expressed as a proportion of these variables measured at ambient CO_2. Responses are highlighted in bold where statistically significant differences were reported. The method of exposure is given; CES (controlled environment chamber), OTC (open top chamber), and FACE (free air CO_2 enrichment).

Experiment	Treatment μmol mol-1	Yield difference from 360 μmol mol^{-1}	Tuber no. difference from 360 μmol mol^{-1}	Above-ground biomass difference from 360 μmol mol^{-1}
Netherlands 1983 Goudriaan & de Ruiter, 1983	700 Glasshouse	↓ −14%		**↓ −23.5%**
United States 1991 Wheeler et al., 1991	1000 CES 1000 CES 1000 CES	**↑ 29%** **↑ 42%** **↑ 43%**		No Change No Change No Change
Italy 1995 Miglietta et al., 1998	460 FACE 560 FACE 660 FACE	**↑ 15%** **↑ 30%** **↑ 45%**	**↑ 13.7%** **↑ 27.4%** **↑ 41.1%**	
Netherlands 1995 Schapendonk et al., 2000	Early 700 OTC Late 700 OTC	**↑ 24%** **↑ 31%**		↑ 2.3% ↓ −16.1%
Netherlands 1996 Schapendonk et al., 2000	Early 700 OTC Late 700 OTC	No Change **↑ 49%**		↓ −13.5% ↑ 9.8%
United States 1997-98 Sicher & Bunce, 1999	530 OTC 700 OTC	↑ 9% **↑ 40%**	↑ 14% **↑ 44%**	
United States 1996 Wheeler & Tibbitts 1997	1000 CES	No Change		**↑ > 100%**
Belgium 1998 Hacour et al., 1998	680 OTC	↑ 2.9%	↑ 3.2%	**↓ −42.5%**
Belgium 1999 Hacour et al., 2000	680 OTC	**↑ 8.9%**	**↑ 14.4%**	↓ −14.7%
England 1998 Lawson et al., 2001	550 OTC 680 OTC	↑ 10.7% ↑ 8.8%	**↑ 17.5%** **↑ 24.1%**	↑ 13.3% ↓ −2%
England 1999 Donnelly et al., 2001	550 OTC 680 OTC	**↑ 26%** **↑ 41%**	↑ 13.5% ↑ 14.6%	**↑ 22.8%** **↑ 28.3%**
Italy 1998 Bindi et al., 1998	550 FACE	**↑ 33.8%**	**↑ 23%**	↑ < 10%
Italy 1999 Bindi et al., 2000	550 FACE	**↑ 54.1%**	**↑ 35%**	**↓ −25.8%**
Ireland 1998 Finnan et al., 2002	680 OTC	**↑ 39.7%**	↑ 1.5%	**↑ 26%**
Ireland 1999 Finnan et al., 2002	680 OTC	**↑ 53.8%**	**↓ −22.4%**	↓ −3.7%
Germany 1998 Fangmeir et al., 1998	550 FACE 550 OTC 680 OTC	↑ 2.8% ↑ 14.0% ↑ 9.7%	↑ 8.6% ↑ 7.8% ↓ −12.3%	↓ −5.5% **↓ −4.5%** **↓ −12.2%**
Germany 1999 Fangmeir et al., 2000	550 FACE 550 OTC 680 OTC	**↑ 8.7%** ↑ 12.6% ↑ 12.1%	↑ 4.4% ↑ 3.1% ↑ 5.5%	↑ 8.2% ↑ 13.9% ↓ −14.6%
Sweden 1998 Pleijel et al., 1998	680 OTC	↓ −7.3%	**↑ 18.2%**	**↓ statistically significant decrease**
Finland 1999 Ojanpera & Vorne, 2000	680 OTC	↑ 24.2%	↑ 16.2%	↓ −10.5%

decrease in tuber yield after exposure to elevated levels of CO_2. Goudriaan and de Ruiter (1983) reported a small negative response of potato to CO_2 enrichment. Potato leaves in this study were reported to have developed a brownish fringe and to contain a very high content of starch and it was suggested that the growing starch grains may have physically damaged the chloroplasts. Pleijel et al. (1998) also reported a small negative effect on tuber yield at elevated CO_2. Plants grown in elevated CO_2 in this experiment produced a significantly larger number of tubers which were, on average, smaller than tubers grown under ambient CO_2, the net effect being a small negative effect on tuber yield. Miglietta et al. (1998) reported that elevated CO_2 substantially increased tuber yield and that the response was essentially linear corresponding to a 10% stimulation in tuber yield for every 100 μmol mol^{-1} increase in ambient CO_2 up to 680 μmol mol^{-1}. Sicher and Bunce (1999) also reported that tuber yield stimulation increased with increasing CO_2 level. Craigon et al. (2002) analysed the yield responses of potato to increased CO_2 recorded in the CHIP experiment and found that CO_2 elevation to 680 μmol mol^{-1} increased the dry weight of marketable sized tubers by approximately 17%. They also concluded that the CO_2 response was at, or approaching, saturation at 550 μmol mol^{-1}. A number of controlled environment studies have investigated the growth and yield of potatoes at a range of elevated CO_2 concentrations up to 10000 μmol mol^{-1}. These studies also reported the existence of a plateau in the response of tuber yield to elevated CO_2 (Wheeler et al., 1994; Cao et al., 1994; Mackowiak & Wheeler, 1996). Tuber yield reached this plateau at 1000 μmol mol^{-1} in these studies.

An overall reduction in above-ground biomass was reported at final harvest in the CHIP experiment, even though tuber yield was stimulated (Craigon et al., 2002). Consequently, increases in harvest index accompany stimulation of tuber yield in potato crops exposed to elevated CO_2. Cao et al. (1994) reported an increase in harvest index in potato plants exposed to elevated CO_2 and Schapendonk et al. (2000) found that assimilates were preferentially exported to the tubers in potato crops exposed to elevated CO_2.

The magnitude of the yield response to elevated CO_2 would not appear to vary with the method of exposure as substantial stimulations of tuber dry matter (>40%) have been reported from controlled environment studies (Wheeler et al., 1991), from open top chamber studies (Schapendonk et al., 2000; Donnelly et al., 2001a; Finnan et al., 2002) as well as from FACE studies (Miglietta et al., 1998; Bindi et al., 1998). However, variation in tuber yield stimulation has been reported both within individual experiments as well as between different experiments using the same method of exposure to elevated CO_2. Contrasting results were reported from two FACE sites participating in the CHIP experiment,with Italian experiments reporting much greater yield responses compared to the German experiment (Fangmeir et al., 1998 and 2000; Bindi et al.,

1998 and 2000). The range of yield response to elevated CO_2 recorded in field studies may not be surprising when it is considered that growing conditions change from year to year and that different potato varieties and agronomic practises were used in separate studies. Schapendonk et al. (2000) exposed an early and a late potato cultivar to elevated CO_2 in open top chambers over a two year period and reported that tuber yield stimulation was greatest for the late cultivar but the yield response was also found to be greater in the temperate year of the study compared to the warm year. Contrasting results were reported from a two year experiment conducted at a site in England as elevated CO_2 had no significant effect on tuber yield in 1998 (Lawson et al., 2001a) but a substantial stimulation in tuber yield was reported in 1999 (Donnelly et al., 2001a). A higher average CO_2 effect was observed in the second year of the CHIP experiment (Craigon et al., 2002), this enhanced CO_2 effect was thought to be at least partially attributable to the reduced planting density used in the second season. Craigon et al. (2002), in a review of the CHIP experiment, reported that the magnitude of the CO_2 effect varied substantially between locations and that there was no obvious relationship between ambient yield and CO_2 effect within any one growing season.

Most studies have reported that an increase in tuber number accompanies a stimulation in tuber yield (Table 1). It would appear that, in general, the tuberization process is enhanced by exposure to elevated CO_2, the net effect being an increase in the number of sinks. Craigon et al. (2002), in an analysis of the final harvest results of the CHIP project, concluded that although tuber number was stimulated by elevated CO_2, the increase was not sufficient to explain the observed increase in tuber yield as tubers also grew larger under elevated CO_2. A number of studies have reported increases in the size of individual tubers (Donnelly et al., 2001a; Finnan et al., 2002). Two studies have reported a decrease in tuber number accompanying an increase in tuber yield (Finnan et al., 2002; Fangmeier et al., 2000). It is unclear in these cases whether exposure to elevate CO_2 altered the hierarchy of tubers causing some to be resorbed, or whether the process of tuberisation was inhibited.

There is some debate as to the physiological reasons underlying tuber yield stimulation by elevated CO_2. Schapendonk et al. (2000) found no change in LAI under elevated CO_2 and concluded that photosynthesis on a leaf area basis is the most important attribute to explain the observed yield increases. Similarly, both Miglietta et al. (1998) and Finnan et al. (2002) reported substantial stimulation in tuber yield without any increase in LAI. An analysis of the results of the CHIP project showed overall tuber yield stimulation (Craigon et al., 2002) with no effect on LAI during the active growing period or at full canopy. In contrast Donnelly et al. (2001a,b) attributed tuber yield stimulation to increases in LAI as elevated CO_2 did not enhance photosynthetic rates. It would thus appear that increases in tuber yield in potato plants can result from either increases in photosynthetic rate per unit leaf area or increases in leaf area and possibly from a combination of both.

Section Summary

Tuber yield is stimulated in potato crops exposed to elevated CO_2. The magnitude of the yield response is determined by interaction with a number of other factors including cultivar, agronomy and growing conditions. Tuber yield stimulation is, in most cases, accompanied by an increase in the number of tubers.

EFFECTS ON TUBER QUALITY

As concluded in the above section, it is now clearly established that elevated concentrations of atmospheric CO_2 will increase yield of potato (Wheeler et al., 1994; Finnan et al., 2002; Donnelly et al., 2001a; Craigon et al., 2002). Potato makes a significant contribution to human nutrition in many parts of the world. For example, 5% of the daily energy requirements of Europeans in conjunction with substantial amounts of carbohydrate, protein and vitamin C comes from potato (Storey & Davies, 1992). Any changes in tuber quality will thus have an effect on human nutrition and these changes will also affect those industrial processes which use potatoes.

In an open-top chamber experiment Donnelly et al., (2001c) reported that elevated atmospheric CO_2 concentrations (680 $\mu mol\ mol^{-1}$) resulted in an increase in tuber starch content as well as in the viscosity of the starch paste. The authors also reported an increase in vitamin C concentrations and a reduction in the nitrogen content of the tubers in response to elevated CO_2. Similarly, Vorne et al. (2002) reported increases in tuber dry matter and starch content and decreases in citric acid and nitrogen concentrations in tubers exposed to elevated CO_2. Increases in vitamin C and in starch content of tubers would enhance the nutritional value of potatoes, increases in starch content would also benefit industries in which potato starch is used as a thickener. The manufacture of processed potato food products such as fries and crisps would benefit from high dry matter content, high dry matter content prevents excessive fat absorption in the manufacture of these products (Storey & Davies, 1992). In contrast to the beneficial effects of elevated CO_2 on tuber starch content, decreases in tuber nitrogen content would not be beneficial as protein uptake would be decreased for consumers of potato. Similarly, decreases in organic acid content would cause discoloration problems with potato products as they prevent the formation of the pigment responsible for after cooking blackening.

Glycoalkaloids are naturally occurring toxic compounds present in tubers which act as a defence mechanism for the plant. The principal glycoalkaloids are α-chaconine and α-solanine. Donnelly et al. (2001c) reported that elevated CO_2 reduced levels of α-chaconine in potato tubers. Similarly, Vorne et al. (2002) reported decreases in total glycoalkaloid content across all elevated CO_2 treatments in the CHIP project. In contrast, Nitithamyong et al. (1999) found an increase in total glycoalkaloid content of tubers exposed to 1000 $\mu mol\ mol^{-1}$ CO_2.

Section Summary

Potato tuber quality will change with rising levels of atmospheric CO_2; increasing the tuber dry matter and starch content whilst reducing tuber nitrogen and glycoalkaloid content.

INTERACTIONS WITH OTHER ELEMENTS OF CLIMATE CHANGE

Increases in the concentration of atmospheric CO_2 is but one aspect of our changing environment and cannot be considered alone if accurate predictions are to be made about crop growth and yield in the future. Concentrations of gases defined as air pollutants have also increased in the atmosphere whilst changes in the composition of the atmosphere are postulated to have an effect on future climate patterns. It is predicted that the climate of Europe is likely to get warmer in the future with drier summers and more variable patterns of rainfall and temperature (IPCC, 2001) and as a consequence responses to increasing atmospheric CO_2 need to be considered in relation to the interaction with these additional changes in the environment. In order to elucidate the interaction between elevated CO_2, solar radiation and increased temperature Cao et al. (1994) exposed potato plants to four CO_2 concentrations combined with two air temperature regimes of 16°C and 20°C in controlled environment chambers. Plant growth was greater at 20°C than at 16°C at all CO_2 concentrations. However, tuber and total dry weights reached a maximum at 1000 $\mu mol\ mol^{-1}$ CO_2 at 16°C but continued to increase up to 2000 $\mu mol\ mol^{-1}$ CO_2 at 20°C. We have not been able to find any evidence from the literature of experiments that have investigated the interactive effects of elevated CO_2 and water deficit. However, Schapendonk et al. (1995) using a crop growth model evaluated the effects of doubled atmospheric CO_2 concentration and elevated temperature on potato crops and concluded that elevated temperature reduced the positive effect of elevated CO_2 except in conditions of optimum water supply when a net positive effect could be expected. Wolf (2000) used the results of the CHIP project experiments at different locations in Europe to calibrate the LPOTCO potato model which he used to simulate the effects of changes in climate on tuber yield of potato (cv. Bintje). The sensitivity of tuber production to separately changed weather variables was investigated first and it was found that irrigated tuber yields increased with temperature rise in northern Europe but remained either unchanged or strongly decreased in central and southern Europe, respectively. In the absence of irrigation, European tuber yields were predicted to be much lower, declining slightly with temperature rise and with increases in radiation, increasing considerably with increasing CO_2 whilst declining slightly with increasing O_3. The model was also used to predict tuber yields in 2050 using two holistic climate change scenarios. The predictions from the model was that irrigated tuber yields in northern Europe would increase by 1-4 tonnes of dry matter/ha whilst tuber yield of unirrigated crops would increase by 1-3 tonnes of dry matter/ha. Yield increases were predicted to be smaller in southern Europe. A subsequent analysis of the impact of climate change in 2050 on tuber yield using two other models (NPOTATO and POTATOS) predicted moderate to strong increases in tuber yield in northern Europe with little or no change in the British Isles, central and southern Europe. This modelling analysis also revealed that the best management strategy to minimise the effects of climate change was to use earlier varieties sown at an advanced planting date (Wolf 2002 a, b).

The most prominent aerial pollutant in rural areas is O_3 (ozone), a secondary pollutant whose concentrations have more than doubled in the last one hundred years (Hough & Derwent, 1990). The CHIP experiments investigated the interactive effects of elevated CO_2 and elevated O_3 on the growth and yield of potato plants. The study concluded that although elevated CO_2 did not protect against O_3 induced yield losses, the positive effects of CO_2 on yield more than compensated for the negative effects of ozone (Craigon et al., 2002). It has also been reported that elevated CO_2 reduces the damaging impact of elevated O_3 on visible foliar damage (Temmerman et al., 2002b; Donnelly et al., 2001b) and chlorophyll content (Donnelly et al., 2001b). Finnan et al. (2002) reported an interaction in which elevated O_3 reduced tuber yield but only at elevated concentrations of CO_2.

Section Summary

It is predicted that the beneficial effects of elevated CO_2 on the growth and yield of potato will be diminished by elevated temperatures and by drought as climate change becomes a reality. The most likely effect of climate change is a stimulation of tuber yield in northern Europe with little or no effect in central and southern Europe. It is possible that elevated CO_2 will provide some protection against the damaging effect of increased concentrations of tropospheric ozone.

CONCLUSIONS

The physiology, growth and harvestable yield of potato plants are all responsive to increases in the concentration of atmospheric CO_2. It is however difficult to generalise on the response of potato to elevated CO_2 as a wide range of responses have been reported by the studies reviewed here. However, common trends are apparent in many of the studies most of which investigated the effect of doubling the current ambient concentration of atmospheric CO_2 on potato. In general, it would appear that twice ambient concentrations of CO_2 stimulate photosynthetic rates initially but that this initial stimulation is followed by photosynthetic acclimation. The stomata in potato leaves partially close in response to elevated CO_2 and a build up of starch in chloroplasts can be expected. Elevated CO_2 will promote an initial growth stimulation above and below ground but the period of active growth will end earlier and accelerated senescence will limit growth towards the end of the growing season. The end result of growing potato crops at approximately twice current ambient CO_2 concentrations will be an increase in tuber yield. However, the magnitude of the yield increase will depend on agronomic practise, cultivar choice and growing conditions. It is likely that the beneficial effects of elevated CO_2 on potato crops may be reduced by other predicted effects of global climate change such as increased temperature and drought stress. Modelling efforts currently suggest that the overall effect of climate change by the middle of this century will be an increase in potato yields in northern Europe with little or no change in central and southern Europe.

RECOMMENDATIONS FOR FUTURE RESEARCH

Much is now understood about the response of potato to elevated CO_2. However, one important aspect of the response, namely photosynthetic acclimation, is poorly understood. Simulated results have indicated that the potential positive effects of elevated CO_2 on tuber yields are reduced by 50% because of this phenomenon. Further research is needed to determine the exact nature and cause of acclimation as an understanding of the effect may show how acclimation might be ameliorated allowing greater stimulation in tuber yield. Further field research is also needed to determine the exact nature of the interaction between drought stress and elevated CO_2. Models used to estimate the effect of climate change on potato have been calibrated using trials in which the effect of drought and the effect of elevated CO_2 have been investigated separately but not in combination. Prediction of the effect of climate change on potato will be enhanced if models can be calibrated using experiments in which this interaction has been quantified. Improved predictions will allow the agricultural industry to prepare for inevitable climate change. Such experiments would also permit predicted increases in the WUE of drought stressed plants exposed to elevated CO_2 to be quantified.

REFERENCES

Bender, J., Herstein, U. and Black, C.R. (1999). Growth and yield responses of spring wheat to increasing carbon dioxide, ozone and physiological stresses: a statistical analysis of 'ESPACE wheat' results. *European Journal of Agronomy*, 10, 185-195.

Bindi, M., Fibbi, L., Frabotta, A., Ottaviani, G. and Magliulo, V. (1998). Annual report for changing climate and potential impacts on potato yield and quality on free air CO2 enrichment of potato (*Solanum tuberosum* L.). In. *Changing Climate and Potential Impacts on Potato Yield and Quality. Progress Report January December 1998).*

Bindi, M., Fibbi, L., Frabotta, A., Chiesci, M., Selvaggi, G. and Magliulo, V. (2000). Report 01.01.1999-30.6.2000 for changing climate and potential impacts on potato yield and quality on free air CO_2 enrichment of potato (*Solanum tuberosum* L.). In. *Changing Climate and Potential Impacts on Potato Yield and Quality: Final Report January 1998-July 2000.*

Bindi, M., Hacour, A., Vandermeiren, K., Craigon, J., Ojanperä, K., Selldén, G., Högy, P., Finnan, J. and Fibbi, L. (2002). Chlorophyll concentration of potatoes grown under elevated carbon dioxide and/or ozone concentrations. *European Journal of Agronomy*, 17, 319-335.

Bunce, J.A. (1992). Light, temperature and nutrients as factors in photosynthetic adjustment to elevated concentrations of CO_2. *Physiologia Plantarum* 111, 501-511.

Bunce, J.A. (2001). Direct and acclamatory responses of stomatal conductance to elevated carbon dioxide in four herbaceous crop species in the field. *Global Change Biology* 7, 323-331.

Cao, W., Tibbetts, T.W. and Wheeler, R.M. (1994). Carbon dioxide interactions with irradiance and temperature in potatoes. *Advances in Space Research* 14, 243-250.

Cure, J.D. and Acock, B. (1986). Crop responses to carbon dioxide doubling: a literature survey. *Agriculture Forest Meteorology*, 38, 127-145.

Collins, W.B. (1976). Effect of carbon dioxide enrichment on growth of the potato plant. *Hort. Science*, 11, 467-469.

Craigon, J., Fangmeier, A., Jones, M., Donnelly, A., Bindi, M., de Temmerman, L., Persson, K. and Ojanperä, K. (2002). Growth and marketable-yield responses of potato to increased CO_2 and ozone. *European Journal of Agronomy*, 17, 273-290.

Cure, J.D. and Acock, B. (1986). Crop responses to carbon dioxide doubling: a literature survey. *Agriculture and Forest Meteorology* 38, 127-145.

De Temmerman, L., Hacour, A. and Guns, M. (2002a). Changing climate and potential impacts on potato yield and quality 'CHIP': introduction, aims and methodology. *European Journal of Agronomy*, 17, 233-242.

De Temmerman, L., Karlsson, G., Donnelly, A. Ojanpera, K., Jager, H.J., Finnan, J. and Ball, G. (2002b). Factors influencing visible ozone injury on potato including the interaction with carbon dioxide. *European Journal of Agronomy*, 17, 291-302.

Donnelly, A., Craigon, J., Black, C.R., Colls, J.J. and Landon, G. (2001a). Does elevated CO_2 ameliorate the impact of O_3 on chlorophyll content and photosynthesis in potato (*Solanum tuberosum*)? *Physiologia Plantarum* 111, 501-511.

Donnelly, A., Craigon, J., Black, C.R., Colls, J.J. and Landon, G. (2001b). Elevated CO_2 increases biomass and tuber yield in potato even at high ozone concentrations. *New Phytologist* 149, 265-274.

Donnelly, A., Lawson, T., Craigon, J., Black, C.R., Colls, J.J. and Landon, G. (2001c). Effects of elevated CO_2 and O_3 on tuber quality in potato (*Solanum tuberosum* L.). *Agriculture, Ecosystems and Environment* 87, 273-258.

Fangmeier, A., De Temmerman, L., Black, C., Persson, K. and Vorne, V. (2002). Effects of elevated CO_2 and/or ozone on nutrient concentrations and nutrient uptake of potatoes. *European Journal of Agronomy*, 17, 353-368.

Fangmeier, A., Jager, H.J. and Hogy, P. (1998). Report of the CHIP experiment 1998 at Giessen, Germany. In. *Changing Climate and Potential Impacts on Potato Yield and Quality. Progress Report January-December 1998.*

Fangmeier, A., Jager, H.J. and Hogy, P. (2000). Changing climate and potential impact on potato yield and quality. Final report (period 1/1/1999-30/6/2000). In. *Changing Climate and Potential Impacts on Potato Yield and Quality: Final Report January 1998-July 2000.*

Farrar, J.F. and Williams, M.L. (1991). The effects of increased atmospheric carbon dioxide and temperature on carbon partitioning, source-sink relations and respiration. *Plant, Cell & Environment*, 14, 819-830.

Farrar, J.F. (1996). Sinks, integral parts of a whole plant. *Journal of Experimental Botany*, 47, 1273-1280.

Finnan, J.M., Donnelly, A., Burke, J.I. and Jones, M.B. (2002). The effects of elevated concentrations of carbon dioxide and ozone on potato (*Solanum tuberosum* L.) yield. *Agriculture, Ecosystems and Environment* 88, 11-22.

Goudriaan, J. and de Ruiter, H.E. (1983). Plant growth in response to CO_2 enrichment, at two levels of nitrogen and phosphorous supply. 1. Dry matter, leaf area and development. *Netherlands Journal of Agricultural Research* 31, 157-169.

Hacour, A., Vandermeiren, K. and De Temmerman, L. (1998). First year report of the CHIP open-top chamber experiment 1998 at Tervuren, Belgium. In. *Changing Climate and Potential Impacts on Potato Yield and Quality. Progress Report January-December 1998.*

Hacour, A., Vandermeiren, K. and De Temmerman, L. (2000). Report of the CHIP open-top chamber experiment 1.1.1999-30.6.2000 Tervuren, Belgium. In. *Changing Climate and Potential Impacts on Potato Yield and Quality: Final Report January 1998-July 2000.*

Hacour, A., Craigon, J., Vandermeiren, K., De Temmerman, Ojanperä, K., Pleijel, H., Danielsson, H., Högy, P., Finnan, J. and Bindi, M. (2002). CO_2 and ozone effects on canopy development of potato crops across Europe. *European Journal of Agronomy*, 17, 257-272.

Hough, A.M. and Derwent, R.G. (1990). Changes in the global concentration of tropospheric ozone due to human activities. *Nature*, 344, 645-648.

Idso, K.B. and Idso, S.B. (1994). Plant response to elevated CO_2 enrichment in the face of environmental constraints; a review of the past 10 years of research. *Agricultural and Forest Meteorology*, 69, 153-203.

IPCC, (2001). *Climate Change 2001:The Scientific Basis.* (Ed. Houghton, J.T. et al.) Cambridge University Press, Cambridge. 881pp.

Kimball, B.A. (1983). Carbon dioxide and agricultural yield: an assemblage and analysis of 430 prior observations. *Agronomy Journal*, 75, 779-788.

Korner, Ch. (2000). Biosphere responses to CO_2 enrichment. *Ecological Applications* 10, 1590-1619.

Korner, C., Pelaez-Riedl, S. and Vanbel, A.J.E. (1995). CO_2 responsiveness of plants. A possible link to phloem loading. *Plant, Cell & Environment*, 18, 595-600.

Komor, E., Orlich, G., Weig, A. and Kockenberger, W. (1996). Phloem loading-not metaphysical, only complex: towards a unified model of phloem loading. J. Exp. Bot., 47, 1155-1164.

Lawlor, D.W. and Mitchell, R.A.C. (1991). The effect of increasing CO_2 on crop photosynthesis and productivity: a review of field studies. *Plant, Cell & Environment.* 14, 807-818.

Lawson, T., Craigon, J., Black, C.R., Colls, J.J., Tulloch, A.-M. and Landon, G. (2001a). Effects of elevated carbon dioxide and ozone on the growth and yield of potatoes (*Solanum tuberosum*) grown in open-top chambers. *Environmental Pollution* 111, 479-491.

Lawson, T., Craigon, J., Tulloch, A.-M., Black, C.R., Colls, J.J. and Landon, G. (2001b). Photosynthetic responses to elevated CO_2 and O_3 in field-grown potato (*Solanum tuberosum*). *Journal of Plant Physiology* 158, 309-323.

Long, S.P. and Drake, B.G. (1992). Photosynthetic CO_2 assimilation and rising atmospheric CO_2 concentrations. In. *Crop Photosynthesis: Spatial and Temporal Determinants* (Eds. Baker, N.R. & Thomas, H.). Elsevier, Amsterdam, 69-101.

Long, S.P., Baker, N.R. and Raines, C.A. (1993). Analysing the response of photosynthetic CO_2 assimilation to long-term elevation of atmospheric CO_2 concentration. *Vegetatio* 104/105, 33-45.

Ludewig, F., Sonnewald, U., Kauder, F., Heineke, D., Geiger, M., Stitt, M., Muller-Röber, B. T., Gillissen, B., Kuhn, C. and Frommer, W.B. (1998). The role of transient starch in acclimation to elevated atmospheric CO_2. *FEBS Letters* 429, 147-151.

Mackowiak, C.L. and Wheeler, R.M. (1996). Growth and stomatal behavior of hydroponically cultured potato (*Solanum tuberosum* L.) at elevated and super-elevated CO_2. *Journal of Plant Physiology* 149, 205-210.

Miglietta, F., Magliulo, V., Bindi, M., Cerio, L., Vaccari, F.P., Loduca, V. and Peressottis, A. (1998). Free Air CO_2 Enrichment of potato (*Solanum tuberosum* L.): development, growth and yield. *Global Change Biology* 4, 163-172.

Miglietta, F., Bindi, M., Vaccari, F.P., Schapendonk, A.C.C.M., Wolf, J. and Butterfield, R.E. (2000). Crop ecosystem responses to climatic change: root and tuberous crops. In. *Climate Change and Global Crop productivity* (Eds. K.R. Reddy & H.F. Hodges). CAB International, pp. 189-289.

Morison, J.I.L. and Gifford, R.M. (1984). Plant growth and water use with limited water supply in high CO_2 concentration. I. Leaf area, water use and transpiration. *Australian Journal of Plant Physiology* 11, 361-384.

Nitithamyong, A., Vonelbe, J.H., Wheeler, R.M. and Tibbits, T.W. (1999). Glycoalkaloids in potato tubers grown under controlled environments. *American Journal of Potato Research* 76, 337-343.

Ojanpera, K. and Vorne, V. (2000). Report of the CHIP open-top chamber experiment 1.1.1999-30.6.2000. In. *Changing Climate and Potential Impacts on Potato Yield and Quality: Final Report January 1998-July 2000.*

Pleijel, H., Persson, K., Danielsson, H., Pihl Karlsson, G., Skarby, L. and Sellden, G. (1998). Changing climate and potential impacts on potato yield and quality: annual report. In. *Changing Climate and Potential Impacts on Potato Yield and Quality. Progress Report January-December 1998.*

Poorter, H. (1993). Interspecific variation in the growth response of plants to an elevated ambient CO_2 concentration. In. *CO_2 and Biosphere* (Eds. Rozema, J., Lambers, A., Van de Geijn, H.S.C.). Klumer Academic, Dordrecht, pp. 77-97.

Poorter, H. (1998). Do slow-growing species and nutrient-stressed plants respond relatively strongly to elevated CO_2? *Global Change Biology* 4, 693-697.

Sage, R., Sharkey, T.D. and Seemann, J.R. (1989). Acclimation of photosynthesis to elevated CO_2 in five C3 species. *Plant Physiology* 89, 590-596.

Schapendonk, A.H.C.M., Pot, C.S. and Goudriaan, J. (1995). Simulated effects of elevated carbon dioxide concentration and temperature on the productivity of potato. In: Haverkort, A.J., MacKerron, D.K.L. (Eds.) *Potato Ecology and Modelling of Crops Under Conditions Limiting Growth.* Klumer Academic Publishers, Amsterdam, pp. 101-113.

Schapendonk, A.H.C.M., van Oijen, M., Dijkstra, P., Pot, C.S, Jordi, W.J.R.M. and Stoopen, G.M. (2000). Effects of elevated CO_2 concentration on photosynthetic acclimation and productivity of two potato cultivars grown in open-top chambers. *Australian Journal of Plant Physiology* 27, 1119-1130.

Sicher, R.C. and Bunce, J.A. (1999). Photosynthetic enhancement and conductance to water vapor of field-grown *Solanum tuberosum* (L.) in response to CO_2 enrichment. *Photosynthesis Research* 62, 153-163.

Sicher, R.C. and Bunce, J.A. (2001). Adjustments of net photosynthesis in *Solanum tuberosum* in response to reciprocal changes in ambient and elevated growth CO_2 parital pressure. *Physiologia Plantarum* 112, 55-61.

Storey, R.M.J. and Davies, H.V. (1992). Tuber quality. In *The Potato Crop.* (ed. P. Harris), pp. 507-570. London: Chapman Hall.

Vandermeiren, K., Black, C., Lawson, T., Casanova, M.A. and Ojanperä, K. (2002). Photosynthetic and stomatal responses of potatoes grown under elevated CO_2 and/or O_3-results from the European CHIP-programme. *European Journal of Agronomy*, 17, 337-352.

Vorne, V., Ojanperä, K., De Temmerman, L., Bindi, M., Högy, P., Jones, M.B., Lawson, T. and Persson, K. (2002). Effect of elevated carbon dioxide and ozone on potato tuber quality in the European multi-site experiment 'CHIP-project.' *European Journal of Agronomy*, 17, 369-381.

Webber, A.N., Nie, G.-Y. and Long, S.P. (1994). Acclimation of photosynthetic proteins to atmospheric CO_2. *Photosynthesis Research* 39, 413-425.

Wheeler, R.M., Mackowiak, C.L., Sager, J.C. and Knott, W.M. (1994). Growth of soybean and potato at high CO_2 partial pressure. *Advances in Space Research* 14, 251-255.

Wheeler, R.M. and Tibbets, T.W. (1997). Influence of changes in daylength and carbon dioxide on growth of potato. *Annals of Botany* 79, 529-533.

Wheeler, R.M., Tibbets, T.W. and Fitzpatrick, A.H. (1991). Carbon dioxide effects on potato growth under different photoperiods and irradiance. *Crop Science* 31, 1209-1213.

Wolf, J. (2000). Analysis of the responses of potato cv. Bintje to elevated CO_2 and O_3 and assessment of future potato yields in Europe. In. *Changing Climate and Potential Impacts on Potato Yield and Quality: Final Report January 1998-July 2000.*

Wolf, J. (2002a) Comparison of two potato simulation models under climate change. 1. Model calibration and sensitivity analyses. *Climate Research*, 21, 173-186.

Wolf, J. (2002b) Comparison of two potato simulation models under climate change. 2. Application of climate change scenarios. *Climate Research*, 21, 187-195.

Crop Responses to Elevated Carbon Dioxide and Interaction with Temperature: Grain Legumes

P. V. Vara Prasad
L. H. Allen, Jr.
K. J. Boote

SUMMARY. Atmospheric carbon dioxide concentration ($[CO_2]$) and other greenhouse gases have risen over the past few decades. If this continues, it could indirectly lead to increases in global temperature. Responses of grain legume crops (soybean, dry bean, peanut and cowpea) to elevated $[CO_2]$ and interactions with temperature are summarized. Our research shows that, in the absence of biotic (pests, diseases and

---

P. V. Vara Prasad and K. J. Boote are affiliated with the Agronomy Department, University of Florida, Gainesville, FL 32611-0500 USA.

L. H. Allen, Jr. is affiliated with the United States Department of Agriculture, Gainesville, FL 32611-0965 USA.

Address correspondence to: P. V. Vara Prasad at the above address (E-mail: vpaga@mail.ifas.ufl.edu or vvpagadala@hotmail.com).

The authors gratefully acknowledge all the scientific, engineering, and technical personnel and students who were involved in designing and executing research in the SPAR units during the last two decades.

The authors acknowledge support by various funding agencies including US Department of Energy, US Environmental Protection Agency, US Department of Agriculture-Agricultural Research Service and University of Florida.

This is a contribution of University of Florida and USDA-ARS at Gainesville, Florida. Florida Agricultural Experiment Station Journal Series No. R-00000.

[Haworth co-indexing entry note]: "Crop Responses to Elevated Carbon Dioxide and Interaction with Temperature: Grain Legumes." Prasad, P. V. Vara, L. H. Allen, Jr., and K. J. Boote. Co-published simultaneously in *Journal of Crop Improvement* (Food Products Press, an imprint of The Haworth Press, Inc.) Vol. 13, No. 1/2 (#25/26), 2005, pp. 113-155; and: *Ecological Responses and Adaptations of Crops to Rising Atmospheric Carbon Dioxide* (ed: Zoltán Tuba) Food Products Press, an imprint of The Haworth Press, Inc., 2005, pp. 113-155. Single or multiple copies of this article are available for a fee from The Haworth Document Delivery Service [1-800-HAWORTH, 9:00 a.m. - 5:00 p.m. (EST). E-mail address: docdelivery@haworthpress.com].

http://www.haworthpress.com/web/JCRIP

Digital Object Identifier: 10.1300/J411v13n01_07

weeds) or abiotic (temperature, water and nutrients) stresses, elevated [CO_2] will increase yield due to increased photosynthesis and growth. However, at above optimum temperatures, the beneficial effects of elevated [CO_2] are more than offset by negative effects of temperature on yield and yield-components, leading to lower seed yield and poor seed quality Future research should focus on developing genetic and agronomic crop management practices to improve crop productivity under changing climates.

KEYWORDS. Carbon dioxide, temperature, climate change, global warming, legumes, photosynthesis, respiration, water use efficiency, growth, yield, seed quality

ABBREVIATIONS. Atmospheric carbon dioxide concentration [CO_2], Soil Plant Atmospheric Research (SPAR), Water Use Efficiency (WUE), Carbon Exchange Rate (CER), Leaf Photosynthetic Assimilation Rate (A), Leaf Area Index (LAI), Total Nonstructural Carbohydrates (TNC), Ribulose Bisphosphate Carboxylase-Oxygenase (Rubisco), Ribulose Bisphosphate (RuBP)

INTRODUCTION

Legumes are important components of cropping systems and are major sources of vegetable oil and protein for human and animal consumption. They provide about 20% of the world's protein for human diet, and one-fourth of the world's fats and oils (Harlan, 1992). Legumes also help to enrich soil through their unique capability to form symbiotic relationships with nitrogen-fixing bacteria that capture atmospheric nitrogen and make it available for crop growth. In addition, legumes are traditionally used as green manure and cover crops in crop rotation to improve soil physical conditions. Legumes belong to the family Leguminosae, subfamily Papilionoideae, and are by far the plant family most utilized by humans in terms of numbers of genera and species. The legume family contains about 650 genera and 18,000 species. This chapter will focus on soybean [*Glycine max* (L.) Merr.], dry bean (*Phaseolus vulgaris* L.), peanut (*Arachis hypogaea* L.), and cowpea [*Vigna unguiculata* (L.) Walp.]. These four grain (seed) legume crops are widely cultivated across the world under a wide range of climatic conditions. Details on origin and production statistics of these crops are given in Table 1. Grain legume crops are commonly referred to as pulse crops in Southeast Asia and Australia.

In the recent years [CO_2] has rapidly increased to about 370 μmol mol^{-1} at present. At present rates of emissions, [CO_2] is predicted to more than double by the end of 21st century (Houghton et al., 2001). These changes in atmospheric [CO_2] along with other greenhouse gases (water vapor, methane, nitrous oxide, sulphur dioxide) can potentially change global climate. At the present rate of CO_2 increase, various atmospheric general circulation models have predicted that the near-surface temperature of the earth could increase as much as 3 to 6°C (Houghton et al., 2001).

Changes in [CO_2] and associated changes in global temperature can cause significant changes in crop production. In this review, we will summarize the responses of soybean, dry bean, peanut and cowpea to elevated [CO_2] and its interaction with temperature. Among all the legume crops, soybean is the most widely studied crop species with reference to responses to elevated [CO_2]. Because of the breadth of available literature, most of the examples are drawn from soybean; however, species differences between these legume crops will be highlighted and discussed as necessary.

CONTROLLED ENVIRONMENT FACILITIES AND EXPERIMENTAL DETAILS

Over the past two decades, considerable research on CO_2 responses of grain crops was conducted in the sunlit, controlled environment chambers known as

TABLE 1. Origin, total world harvested area and production during 2001 (FAO, 2001) and important countries producing soybean, dry bean, peanut and cowpea.

Crop	Origin	Harvested area (Million ha)	Total production (Million tons)	Important countries
Soybean	Northern China	75. 5	176.6	USA, China, Argentina, Brazil and India
Dry bean	Central and/or South America	23.2	16.8	India, Brazil, China, Myanmar, Mexico
Peanut	Argentina and/or Bolivia	25.5	35.1	India, China, Nigeria, Senegal, USA
Cowpea	West Africa and/or India	9.9	3.0	Nigeria, Niger, Myanmar

Soil-Plant-Atmosphere-Research (SPAR) chambers at the University of Florida, Gainesville, Florida, USA. A list of grain legumes experiments conducted in the SPAR facility is given in Table 2. Details of the chamber characteristics and functions, methods and quality of specific chamber environment, control are described in several publications (Jones et al., 1984; Baker et al., 1989; Pickering et al., 1994; Prasad et al., 2002). In brief, SPAR growth chambers are located outdoors and use solar radiation as light source. Each chamber has two parts, the upper chamber measures 1 m wide, 2 m long and 1.5 m high and is made of aluminum frame work with polyethylene telephtalate "six light" film walls. The bottom rooting-chamber (aluminum lysimeter) has the same cross section and is 0.6 m deep to accommodate the rooting medium. Each chamber has the capability to control air temperature and dewpoint temperature and [CO_2] at given set points. Air is circulated from top to bottom of the canopy chamber using fans located in external ductwork. The air handling system consists of a chilled water heat exchanger and electrical resistive heater coils.

Dry bulb air temperatures were measured with aspirated and shielded copper-constantan thermocouples and controlled by turning on and off the electrical resistance proportional heaters located in the exits ductwork. Air dewpoint temperatures were measured using hygrometers and controlled by opening and closing bypass valves that determine the flow of chilled water through the cooling coils. The condensate from each cooling coil was collected by a tip-

TABLE 2. Treatment details of experiments on grain legumes conducted in SPAR growth chambers at the University of Florida, Gainesville, USA.

Year	Crop	Carbon dioxide concentration (μmol mol^{-1})	Diurnal temperature (maximum/minimum °C)
1980	Soybean	320, 640	31/23
1981	Soybean	330, 450, 600, 800	31/23
1982	Soybean	330, 450, 600, 800	31/23
1983	Soybean	330, 660	31/23
1984	Soybean	160, 220, 280, 330, 660, 990	31/23
1985	Soybean	330, 660	31/23
1985	Soybean	330, 660	26/19, 31/24, 36/29
1993	Soybean	350, 700	28/18, 32/22, 36/26, 40/30
1994	Soybean	350, 700	28/18, 32/22, 36/26, 40/30,44/34, 48/38
2000	Dry bean	350, 700	28/18, 31/21, 34/24, 37/27, 40/30
2002	Peanut	350, 700	32/22, 34/24, 40/30, 44/34

ping bucket rain-gauge to estimate evapotranspiration at 5 min intervals. In most of the experiments, plants were grown throughout the season at a range of pre-set [CO_2] and/or temperature regimes. Air temperature and dewpoint temperature were typically controlled according to a sinusoidal wave function during the day and a decay function during the night. Dewpoint temperatures were maintained 10°C below target dry bulb air temperature. The temperatures indicated in this chapter are the daytime maximum and nighttime minimum temperature, unless stated otherwise.

Daytime [CO_2] was measured by drawing samples from the chamber every 10 s and controlled at given set points by injecting pure CO_2 from a high pressure cylinder through a mass flow controller. [CO_2] was not controlled during nighttime and vents were used to remove excess CO_2 due to respiration every hour. The carbon exchange rates were estimated from CO_2 injection rate and the change in [CO_2] after correcting for leakage in the chamber. Incoming solar radiation was measured using a pyranometer or solar cells. Canopy temperature was measured using infrared sensors. Measurements of all parameters were taken every 10 s, means were stored at 5-min intervals, and all data were downloaded automatically to a computer every night.

Leaf-level photosynthesis, stomatal conductance, and transpiration were measured on individual attached leaves using a LICOR 6200 photosynthesis system. Phenology, growth and dry matter production data were collected at frequent time intervals. At each harvest, plant height, vegetative stage, leaf area, and dry weights of components parts (leaf, stem, root, pod and seed) were recorded. At maturity, final yield, biomass production, partitioning, and yield component were measured either from a sub-sample of plants or on a land area basis. For detailed methods see individual references.

In most experiments, temperature and [CO_2] treatments were started from emergence to maturity, unless indicated otherwise. Crops were grown under optimal crop management practices with no water or nutrient stress. Irrigation was provided by subsurface irrigation and/or overhead sprinklers as necessary. Crops were kept weed free by regular hand weeding and insect and disease problems were controlled by appropriate chemical sprays as necessary.

EFFECTS OF [CO_2]

Carbon and Water Use Efficiency

Photosynthesis

Current atmospheric [CO_2] is sub-optimal for photosynthesis of C_3 plant species including soybean, dry bean, peanut and cowpea. In C_3 species, the primary acceptor of CO_2 is ribulose bisphosphate (RuBP) and the enzyme that catalyzes this reaction is ribulose bisphosphate carboxylase-oxygenase (Rubisco). As Rubisco catalyzes both carboxylation and oxygenase reactions, CO_2 and O_2 compete for the same site on Rubisco. The oxygenase activity of Rubisco (photorespiration) leads to = 25% (20 to 60%) loss of carbon in C_3 species (Bowes, 1996). Due to competitive interaction, increase in [CO_2] will diminish the oxygenase activity of O_2. Therefore, doubling the current [CO_2] will increase carboxylation, decrease oxygenation and photorespiration, and thus will increase the rate of net photosynthesis of C_3 species by about 30 to 40% (Bowes, 1996).

Photosynthetic rates of soybean were increased by elevated [CO_2] at both leaf (Vu et al., 1997, 2001) and canopy levels (Jones et al., 1984a, 1985a; Allen et al., 1987). Increased photosynthetic rates were observed with both short-term and long-term exposures to elevated [CO_2] (Valle et al., 1985a; Campbell et al., 1990). Maximum midday photosynthetic rates of soybean canopies during periods of peak irradiance (1,100 to 1,370 μmol photons m^{-2} s^{-1}), showed a strong positive response to growth [CO_2] (Figure 1). Furthermore, photosynthetic rates of soybean plants grown at high [CO_2] were greater compared to those at 330 μmol mol^{-1} at all levels of irradiance ranging from 100 to 1200 μmol photons m^{-2} s^{-1} (Campbell et al., 1988; 1990). Increased photosynthetic rates of soybean leaves and/or canopies with elevated [CO_2] have been reported in several other studies (Huber et al., 1984; Hesketh et al.,

FIGURE 1. Maximum midday photosynthetic rates (measured when solar irradiance was between 1,100 and 1,370 μmol photons m^{-2} s^{-1}) on land area basis of soybean canopies grown and measured at different [CO_2]. (Adapted from Campbell et al., 1990.)

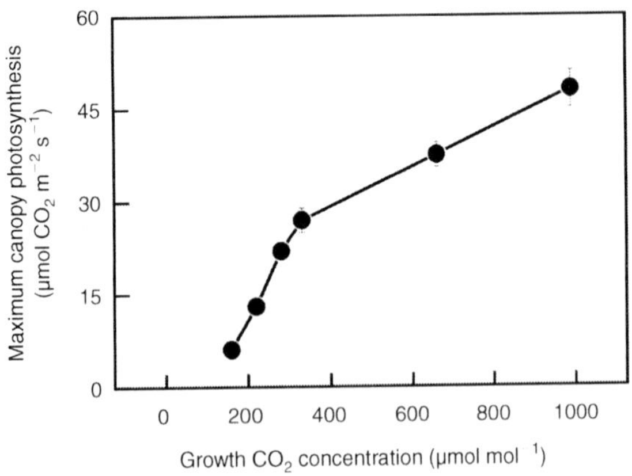

1984; Sionit et al., 1984; Chen et al., 1995; Ferris et al., 1998; Ziska et al., 2001). An increase in daytime canopy photosynthesis rates at elevated [CO_2] was observed at both vegetative and reproductive stages (Jones et al., 1985a and b). However, crop canopy photosynthetic rates can vary throughout the growing season mainly due to different requirements of photo-assimilates (Acock et al., 1985).

Campbell et al. (1988) measured Rubisco activity and amount in soybean leaves grown at a range of sub and super-ambient [CO_2] and found that Rubisco activity was almost constant (1.0 μmol CO_2 min^{-1} g^{-1} soluble protein) mainly because leaf soluble protein was nearly constant (2.4 g m^{-2}) with 55% being Rubisco protein. In contrast, Rubisco activity expressed on a leaf dry weight basis decreased because of an increase in specific leaf weight with an increase in [CO_2]. Campbell et al. (1990) also showed that photosynthetic capacity of soybean canopies grown at 330 and 660 μmol mol^{-1} of [CO_2] were similar for short exposures at [CO_2] below 550 μmol mol^{-1}, but above this value, canopies grown at 660 μmol mol^{-1} had a higher photosynthetic capacity. Vu et al. (1989) suggested that an increase in starch storage sites due to an increased number of palisade cells, together with the maintenance of Rubico activity, enables soybean to continue to exhibit high photosynthetic rates throughout the growth period under high [CO_2]. This suggests that photosynthetic capacity of soybean was not down regulated at elevated [CO_2]. Similar responses were observed in most of our studies on soybean. Allen et al. (1996), attributed this response to the typical characteristics of the soybean crop, which include: (i) high symbiotic N_2 fixation capability; (ii) the capacity to form an additional layer of palisade cells in the leaf tissue; (iii) the capacity to shunt much of the photoassimilates into relatively inert starch rather than soluble sugars during photosynthesis; (iv) a relatively strong leaf and stem sink during vegetative development; and (v) a strong seed sink during reproductive development. Plants that lack these capacities, either inherently or because of growth in limiting environments, are more likely to demonstrate some degree of down regulation of photosynthesis (Allen, 1994).

In dry bean, growth at elevated [CO_2] increased leaf photosynthetic rates (Figure 2; Prasad et al., 2002). This agrees with Sage (1994) who reported an increase in net CO_2 assimilation rates by about 30%. Similarly, growth at sub-ambient [CO_2] (200 μmol mol^{-1}) compared with near ambient [CO_2] (380 μmol mol^{-1}) decreased net assimilation rates by 40% at 25/20°C (Cowling and Sage, 1998). In contrast, Socias et al. (1993) found that photosynthetic rates of dry bean plants grown at 350 and 650 μmol mol^{-1} of [CO_2] were similar, when measured at the growth [CO_2]. Furthermore, the rate of RuBP consumption was lower in plants grown at elevated [CO_2], but the [CO_2] at which maximum consumption occurred was higher in plants grown in high [CO_2], indicating acclimation to high [CO_2]. This acclimation of RuBP consumption

FIGURE 2. Carbon dioxide enhancement (%) in leaf photosynthetic rates due to twice-ambient [CO_2] (700 μmol mol^{-1}) of soybean and dry bean leaves grown and measured at different growth temperatures. (Redrawn from Vu et al., 1997 and Prasad et al., 2002.)

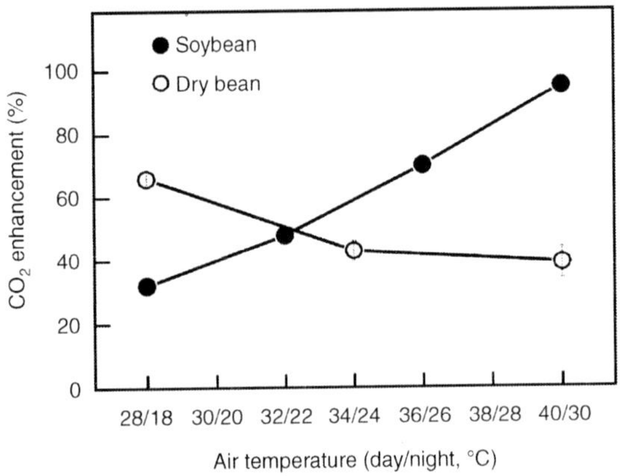

was related to a decrease in Rubisco activity resulting from decreased carbamylation, not a loss of Rubisco protein. Similarly, leaves of dry bean plants grown in enriched CO_2 (1,400 μmol mol^{-1}) showed a decrease in CO_2 exchange capacity compared with leaves grown in ambient [CO_2] (340 μmol mol^{-1}), when measured at the same [CO_2] (Ehret and Jolliffe, 1985). Acclimation of photosynthesis to elevated [CO_2] with decreases in Rubisco activity was reported by Sage et al. (1989). These differential responses could be related to growth conditions and nutrient stress. In the later two studies, dry bean plants were grown in small pots, i.e., 3 L (Socias et al., 1993) and 0.85 L (Ehret and Jolliffe, 1985), while those by Prasad et al. (2002) were grown in 2 m long, 1 m wide and 0.6 m deep lysimeter and by Sage (1994) in 20 L pots. The effects of restricted pot volume on photosynthetic acclimation are well documented (Arp, 1991). Sage (1994) compared the ratios of net CO_2 assimilation rates of plants grown at elevated and ambient [CO_2] obtained from four independent studies on dry bean and soybean, where plants were grown in different size pots. In both species, pot size had pronounced effects on net CO_2 assimilation rates (A). Plants grown in small pots (<5 L) exhibited carbon assimilation ratios (i.e., A at elevated CO_2/A at ambient CO_2) less than one across a wide range of intercellular [CO_2] (Ci) between 100 to 900 μmol mol^{-1}. In contrast, the carbon assimilation ratio of plants grown in large 20 L pots or in field experiments was not affected or enhanced by CO_2 enrichment across a range of Ci. Radogalou et al. (1992) found that the net photosynthetic rate of dry bean leaves grown in high nutrient levels did not show a down-regulation under elevated [CO_2]. Mjwara et al. (1996) reported that CO_2 enrichment (700 μmol mol^{-1}) compared to ambient (360 μmol mol^{-1}) increased carbon assimilation rates of bean early in the season, but from 20 days after germination there was a steady decline to almost similar levels to those measured in plants grown under ambient [CO_2]. This clearly suggests that acclimation response varies with the growth environment, age, nutrient status of the soil, and ability of plant roots to grow in an unrestricted volume.

The possible mechanisms for differential responses to elevated [CO_2] in small pots could be attributed to the following: (i) accumulation of greater nonstructural carbohydrates in plants grown in small pots or under nutrient deficiency (Robbins and Pharr, 1988) which in turn can lead to feedback inhibition of photosynthesis (Stitt, 1991; Bowes, 1996); (ii) plants exposed to elevated [CO_2] have greater initial growth rates and become nutrient-limited earlier than those grown in ambient [CO_2] (Coleman et al., 1993); (iii) root crowding and resistance to root expansion in small pots leads to release of abscisic acid and other signals that inhibit photosynthesis (Tardieu and Davis, 1993); and (iv) root tip damage due to desiccation at the bottom of the pots, which can trigger the release of large amounts of inhibitory chemicals which reduce stomatal conductance and lead to hormonally-controlled restrictions of shoot expansion, and lower sink capacity of leaves (Zhang and Davies, 1987; Davies et al., 1990). These mechanisms are discussed in detail by Sage (1994).

Elevated [CO_2] increased net photosynthesis and light saturated carbon assimilation rates of peanut under both irrigated and drought conditions (Clifford et al., 1993, 2000; Stronach et al., 1994). Clifford et al. (1993) reported a 40% increase in maximum leaf photosynthetic rate in peanut. Chen and Sung (1990) observed that when peanut plants were exposed to elevated [CO_2] (1,000 μmol mol^{-1}) starting from pod formation to maturity, there was an increase in soluble protein content and Rubisco, and plants even showed upregulation of photosynthesis and internal [CO_2] response (A:Ci) curve. Valle et al. (1985a) reported that soybean leaves grown at elevated [CO_2] had higher A_{max} with respect to saturating [CO_2]. Such responses are likely to be associated with increases in specific leaf weight under elevated [CO_2] (Vu et al., 1989). Increases in leaf photosynthetic rates were also reported in cowpea (Lopez et al., 1987; Overdieck et al., 1988; Ahmed et al., 1993).

Respiration

Amthor (1991, 1997) suggested that the response of respiration to CO_2 enrichment can be described in terms of effects of instantaneous or current [CO_2] (direct effects) or those related to CO_2-history of plants (indirect effects). Examples of indirect effects include differences in respiration caused by CO_2-induced changes in carbohydrate or protein concentration of the plant. Thomas and Griffin (1994) studied soybean leaf respiration in response to [CO_2] and found that effects were variable and dependent on both the basis on which the respiration was expressed and the extent of leaf expansion (leaf age). Respiration rates of expanding leaves were not affected by growth [CO_2], whereas CO_2 enrichment significantly increased respiration in fully expanded leaves when rates were expressed on either a leaf area basis (+ 27%), total non-structural carbohydrate basis (+ 35%) or leaf N basis (+ 42%). Contrasting results were found when respiration was expressed on a leaf dry mass basis. Respiration rates of fully expanded leaves were not significantly affected by growth [CO_2], but expanding leaves had a 19% lower respiration rate under elevated [CO_2] than under ambient [CO_2]. Similar responses were observed in several plant species (Gifford et al., 1985; Bunce and Caulfield, 1991). Thomas and Griffin (1994) partitioned total respiration into the components contributing to the formation of new tissues (growth respiration) and to the maintenance of existing tissues (maintenance respiration). They found that elevated [CO_2] did not affect growth respiration but increased maintenance respiration (+ 34%), which was correlated to with high leaf carbohydrate levels (+ 33%). Boote et al. (1997) suggested that the lower respiration rates expressed on unit tissue mass (specific respiration rate) could be attributed to indirect effects of elevated [CO_2] associated with: (i) shifts in composition with more carbohydrate and less protein causing lower synthesis cost, (ii) increase in structure-support components that are lower in synthesis and maintenance cost, and (iii) lower protein causing lower maintenance respiration.

Results of respiration studies differ with the experimental methodology used to measure CO_2 efflux, i.e., closed or open system (Long and Hallgren, 1993; Drake et al., 1999; Amthor, 1997). Preliminary studies conducted by Jahnke (2001) on dry bean showed that inhibition of dark CO_2 efflux by elevated [CO_2] was found using an open gas exchange system; but when the same leaf was investigated in a closed system, no such inhibition was detected. Jahnke (2001) conducted detailed experiments on dry bean in a specialized gas exchange system in three different ways: (i) as a closed system in which [CO_2] increased continuously from 200 to 4,200 $\mu mol\ mol^{-1}$ due to respiration of the enclosed leaf; (ii) as an intermittently closed system that was repeatedly closed and opened during [CO_2] period of either 350 or 2,000 $\mu mol\ mol^{-1}$; and (iii) as an open system in which [CO_2] varied between 350 and 2,000 $\mu mol\ mol^{-1}$. As a control, respective system characteristics were evaluated carefully in an empty leaf chamber. It was observed that when all the relevant system characteristics were taken into account, no effects of short-term changes in CO_2 on dark CO_2 efflux of dry bean leaves were found, even when [CO_2] increased to 4,200 $\mu mol\ mol^{-1}$. It was concluded that the leaf respiration of dry bean is not directly inhibited by elevated [CO_2] (Jahnke, 2001). Jannke and Krewitt (2002) reported older leaves of tobacco which has homobaric leaf anatomy (intercellular air spaces are not compartmented and have a continuous system of open pores) can lead to internal leakage and give a false indication of leaf respiration rates being decreased in elevated [CO_2] chambers. When internal leakage was experimentally restricted, the CO_2 inhibition on CO_2 efflux was no longer observed. They concluded that a true direct effect of [CO_2] on leaf respiration does not exist and direct effects of CO_2 on dark respiration reported in the literature might be artifactual (Jahnke and Krewitt, 2002). However, further research is needed across other crop species to confirm these. Clearly Thus, the variable response of respiration to elevated [CO_2] may be attributed to age of the leaf, period of exposure, method of measurement, units of measurement and expression and defects if measurement techniques. Further research is needed to understand the effects of elevated [CO_2] on growth and maintenance respiration of various component parts including leaves and roots.

Stomatal Conductance, Transpiration, and Water Use Efficiency

One of the most important phenomena responsive/sensitive to increasing [CO_2] levels is partial stomatal closure. A recent review on soybean showed that stomatal conductance was decreased by 31% at 450-550 μmol mol^{-1} [CO_2], 36% at 600-800 μmol mol^{-1}, and 51% at > 850 μmol mol^{-1} of [CO_2] with respect to ambient 330-360 μmol mol^{-1} [CO_2] (Ainsworth et al., 2002). Similarly, a previous review of soybean reported a 30% decrease in stomatal conductance with a doubling of [CO_2] (Cure, 1985; Cure and Acock, 1986). In addition to decreased stomatal conductance, partial stomatal closure increases leaf resistance to transpirational water loss resulting in lower leaf transpiration rates. This may lead to an increase in leaf or tissue temperatures. In controlled environments, foliage temperatures of soybean grown at 700 μmol CO_2 mol^{-1} were 1 to 2°C warmer compared to those grown at 350 μmol mol^{-1} (Campbell et al., 1990). Similar increases were observed in dry bean (Prasad et al., 2002). As leaf temperature increases, the vapor pressure inside the leaves will increase leading to a higher leaf-to-air vapor pressure difference, which is the driving force for transpiration leading to increased transpiration rate per unit leaf area and thus maintain those rates at slightly lower values than would exist at ambient [CO_2].

Valle et al. (1985b) reported that diurnal transpiration rates of soybean leaflets grown and measured at high (660 μmol mol^{-1}) and ambient (330 μmol mol^{-1}) [CO_2] were almost similar. This is mainly because the increases in leaf resistance caused by high [CO_2] were partially offset by increases in vapor pressure gradient between leaves to air caused by increases in leaf temperatures (Table 3). Effects of changes in leaf conductance to elevated [CO_2] on single leaf transpiration cannot be directly transferred to transpiration from

TABLE 3. Carbon exchange rate (CER), transpiration rate (TR), leaf temperature (TL), leaf resistance (rL), vapor pressure gradient (VPG), water-use efficiency (WUE), air temperature (Ta) and air vapor pressure (VP_a) at midday (1130-1230 h EST) of soybean leaves grown and measured at ambient (330 μmol mol^{-1}) and high (660 μmol mol^{-1}) [CO_2]. (Adapted from Valle et al., 1985b.)

	Carbon dioxide concentration (μmol mol^{-1})	
Parameter	330	660
CER (μmol m^{-2} s^{-1})	21.2	41.1*
TR (mmol m^{-2} s^{-1})	5.46 ± 0.19†	5.17 ± 0.11
TL (°C)	30.7 ± 0.30	32.5 ± 0.50
rL (s m^{-2})	146 ± 6.50	187 ± 5.00*
VPG (10^2 Pa)	19.9 ± 0.80	24.3 ± 1.60
WUE (mmol mol^{-1})	3.9 ± 0.10	7.9 ± 0.20*
Ta (°C)	29.3 ± 0.13	29.9 ± 0.23
VPa (10^2 Pa)	20.7 ± 0.17	20.9 ± 0.24

* Significant at $P < 0.05$ t-test; comparison within rows
† Average of four leaves ± standard error; n = 24 for 330 and 21 for 660 μmol mol^{-1}

whole crop canopies under field conditions. Comparison of transpiration rates of soybean canopies grown at high (800 μmol mol^{-1}) versus ambient (330 μmol mol^{-1}) [CO_2] (corresponding LAI values were 6.0 to 3.3, respectively) showed that changes in transpiration rates were in the range of −2% (Jones et al., 1985a) to +11% (Jones et al., 1985b). In another experiment, where soybean canopies grown at ambient (330 μmol mol^{-1}) and high (700 μmol mol^{-1}) [CO_2] had similar LAI (about 3.4), the seasonal cumulative water use was decreased by 12% at high [CO_2] (Jones et al., 1985c).

In some studies, total canopy water use of soybean was reduced by [CO_2] enrichment (Jones et al., 1985 c; Rogers et al., 1984), while in others it was unaffected (Jones et al., 1985a; Kimball et al., 1984). Albeit to partial stomatal closure, water use per unit leaf area is reduced, much of this reduction in transpiration is offset due to greater LAI increases at elevated [CO_2]. Even at similar LAI, the canopy energy balance would cause foliage temperature to rise and prevent full expression of reduction effects of stomatal conductance on transpiration as indicated above.

Water use efficiency (WUE) is defined as the ratio of the quantity of plant production to the quantity of the water transpired, evaporated or consumed. It can be described in terms of dry matter production, grain yield, or CO_2 uptake. For example, photosynthetic WUE is defined as a ratio of the photosynthetic carbon exchange rate to transpiration and can be expressed on a leaf area, whole-canopy, or ground area basis. Similarly, whole season WUE is best defined as the ratio of total dry matter (or seed yield) produced to total amount of water used by the crop. Based on the above definitions an increase in WUE can be obtained from increased carbon exchange rate and dry matter production, or from decreased transpiration and total water use.

Allen et al. (1985) compared WUE of soybean canopies grown at 330 and 800 μmol CO_2 mol^{-1}, with an LAI of 3.3 and 6.0, respectively. In each of the two treatments, [CO_2] levels were cross-switched for one day. The ratio of the WUE (i.e., WUE at 800 ppm [CO_2] divided by the WUE at 330 μmol mol^{-1}) averaged 2.33. The relative contributions of photosynthesis and transpiration to the ratio of WUE were 73 and 27%, respectively. These comparisons are valid as the same canopy was used for both [CO_2] concentrations. Exposure to elevated [CO_2] increased water use efficiency, mainly due to increased carbon exchange rates (Table 3).

Clifford et al. (2000) observed that stomatal conductance and transpiration rate of peanut were decreased under elevated [CO_2] in both irrigated and drought conditions. These reductions in combination with a relative delay in decline in relative water content, leaf water potential and turgor potential can lead to an increase in seasonal water use efficiency in peanut (Stronach et al., 1994). Decreases in stomatal conductivity and transpiration rates at elevated [CO_2] were reported in other studies on peanut (Chen and Sung, 1990; Clifford et al., 1995), dry bean (Radoglou et al., 1992; Prasad et al., 2002) and cowpea (Morison and Gifford, 1984).

Allen (1999) concluded that although stomatal conductance may be decreased by about 40% for doubled [CO_2], water use by C_3 crops under field conditions will probably be decreased only by up to 12%. If increases in leaf area due to doubled [CO_2] are small (or can be controlled), then the transpiration reductions would be meaningful, albeit small. If increases in leaf area due to doubled [CO_2] are large, then no reductions in transpiration would be expected, and even small increases might be possible.

Phenology, Growth, Yield, and Seed Quality

Phenology

Allen et al. (1990) studied the effects of [CO_2] on soybean developmental stages and expansive growth over four years and concluded that vegetative stages progressed slightly faster and the final number of nodes was slightly greater with increased [CO_2]. Time to reproductive stages was similar across the range of [CO_2] (160 to 990 μmol mol^{-1}), however, plants grown at 160 μmol mol^{-1} [CO_2] lagged behind plants grown at other [CO_2]'s at the beginning of flowering, but after 52 days, plants in all treatments had reached the same reproductive stage (Allen et al., 1991). In contrast, earlier studies showed that above-ambient [CO_2] either had no effect (Allen et al., 1988), decreased (Baker et al., 1989; Allen et al., 1990), or increased time to flowering (Rogers et al., 1984; Ellis et al., 1995). There was no effect of elevated [CO_2] on time to flowering or maturity for dry bean (Prasad et al., 2002). In cowpea, Bhattacharya et al. (1985) and Ellis et al. (1995) reported hastening of flowering by CO_2 enrichment, while Ahmed et al. (1993) reported no effect of CO_2 enrichment on time to first flowering.

Organ Growth and Partitioning

Allen et al. (1991) grew soybean from planting to maturity in a wide range of sub-ambient and super-ambient [CO_2] and acquired time series data on dry matter accumulation and leaf area. Responses of dry matter accumulation to [CO_2] during the linear phase were fitted by an empirical nonlinear regression model. A rectangular hyperbola of a modified Michaelis-Menten equation given below was adapted to model the response as a function of [CO_2] (Allen et al., 1987).

$$Y = \beta_0 + \beta_1 \,(DAP - D)(C - G)/(C - Kc)$$

where Y = dependent growth variable (dry weight or leaf area); β_0 = intercept on Y-axis when second term is zero; β_1 = maximum asymptotic value of dry matter or leaf area growth rate when [CO_2] is no longer limiting; D = days after planting (DAP) offset parameter to remove the time lag between seedling emergence and nonlinear early plant growth; C = [CO_2]; G = apparent [CO_2] compensation point for the dependent variable; and Kc = hyperbolic function shape factor, related to the apparent Michaelis constant. Allen et al. (1991) concluded that growth rates of all plant parts during the linear phase of growth (24 to 66 DAP) were stimulated by elevated [CO_2] (Table 4). This resulted in greater dry weights of all of the various component parts (Figure 3). The relative ranking of the [CO_2] treatment effect on growth rates among different plant parts was stem > petiole > leaf biomass > leaf area (Table 4).

Initiation and expansion of leaves on both the main stem and branches were faster in soybean plants grown at elevated [CO_2] compared to ambient [CO_2] (Cure et al., 1988). Allen et al. (1990) reported that expansive soybean growth increased with [CO_2], leading to greater mainstem height, branch lengths, stem diameter, individual leaf area and leaf area per plant. An increase in mainstem height was attributed to increased node number and stimulation of internode elongation (Rogers et al., 1984; Allen et al., 1990). Furthermore, in-

TABLE 4. Growth rates of different plant component during the linear phase of growth (24 to 66 days after planting) of soybean plants grown at sub-ambient, ambient and super-ambient [CO_2]. (Adapted from Allen et al., 1991.)

Parameter	Carbon dioxide concentration (μmol mol^{-1}) 160	220	280	330	660	990
Total shoot dry weight (g m^{-2} day^{-1})	5.0	8.4	10.9	12.5	18.2	20.7
Stem dry weight (g m^{-2} day^{-1})	0.8	1.7	2.4	2.9	4.7	5.5
Petiole dry weight (g m^{-2} day^{-1})	0.5	1.0	1.4	1.6	2.5	2.9
Leaf dry weight (g m^{-2} day^{-1})	1.7	2.7	3.4	3.9	5.3	5.8
Leaf area (m^2 m^{-2} day^{-1})	0.07	0.12	0.14	0.15	0.19	0.20

FIGURE 3. Component dry weights of different component parts of soybean plants grown at different [CO_2] during the linear phase from 13 to 66 days after planting. (Redrawn from Allen et al., 1991.)

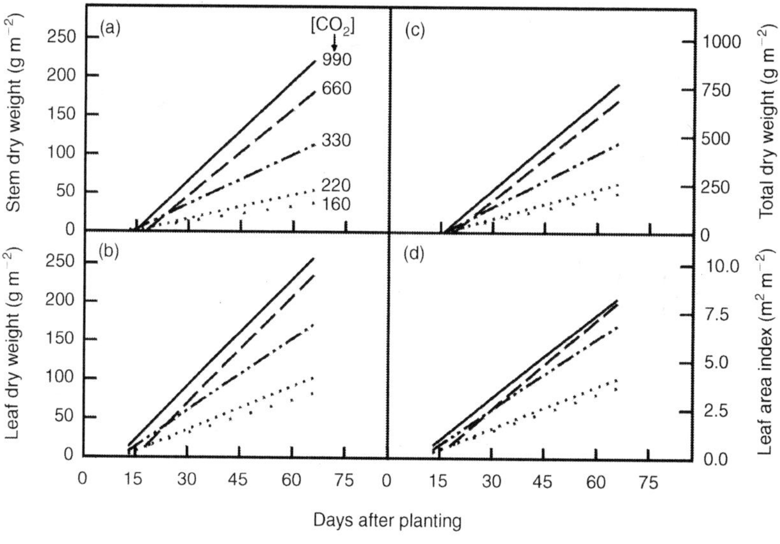

creases in extension rates of branches were greater than of mainstems under elevated [CO_2] (Allen et al., 1990).

Partitioning of photoassimilates to various plant organs of soybean over time was influenced by [CO_2]. Plants grown at sub-ambient [CO_2] partitioned carbon preferably to leaves, which served to increase carbon fixation capability, while plants grown at super-ambient [CO_2] partitioned more carbon to supporting structures such as stems, petioles and roots than to leaf laminae, especially during the vegetative stages (Allen et al., 1991). This mechanism could be for balancing the supply and demand for carbohydrates needed by the whole plant for balanced growth (Acock and Allen 1985). Alternatively, larger plants normally produce more supporting structure, so the partitioning to these structures may have been a consequence of plant size and hence an indirect effect rather than the direct consequence of super-ambient [CO_2]. Growth at elevated [CO_2] increased stem, petiole and leaf dry matter production as well as number of branches (Tables 5 and 6). Elevated [CO_2] increased the total dry matter production in soybean (Allen et al., 1987, 1991; Baker et al., 1989; Pan, 1996), dry bean (Prasad et al., 2002), peanut (Clifford et al. 1993, 2000), and cowpea (Bhattacharya et al., 1985; Overdieck et al., 1988; Ahmed et al., 1993; Ellis et al., 1995).

Elevated [CO_2] increased specific leaf weight of soybean (Leadley and Reynolds, 1988; Campbell et al., 1988; Vu et al., 2001). This could be attributed to increased carbohydrate accumulation and/or changes in leaf thickness. In soybean, exposure to elevated [CO_2] increased the number of palisade cells in an extra layer of cells, which contributed to leaf thickness (Thomas and Harvey, 1983). Plants dependent upon nitrate and those dependent on symbiotic N_2 fixation both responded in a similar manner.

Root Growth and Nitrogen Fixation

Exposure to elevated growth [CO_2] increased total root dry matter, root volume, and root length (Rogers et al., 1992). There were also significant changes

TABLE 5. Dry weights of component plant parts at final harvest of soybean plants grown at sub-ambient, ambient and super-ambient [CO_2]. (Adapted from Allen et al., 1991.)

	Carbon dioxide concentration (μmol mol^{-1})		
Parameter	160	330	660
Leaf dry weight (g plant^{-1})	2.30	6.09	5.43
Petiole dry weight (g plant^{-1})	0.66	2.19	2.63
Stem dry weight (g plant^{-1})	1.48	4.63	6.42
Pod dry weight (g plant^{-1})	7.98	20.12	24.64
Seed dry weight (g plant^{-1})	5.77	14.86	17.85
Total dry weight (g plant^{-1})	12.88	33.03	39.12
Seed weight (g seed^{-1})	0.128	0.176	0.167

TABLE 6. Numbers of main stem nodes, branches, pods, seeds and seeds per pod at final harvest of soybean plants grown at sub-ambient and super-ambient [CO_2]. (Adapted from Allen et al., 1991.)

	Carbon dioxide concentration ($\mu mol\ mol^{-1}$)		
Parameter	160	330	660
Main stem nodes ($plant^{-1}$)	11	14	14
Number of branches ($plant^{-1}$)	4.4	5.6	6.7
Pods on main stem ($plant^{-1}$)	11.3	17.8	15.3
Pods on branches ($plant^{-1}$)	12.8	18.5	34.9
Total pod number ($plant^{-1}$)	24.0	36.3	50.2
Total seed number ($plant^{-1}$)	45.1	84.3	106.7
Number of seeds per pod	1.88	2.32	2.13

in root anatomy with increases in the diameter of the central stele and width of the cortex. However, there were no effects of CO_2 enrichment on total number of roots or number of roots at specific depths. Similar increases in root weight, length and volume were reported for cotton (Prior et al., 1994; Reddy et al., 1997).

A direct effect of increased root growth and mass due to elevated [CO_2] in legumes would be on nodulation and microbial activity and thus on biological nitrogen fixation. Increased photosynthetic carbon fixation under elevated [CO_2] will provide more energy for microbial activity and facilitate nitrogen fixation in legumes. Symbiosis between legume and bacteria was greater under elevated [CO_2] (Reddy et al., 1989; Reardon et al., 1990). Rogers et al. (1994, 1997) reported that elevated [CO_2] increased nodule development (i.e., increased nodule dry weight and increased nodule number per plant) and/or nodule function (i.e., increased nitrogenase activity, increased N_2 fixation, and increased whole plant N). Doubling of [CO_2] resulted in substantial increases in biological nitrogen fixing ability of soybean (Hardman and Brun, 1971; Hardy and Havelka 1973, 1975; Williams et al., 1981). Exposure to elevated [CO_2] for two weeks increased the number and mass of nodules and nitrogenase activity (Finn and Brun, 1982). However, there was no effect of short-term exposure (48 h) on nitrogenase activity (Finn and Brun, 1982; Williams et al., 1981). Phillips et al. (1976) indicated that in pea long-term CO_2 enrichment promoted N_2 fixation by enhancing nodule development, while short-term increases in [CO_2] increased N_2 fixation by affecting nodule function. Exposure to high [CO_2] increased nitrogenase activity in dry bean (Ortega et al., 1992). Kimball (1985), suggested that leguminous crops will probably not be nitrogen limited in future climates with elevated [CO_2] any more than they are at present [CO_2].

Carbohydrate Accumulation

Elevated [CO_2] increased accumulation of carbohydrates in soybean (Allen et al., 1988; Vu et al., 1989, 1997, 2001), dry bean (Sharkey et al., 1985; Ehret and Jolliffe, 1985), and cowpea (Bhattacharya et al., 1987; Ahmed et al., 1993). Vu et al. (1989) reported linear increases in carbohydrate content and protein concentration of soybean leaves grown across four different [CO_2] treatments (330, 450, 600 and 800 μmol mol^{-1}). Growth at 800 μmol CO_2 mol^{-1} significantly increased starch, sucrose, reducing sugars content and concentration chlorophyll and soluble protein compared to plants grown at 330 μmol mol^{-1} of [CO_2].

Increases in concentrations of total non-structural carbohydrates (TNC) in vegetative tissues due to increased [CO_2] were also reported at all stages of soybean development (Allen et al., 1998). Linear regression analyses indicated that both morning (a.m.) and evening (p.m.) leaf, petiole and stem TNC concentrations increased with increasing [CO_2] (Figure 4; Allen et al., 1998). In contrast, there was no significant effect of [CO_2] on TNC concentrations in pod walls and seeds at any stages of development a feature also confirmed by Thomas (2001). This suggests that individual pods and seeds, once set, are well-buffered and have a limited capacity to respond to [CO_2]. Thus, seed yield response to elevated [CO_2] is related to other traits such as increase in number of pods (Allen et al., 1991). The fact that plants grown at elevated [CO_2] contained large amounts of TNC in vegetative parts at maturity indicates that not all the reserves were utilized by seed yield. Therefore, soybean cultivars in the future should be designed to utilize carbohydrates more readily for seed production so that a greater benefit can be realized from rising atmospheric [CO_2] (Allen, 1994; Allen et al., 1998).

Yield and Yield Components

Several reports showed that increases in [CO_2] increased seed yields in soybean (Huber et al., 1984; Roger et al., 1986; Cure et al., 1988; Allen et al., 1991), dry bean (Prasad et al., 2002), peanut (Clifford et al., 1993; 2000; Mortley et al., 1997), and cowpea (Ahmed et al., 1993). Ainsworth et al. (2002) summarized the effects of doubled [CO_2] on soybean yield from 57 different data sets and reported an average increase of 24%. Allen et al. (1987) modeled the response of seed yield to [CO_2] from experiments conducted at different locations and reported an increase of 32% for doubling of [CO_2] from 315 to 630 μmol mol^{-1}. Similarly, effects of elevated [CO_2] across five different experiments in Gainesville, Florida showed that doubling of [CO_2] in-

FIGURE 4. Linear regression of whole-season total nonstructural carbohydrates in (a) leaf blades, (b) petioles, and (c) stems of soybean plants grown at different [CO_2] at 0800 (a.m.) and 1700 (p.m.) h EST. (Adapted from Allen et al., 1998.)

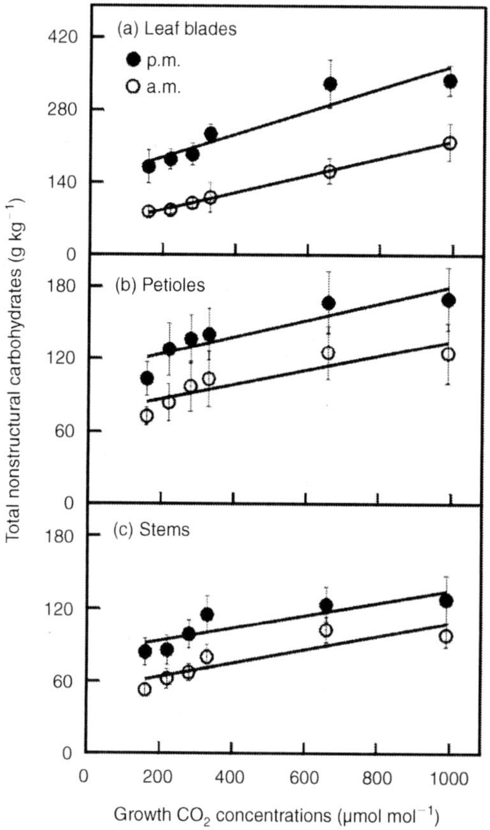

creased seed yield of soybean by about 35% (Allen and Boote, 2000). Higher seed yields due to elevated [CO_2] have been shown in other studies (Kimball, 1983; Cure, 1985; Rogers et al., 1986; Ferris et al., 1999; Ziska et al., 2001; Ainsworth et al., 2002). Increases in seed yield at elevated [CO_2] were mainly attributed to increases in number of pods and seeds from improved branching and greater number of pods on branches (Tables 5 and 6). Harvest index (ratio of seed to total dry matter) decreased at elevated [CO_2] (Baker et al., 1989; Allen et al., 1991), mainly because CO_2 enrichment causes greater increases in vegetative biomass than in seed yield.

Total dry matter production and pod yields of dry bean increased linearly as $[CO_2]$ increased from 340 to 1,200 μmol mol^{-1}, at which the maximum increase was observed, and further enrichment to 2,000 or 3,000 μmol mol^{-1} had no additional effect (Jolliffe and Ehret, 1985). The percentage of flowers setting seed was lower at elevated $[CO_2]$ (Figure 5) in spite of greater photosynthetic rates (Figure 2; Prasad et al., 2002). This clearly suggests that reduced seed-set was not a result of decreased availability of photosynthates. The lower seed-set at elevated $[CO_2]$ may be the result of a greater number of flowers produced due to improved vegetative growth and branching (Jolliffe and Ehret, 1985) or a small shift in tissue temperature at elevated $[CO_2]$ due to reduced stomatal conductance and transpiration (Matsui et al., 1997; Prasad et al., 2002). CO_2 enrichment increased seed yields of dry bean but did not affect individual seed weight and number of seed per pod at maturity (Figures 6 and 7; Prasad et al., 2002).

CO_2 enrichment increased total dry matter production and pod yield of peanut (Chen and Sung, 1990; Clifford et al., 1993; 2000; Mortley et al., 1997; Stanciel et al., 2000); however, there was no effect of CO_2 enrichment on harvest index (Clifford et al., 1993; Stanciel et al., 2000). Increased seed yield and number of seeds (Bhattacharya et al., 1985; Ahmed et al., 1993) and harvest index (Biswas and Hileman, 1985) were reported in cowpea plants grown under elevated $[CO_2]$.

FIGURE 5. Effect of temperature on seed-set (proportion of flowers setting seed) of dry bean plants grown at (a) ambient and (b) elevated $[CO_2]$. Fitted regressions: (a) $Y = 263 + 6.5X$, $r^2 = 0.93$; and (b) $Y = 250 + 6.5X$, $r^2 = 0.93$. (Adapted from Prasad et al., 2002.)

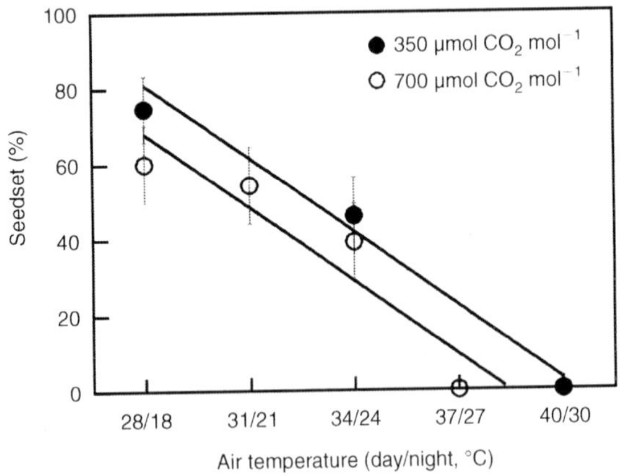

FIGURE 6. Relations between temperature and (a) number of seeds per plant, (b) seed yield per plant, and (c) total dry weight per plant of dry bean plants grown at ambient and elevated [CO_2]. Fitted regressions for sloping lines: (a) ambient [CO_2], Y = 87.2 − 2.3X, r^2 = 0.99: elevated [CO_2], Y = 120.7 − 3.2X, r^2= 0.88; (b) ambient [CO_2], Y = 42.5 − 1.2X, r^2 = 0.99: elevated [CO_2], Y = 56.7 − 1.5X, r^2 = 0.88; and (c) ambient [CO_2], Y = 68.3 − 1.6X, r^2 = 0.95: elevated [CO_2], Y = 75.5 − 1.6X, r^2 = 0.88. (Adapted from Prasad et al., 2002.)

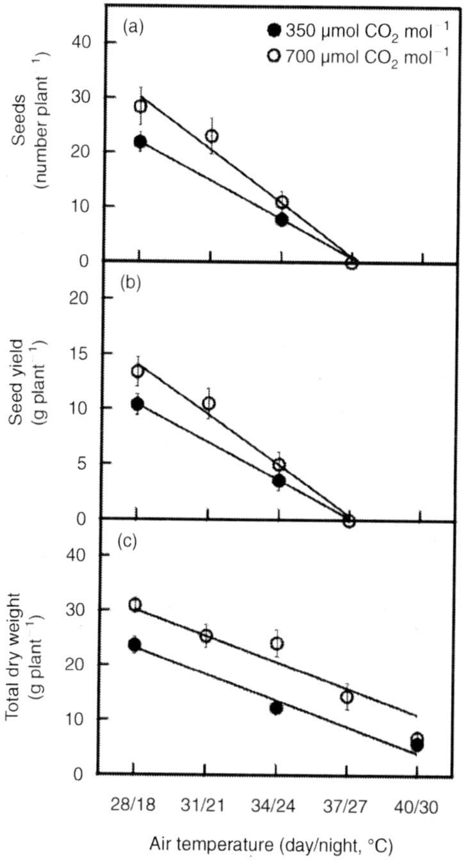

Thus, elevated [CO_2] increases seed yield of soybean, dry bean, peanut and cowpea under optimum temperature with no nutrient deficiency and water deficit. Greater seed yields under CO_2 enrichment were attributed to increases in seed numbers due to increased branching, but not due to increases in individual seed weight or number of seeds per pod.

FIGURE 7. Relations between temperature and (a) number of seeds per pod and (b) weight per seed (seed size) at maturity, of dry bean plants grown at ambient and elevated [CO_2] in dry bean. Fitted regression: (a) Y = 13.2 − 3.4X, r^2 = 0.94; and (b) Y = 2.66 − 0.67X, r^2 = 0.86. (Adapted from Prasad et al., 2002.)

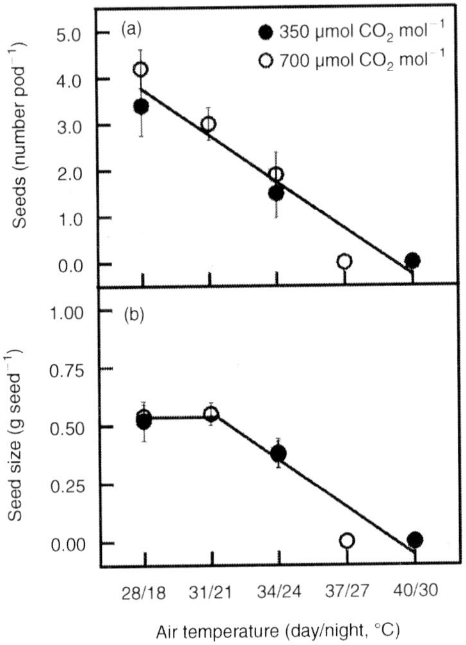

Seed Quality

Three important aspects of seed quality include: (i) individual seed size; (ii) composition or nutritional quality of seed; and (iii) ability of seed to germinate and grow. Although several studies have documented the effects of elevated [CO_2] on seed size, very few have examined composition, nutritional quality and germination of seeds produced at high [CO_2]. Despite the importance of seed composition for human and animal nutrition, research on seed quality aspects are however, limited (Murray, 1997).

Elevated [CO_2] did not influence seed size in soybean (Baker et al., 1989; Pan, 1996; Thomas, 2001) and dry bean (Prasad et al., 2002). As for seed composition, Allen et al. (1988) reported no effect of elevated [CO_2] on total seed N concentration in plants dependent on symbiotic N fixation. Rogers et al. (1984) reported that seed protein as a percentage of seed dry weight decreased as [CO_2] increased from 331 to 910 μmol mol^{-1}, although total seed protein per plant increased. Since Rogers et al. (1984) research was conducted in open-top field chambers; it is unlikely that the decline in seed protein was caused by constraints to the root system.

Thomas et al. (2003) studied effects of elevated [CO_2] on composition of mature soybean seed at different temperature regimes and concluded that there was no effect of elevated [CO_2] (700 μmol mol^{-1}) on N, P, starch, total oil, fatty acids, and total nonstructural carbohydrates compared to ambient [CO_2] across all temperatures ranging from 28/18 through 44/34°C. There were no effects of elevated [CO_2] (700 and 1,400 μmol mol^{-1}) on nutrient composition and mineral nutrient in peanut kernels (Wu et al., 1997). Effects of elevated [CO_2] on composition and nutritional quality of dry bean and cowpea are known. These crop species could respond differently than soybean and peanut because of their variable ability to fix atmospheric nitrogen, especially dry bean, which is less efficient in nitrogen fixation.

Idso and Idso (2001) reviewed effects of [CO_2] on plant constituents related to animal and human health, and argued that although elevated [CO_2] decreased nitrogen and protein concentration in some species, these decreases were not observed when plants were supplied with sufficient N. Decreases in mineral concentration were reported in dry bean (Porter and Grodzinski, 1984). Recently, Jablonski et al. (2002) conducted meta analysis of 79 crops species including legumes, non-legumes, C_3 and C_4 species and concluded that seed N concentration was unaffected by growth at high [CO_2] (500 to 800 μmol mol^{-1}) in legumes, while in non legume C_3 plants it was reduced by 15%. Legumes growing symbiotically with *Rhizobium* have an enhanced root system (Finn and Brun, 1982; Bottomely and Rogers, 1993) and may be able to support the increased N_2 fixation necessary to maintain the N concentration for an increased number of seeds per plant under elevated [CO_2] (Murray, 1995). Similarly, Allen and Boote (2000) opined that legumes are often able to use increased carbon gain under elevated [CO_2] for increased N_2 fixation and thus may be able to increase seed number and/or mass without a loss in seed N.

INTERACTION EFFECTS OF [CO_2] AND TEMPERATURE

To properly quantify the effects of potential climate change on crops, studies on interactions of [CO_2] and temperature are more important than effects of [CO_2] alone, because in the future climate both of these factors are likely to increase. Therefore, in this section the interaction effect of [CO_2] and temperature on carbon and water use, phenology, growth, carbohydrate accumulation, yield components, yield, and seed quality is discussed.

Carbon and Water Use Efficiency

Photosynthesis

Elevated CO_2 enhancements of leaf photosynthesis increased in a linear manner by 32 to 95% with increasing growth temperatures from 28/22 to 40/30°C (Figure 2; Vu et al., 1997). Furthermore, it was observed that at ambient [CO_2], photosynthetic rates of soybean decreased above 32/22°C, while at elevated [CO_2] photosynthetic rates at 32/22 and 36/26°C were similar and a further increase in temperature to 40/30°C decreased leaf photosynthetic rates. This clearly suggests a greater and wider temperature optima for photosynthesis of plants grown at elevated [CO_2]. Similar responses were observed for cotton (Reddy et al., 1998) and in the mechanistic model predictions by Long (1991), where the temperature optimum for light saturated photosynthesis increased by 5°C with increases in [CO_2] from 350 to 700 μmol mol^{-1}.

There was no effect of elevated [CO_2] and/or temperature up to 40/30°C on midday total Rubisco activity or protein concentration, but both elevated temperature and [CO_2] decreased initial Rubisco activities and Rubisco activation (Vu et al., 1997). A further increase in temperature to 44/34 or 48/38°C significantly decreased midday Rubisco activity and Rubisco protein concentration (Vu et al., 2001).

Interaction studies on dry bean showed negative effects of temperature on CO_2 enhancement of leaf photosynthesis (Prasad et al., 2002). For example, CO_2 enrichment increased leaf photosynthetic rate by 66, 43, and 39% at temperature regimes of 28/18, 34/24, and 40/30°C, respectively (Figure 2). Two main reasons for a greater response of soybean photosynthesis at high temperature, as suggested by Morrison and Lawlor (1999) include (i) a decreased ratio of photosynthesis to photorespiration; and (ii) a decreased ratio of gross photosynthesis to dark respiration in warmer conditions. As discussed earlier, an increase in [CO_2] decreases oxygenase activity of Rubisco and the loss of carbon through photorespiration. On the other hand, increasing temperature would increase photorespiration by lowering solubility of [CO_2] compared to O_2 and decreasing specificity of Rubisco at higher temperature. Thus, the largest effect of increased [CO_2] on photosynthetic carbon fixation is observed at high temperature (Lawlor and Keys, 1993). However, this phenomenon is not observed in all C_3 species, for example in dry bean the percentage increase in photosynthesis was greater at 28/18°C compared to that at 40/30°C (Prasad et al., 2002), while in rice, percentage enhancement was not altered over a range of temperatures between 32/23 and 38/29°C (Vu et al., 1997). These species differences could partially be attributed to different temperature optima for photosynthesis and differential effects of temperature and [CO_2] on Rubisco (Sage et al., 1989; Long, 1991; Vu et al., 1997; Bunce, 1998).

Respiration

Bunce and Ziska (1996) studied the response of three soybean cultivars from three different maturity groups at three different temperatures (20/15, 25/20, and 31/26°C) and [CO_2] (370, 555 and 740 μmol mol^{-1}) and reported that at all temperatures, growth at elevated [CO_2] decreased total respiration when measured at the growth concentration. Growth respiration as estimated from elemental composition was not affected by [CO_2] at any temperature, but increased with temperature. Pan (1996) measured soybean respiration at five temperatures (28/18, 32/22, 36/26, 40/30, and 44/34°C) at elevated [CO_2] (700 μmol mol^{-1}) and at two temperatures (28/18, and 40/30°C) at ambient [CO_2] (350 μmol mol^{-1}). He reported no interaction between [CO_2] and temperature. Plants grown and measured at higher temperature had higher specific respiration rate, regardless of [CO_2].

Stomatal Conductance, Transpiration, and Water Use Efficiency

Studies on combined effects of elevated [CO_2] and temperature on water use are scarce. High [CO_2] decreases stomatal conductance, while high temperature increases stomatal conductivity and transpiration due an increased vapor pressure gradient between leaf surface and atmosphere. Therefore, a combination of elevated [CO_2] and temperature could give variable effects. Transpiration of soybean plants increased by about 4% per °C increase in temperature in the range of 28/23 to 35/23°C (Jones et al., 1985c). They also found that photosynthetic WUE of soybean plants was decreased at high temperature due to increased transpiration caused by greater evaporative demand. Similar observations were made in peanut canopies (Stronach et al., 1994).

Pan (1996) studied the interaction effects of high [CO_2] and temperature on soybean crop water use and concluded that the extent of decreases in evapotranspiration by elevated [CO_2] were offset by elevated temperature (>36°C). Thus, more water would be needed at high temperatures. These results confirm the earlier suggestion that both increased leaf area and increased foliage temperature at doubled [CO_2] can offset reductions in crop transpiration caused by elevated [CO_2] (Allen et al., 1985). Increased WUE due to decreased stomatal conductance at high [CO_2] were offset by increased temperature (Jones et al., 1985c; Allen et al., 1985; Allen, 1999). Polley (2002) concluded that yield benefits of improved water use efficiency at high [CO_2] are eroded due to decreased harvest index at high temperatures. Therefore, if global temperatures rise with increasing [CO_2], future crop water use will increase especially in arid subtropical and tropical regions where air temperatures are already high.

Phenology, Growth, Yield, and Seed Quality

Phenology

Studies have shown that both increasing temperature and increasing [CO_2] decreased the plastochron interval of soybean (Baker et al., 1989). However, the effects of temperature on phenology and development are much more prominent than those of [CO_2]. Temperature is the main environmental factor affecting phenology apart from photoperiod. Plants require a certain amount of heat units (thermal time) to reach each physiological development stage. At warmer temperatures, plants achieve these requirements earlier, thus enhancing physiological development in most plants. Increase in temperature within the optimal range shortens time to flowering in soybean (Baker et al., 1989), dry bean (Prasad et al., 2002), peanut (Bell et al., 1991; Awal and Ikeda, 2002), and cowpea (Craufurd et al., 1996). Long-term exposure to elevated [CO_2] in soybean (Valle et al., 1985b) and dry bean (Prasad et al., 2002) increased tissue (leaf) temperature by about 2°C compared to those at ambient [CO_2] at temperatures ranging from 28/18 to 40/30°C. This was mainly due to decreased transpirational cooling caused by partial stomatal closure. These increases in tissue temperature may explain the hastening of time to flowering in soybean grown at elevated [CO_2] as reported by Sionit et al. (1987a, b), Allen et al. (1988), and Baker et al. (1989). Exposure to excessively high temperature delayed time to seed initiation in soybean at ambient or elevated [CO_2] (Pan, 1996) and heat susceptible peanut genotype grown under ambient [CO_2] (Wheeler et al., 1997).

Organ Growth and Partitioning

Although the effects of elevated [CO_2] and temperature are well documented, few studies have focused on interaction between elevated [CO_2] and temperature (Sionit et al., 1987a and b; Baker et al., 1989; Ahmed et al., 1993; Clifford et al., 2000; Prasad et al., 2002). Sionit et al. (1987a) studied the effects of three temperatures (18/12, 22/16, and 26/20°C) at three [CO_2]'s (350, 675, and 1000 $\mu mol\ mol^{-1}$) on growth and dry matter production of soybean. They reported that increasing [CO_2] and temperature increased stem height, number of nodes, number of branches, number of leaves, leaf weight, stem weight and root weight and leaf area per plant. Similarly, increases in temperature from 26/19 to 36/29°C increased above ground dry matter and leaf area at both ambient and elevated [CO_2] (Baker et al., 1989). Studies of Sionit et al. (1987a, b) were at the cool end of optimum temperature range, while those of Baker et al. (1989) were slightly higher in the optimum range. Pan (1996) studied the effects of elevated [CO_2] at optimal and super-optimal temperatures (28/18, 32/22, 36/26, 40/30, and 44/34°C). It was observed that the optimum temperature for leaf area growth during the vegetative phase was 36/26°C, whereas lower or higher temperature decreased leaf area growth. At later stages, leaf area of plants grown at 44/34°C was greater than other treatments; this was because flowering and pod formation were delayed and decreased at high temperatures and available assimilates were used for continued leaf growth. There were no interactions between temperatures and [CO_2], and doubling of [CO_2] increased total dry matter at all temperatures. Elevated temperature increased dry matter partitioning to stems and leaves, while doubling of [CO_2] decreased partitioning to leaves and pods, and increased dry matter partitioning to stems.

In dry bean, Prasad et al. (2002) reported no interaction between elevated [CO_2] and temperatures on dry matter production and seed yield. As temperature increased from 28/18 to 40/30°C, total dry matter production decreased linearly by 1.6 g $plant^{-1}$ $°C^{-1}$ at both ambient and elevated [CO_2] (Figure 6).

Root Growth and Nitrogen Fixation

Interaction effects of elevated [CO_2] and temperature on root growth and nodule function of grain legumes have not been well studied. Separate studies have shown a significant influence of elevated [CO_2] (Rogers et al., 1992; Hardy and Havelka 1975) or soil and/or air temperature (Prasad et al., 2000) on root growth and nodule dry weights. As mentioned earlier, elevated [CO_2] increased root growth mass, growth and distribution in soybean. Conversely, high soil temperature significantly decreased root to shoot ratio, root growth, nodule weight and nodule numbers in peanut (Prasad et al., 2000, 2001a). Furthermore, it was noticed that decreases in number of nodules and nodule dry weights were more sensitive to high soil temperature than high air temperature (Prasad et al., 2001b). Various aspects of rhizobium-legume symbiosis are known to be affected by high soil temperatures including growth and survival of rhizobia, formation of root hairs, formation of infection threads, structural development of root nodules, and activity of nitrogenase enzyme, resulting in reduced nodule numbers, and nodule mass per plant (Giller and Willson, 1991). However, it is not known if deleterious effects of high temperature on root growth and nodule function could be overcome by elevated [CO_2].

Serraj et al. (1998) studied the combined effects of elevated [CO_2] and drought on nodulation and N_2 fixation of soybean. They concluded that under drought conditions, elevated [CO_2] delayed decreases in N_2 fixation rates associated with drying. Elevated [CO_2] also allowed the plants under drought to sustain significant increases in nodule number and mass relative to those under ambient [CO_2]. Thus, it is possible that future increases in [CO_2] will benefit soybean production by increasing the drought tolerance of N_2 fixation, increasing carbon assimilation, and reducing water requirement. Similar interaction studies with [CO_2] and elevated temperature need further attention.

Carbohydrate Accumulation

Carbohydrate metabolism in soybean leaves grown at elevated [CO_2] remained relatively high compared to leaves grown at ambient [CO_2] at temperatures ranging between 28/18 and 44/34°C (Vu et al., 2001). However, as temperature increased from 28/18 to 44/34°C, predawn and midday starch were decreased at both ambient and elevated [CO_2]. The net increase in starch and sugar accumulation (i.e., difference between predawn and midday values) was highest at 40/30°C, and decreased with a further increase or decrease in temperature to 28/18 or 48/38°C. Similarly the midday activity of Rubisco was decreased at temperatures above 32/23°C in both ambient and elevated [CO_2] (Vu et al., 1997). It was observed that even at a growth temperature of 44/34°C, photosynthesis and carbohydrate metabolism under elevated [CO_2] were greater than those grown at 28/18°C under ambient [CO_2] (Vu et al., 1997, 2001). They concluded that soybean carbohydrate metabolism and photosynthesis would not be hindered by predicted changes in elevated [CO_2] and/or temperatures.

Yield and Yield Components

Beneficial effects of elevated [CO_2] on photosynthesis, carbohydrate metabolism and vegetative growth are not always reflected in seed yields, because vegetative and reproductive processes have different responses to temperature and therefore show different stimulations by [CO_2] at elevated temperatures (Boote et al., 1997). In addition, reproductive development is more sensitive to high temperatures than overall plant biomass production, vegetative processes or photosynthesis. Temperatures that stimulate photosynthesis and vegetative growth are often detrimental to reproductive growth, resulting in lower yield. Lower seed yields at super-optimal temperatures are due mainly to a decreased number of fruits and a smaller seed size (Baker et al., 1989; Pan, 1996; Prasad et al., 2002). Lower fruit numbers at super-optimal temperatures are caused by decreased fruit-set, due mainly to decreased pollen production and poor pollen viability (Hall, 1992; Gross and Kigel, 1994; Prasad et al., 1999, 2000, 2002).

There were no interactions between temperature and [CO_2] on seed yields or harvest index. Seed yields of soybean were decreased at temperature >36/26°C, due to lower seed numbers per plant, lower numbers of seed per pod and smaller seed size (Pan, 1996; Thomas, 2001). Harvest index of soybean progressively decreased with increasing day temperature from 28/18 to 36/26°C in both elevated and ambient [CO_2] (Baker et al., 1989). Similarly, Prasad et al. (2002) reported no interaction effect between [CO_2] and temperature on seed-set in dry bean. As temperature increased from 28/18 to 40/30°C, seed-set was reduced by 6% per °C rise in temperature at both ambient and elevated [CO_2] (Figure 5). Furthermore, the fact that the ceiling temperature for seed set was 2°C lower at elevated [CO_2] suggests that yield losses associated with high temperature will increase with elevated [CO_2]. Similar effects were observed in rice (Matsui et al., 1997), where the critical temperature for pollen fertility was lower by 1.5 to 2°C in plants grown at elevated [CO_2] compared to those at ambient [CO_2]. Seed yields of rice decreased 10% per °C rise at temperatures from 26 to 36°C under both ambient and elevated [CO_2] (Baker and Allen, 1993). Both short-term and long-term exposure to super-optimal temperature decreased seed yields of soybean and peanut (Wheeler et al., 2000). Similar yield decreases due to super-optimal temperature were observed in several other crops (Reddy and Hodges, 2000). Exposure to temperatures > 28/18°C decreased seed number, seed yield per plant, number of seeds per pod, and seed size of dry bean at both ambient and elevated [CO_2] (Figures 6 and 7; Prasad et al., 2002).

Clearly, there are no beneficial interactions of [CO_2] at higher temperatures and damage caused by high temperature on reproductive growth and seed yield of soybean (Baker et al., 1989; Pan 1996; Ferris et al., 1999; Wheeler et al., 2000; Thomas, 2001), dry bean (Prasad et al., 2002), and cowpea (Hall, 1992) were not overcome by elevated [CO_2]. It is therefore concluded that greater [CO_2] enhancements with increasing temperature for photosynthesis and vegetative dry matter production are not applicable for reproductive growth, yield components, and seed yield.

The optimum temperature for seed yield and harvest index of soybean (Pan, 1996), peanut (Hammer et al., 1995) and cowpea is close to 25°C and even lower for dry bean (Prasad et al., 2002). Average temperatures of 25°C are very common in subtropical and tropical regions where these crops are currently grown. For example, mean air temperatures in Florida are about 26°C in the normal summer season, with up to 27°C in hot summers (Boote et al., 1997). Therefore, any warming of climate will decrease seed yields of these crops in regions where present temperatures are close to optimum irrespective of changes in [CO_2]. However in regions where temperatures are cool seed yield would be benefited by small increases in temperatures as long as they do not exceed optimum temperatures.

Seed Quality

There were no effects of [CO_2] or interaction between [CO_2] and temperature on seed size and percent shriveled seeds in soybean (Pan, 1996; Thomas,

2001). Increase in growth temperature from 28/18 to 40/30°C decreased seed size and increased the percentage of shriveled seed under both ambient and elevated [CO_2]. In a similar study on dry bean, Prasad et al. (2002) reported that seed size was significantly decreased with an increase in temperature above 31/21°C at both ambient and elevated [CO_2]. There was no effect of elevated [CO_2] or interaction between [CO_2] and temperature on number of seeds per pod and seed size (Figure 7).

Seed composition and transcript abundance were significantly affected by temperature, but not by [CO_2] or interaction of [CO_2] and temperature (Thomas et al., 2003). It was reported that oil concentration increased with increasing temperature with an optimum at 25 to 28°C, above which the oil concentration declined (Dornbos and Mullen, 1992; Gibson and Mullen, 1996a; Piper and Boote, 1999). In contrast, seed protein concentration of soybean was constant or slightly increased at temperatures between 16 and 25°C, but increased at temperatures above 25°C (Wolf et al., 1982; Dornbos and Mullen, 1992; Gibson and Mullen, 1996a; Piper and Boote, 1999).

Studies on soybean have shown that seeds obtained from plants grown at high day (35°C) and high night (30°C) temperatures had reduced subsequent seed germination and seedling vigor (Gibson and Mullen, 1996b). The extent of reductions in seed germination and vigor by high temperature was influenced by duration of temperature and the phase of reproductive growth. Greater reductions were observed with longer duration of exposure to high temperatures especially during seed fill and maturation. However, further research is required across other cultivars and crop species to clearly understand effects of super-optimal temperature and elevated [CO_2] on seed quality.

RESPONSE OF CULTIVARS TO [CO_2] AND/OR TEMPERATURE

Cultivars within the same crop species are reported to differ in their responses to [CO_2] and temperature and their interactions. Ziska et al. (2001) tested nine soybean cultivars differing in determinacy, maturity group and morphology in response to elevated [CO_2]. They found that although all cultivars showed a significant increase in seed yield with elevated [CO_2] (approximately 40%), Mandarin, an ancestral indeterminate cultivar, showed a greater relative response of seed yield to elevated [CO_2] (approximately 80%). The greater response of Mandarin was not related to any vegetative parameters, but to a greater pod weight per plant and seed weight from branches.

Ahmed et al. (1993) studied the response of three cowpea cultivars with varying heat tolerance to high night temperatures under ambient and elevated [CO_2] (Table 7). Results showed that under high night temperatures (30°C), the heat sensitive cultivars did not produce any flowers (cv. CB 5) and pods

TABLE 7. Reproductive responses of contrasting cowpea lines grown at two [CO_2] concentrations and two day/night temperature regimes. Values are means ± SEM. (Adapted from Ahmed et al., 1993.)

		Carbon dioxide concentration (μmol mol^{-1})							
Genotype	Temperature (day/night, °C)	350	700	350	700	350	700	350	700
		Period to first flower (d)		Flowers (no. $plant^{-1}$)		Pods (no. $plant^{-1}$)		Total shoot biomass (g $plant^{-1}$)	
CB 5	33/20	30.3 ± 0.3	30.8 ± 0.3	74 ± 6	84 ± 12	15 ± 1	26 ± 2	91 ± 3	109 ± 5
	33/30	++	++	0	0	-	-	76 ± 5	82 ± 10
7964	33/20	29.5 ± 0.3	30.3 ±0 .3	65 ± 6	64 ± 2	23 ± 2	27 ± 2	66 ± 3	89 ± 4
	33/30	24.8 ± 0.3	25.3 ± 0.2	100 ± 8	104 ± 8	0	0	58 ± 7	63 ± 3
518	33/20	30.5 ± 0.3	30.3 ± 0.3	70 ± 4	81 ± 12	28 ± 1	42 ± 5	89 ± 4	118 ± 5
	33/30	24.5 ± 0.3	25.6 ± 0.2	75 ± 3	70 ± 4	28 ± 1	40 ± 2	91 ± 5	103 ± 5

++: Did not flower during the experiment which lasted 6 weeks.

(cv. 7964) under both ambient and elevated [CO_2]. In contrast, the heat tolerant cultivar (cv. 518) produced a greater number of pods and a higher pod yield under high temperatures and responded strongly to CO_2 enrichment at high temperatures. Increased pod numbers in the heat tolerant cultivar were a result of greater fruit-set.

Cultivars of soybean (Sapra and Anaele, 1991; Srinivasan et al., 1996), peanut (Ntare et al., 2001; Craufurd et al., 2002, 2003; Kakani et al., 2002), dry bean (Agtunong et al., 1992), and cowpea (Ahmed et al., 1993; Ehlers and Hall, 1996) are known to vary in their response to high temperatures. However, further research is necessary to test if heat tolerant cultivars of legumes and other seed producing crops are more or less responsive to elevated [CO_2].

Cultivars with heat tolerance during reproductive development, high harvest index, high photosynthetic capacity per unit leaf area, small leaves and low leaf area per unit ground area, under present levels of [CO_2] will be most responsive to both high and intermediate temperatures (Hall and Allen, 1993). Therefore, research should be focused on searching for cultivars that show these characters and testing them under controlled and field environments. In addition, farmers may also make adjustments in planting dates, selection of crop, and selection of cultivars, so as to avoid severe yield losses. Future research should focus on designing suitable crop management practices for future climates, which will not only have higher [CO_2] and mean temperatures, but also more short episodes of extreme temperature events.

CONCLUSIONS

Photosynthetic rates of soybean, dry bean, peanut, and cowpea were enhanced when grown at high [CO_2]; however, the extent of stimulation varied with temperature. There were species differences in these responses, with soybean responding positively and dry bean responding negatively at higher temperatures. Decreases in stomatal conductivity and transpiration caused by partial stomatal closure under elevated [CO_2] were more than offset by increased leaf area and vapor pressure deficits caused by increased leaf temperature. These changes in energy balance of leaves resulted in similar total water use by crop canopies when expressed on a land area basis. There was very little or no effect of elevated [CO_2] on major phenostages such as flowering of these legumes, which were driven mainly by temperature and photoperiod. Elevated [CO_2] increased total biomass production due to larger total leaf area, greater stem growth, and greater root growth. Seed yields increased at elevated [CO_2] under optimal temperature and irrigated conditions due to greater numbers of pods and seeds caused by increased branching. However, at super-optimal temperature, fruit-set, number of seeds, seed growth rates, individual seed weight, and harvest index decreased under both ambient and elevated [CO_2]. Adverse effects of super-optimal temperature on seed yield and yield components were not offset by stimulation of photosynthesis or growth under elevated [CO_2]. In the future, if increases in [CO_2] are associated with increases in temperature, seed yield will decrease in regions where present temperatures are already close to optimum. Future research should be aimed at identifying heat tolerant cultivars that can produce more seeds, sustain them to maturity, and utilize available carbohydrates more efficiently to improve crop productivity under hostile climatic conditions.

REFERENCES

Acock, B. and L.H. Allen, Jr. (1985). Crop responses to elevated carbon dioxide concentration. pp 53-97. In: B.R. Strain and J.D. Cure (eds.), *Direct Effects of Increasing Carbon Dioxide on Vegetation*. DOE/ER-0238. U.S. Department of Energy, Carbon Dioxide Research Division, Washington, DC.

Acock, B., V.R. Reddy, H.F. Hodges, D.N. Baker and J.M. McKinion (1985). Photosynthetic response of soybean canopies to full season carbon dioxide enrichment. *Agronomy Journal* 77: 942-947.

Agtunong, T.P., R. Redden, M.A. Menggenang, C. Searle and S. Fukai. (1992). Genotypic variation in response to high-temperature at flowering in common bean (*Phaseolus vulgaris* L). *Australian Journal of Experimental Agriculture* 32:1135-1140.

Ahmed, F.E., A.E. Hall and M.A. Madore. (1993). Interactive effects of high-temperature and elevated carbon dioxide concentration on cowpea [*Vigna unguiculata* (L.) Walp]. *Plant Cell and Environment* 16:835-842.

Ainsworth, E.A., P.A. Davey, C.J. Bernacchi, O.C. Dermody, E.A. Heaton, D.J. Moore, P.B. Morgan, S.L. Naidu, H.S.Y. Ra, X.G. Zhu, P.S. Curtis and S.P. Long. (2002). A meta-analysis of elevated [CO_2] effects on soybean (*Glycine max*) physiology, growth and yield. *Global Change Biology* 8:695-709.

Allen, L.H., Jr., J.T. Baker and K.J. Boote. (1996). The CO_2 fertilization effect: higher carbohydrate production and retention as biomass and seed yield. pp. 65-100. In: F. Bazzaz and W. Sombroek (eds.), *Global Climate Change and Agricultural Production*. FAO, Rome, Italy.

Allen, L.H., Jr. (1999). Evapotranspiration responses of plant and crops to carbon dioxide and temperature. *Journal of Crop Production* 2: 37-70.

Allen, L.H., Jr. and K.J. Boote. (2000). Crop ecosystems responses to climate change: soybean. pp, 133-160. In: K.R. Reddy and H.F. Hodges (eds.), *Climate Change and Global Crop Productivity*. CABI Publishing, Oxon, UK.

Allen, L.H., Jr., E.C. Bisbal and K.J. Boote. (1998). Nonstructural carbohydrates of soybean plants grown in subambient and superambient levels of CO_2. *Photosynthesis Research* 56:143-155.

Allen, L.H., Jr., E.C. Bisbal, K.J. Boote and P.H. Jones. (1991). Soybean dry matter allocation under subambient and superambient levels of carbon dioxide. *Agronomy Journal* 83: 875-883.

Allen, L.H., Jr., E.C. Bisbal, W.J. Campbell and K.J. Boote. (1990). Carbon dioxide effects on soybean developmental stages and expansive growth. *Soil and Crop Science Society of Florida Proceedings* 49:124-131.

Allen, L.H., Jr., J.C.V. Vu, R.R. Valle, K.J. Boote and P.H. Jones. (1988). Nonstructural carbohydrates and nitrogen of soybean grown under carbon dioxide enrichment. *Crop Science* 28: 88-94.

Allen, L.H., Jr., P.H. Jones and J.W. Jones. (1985). Rising atmospheric CO_2 and evapotranspiration. pp. 13-17. In: *Advances in Evapotranspiration. ASAE Publication* no. 14-16. American Society of Agricultural Engineers, St. Joseph, Michigan.

Allen, L.J., Jr. (1994). Carbon dioxide increase: direct impacts on crops and indirect effects mediated through anticipated climate changes. pp. 425-459. In: K.J. Boote, J.M. Bennett, T.R. Sinclair and G.M Paulsen (eds.), *Physiology and Determination of Crop Yield*. ASA, CSSA and SSSA, Madison, Wisconsin.

Allen. L.H., Jr., E.C. Bisbal, W.J. Campbell and K.J. Boote. (1987). Responses of vegetation to rising CO_2: Photosynthesis, biomass and seed yield of soybean. *Global Biochemical Cycles* 1: 1-14.

Amthor, J.S. (1991). Respiration in future, higher CO_2 world. *Plant Cell and Environment* 14: 13-20.

Amthor, J.S. (1997). Plant respiratory responses to elevated carbon dioxide partial pressure. pp. 35-77. In: L.H. Allen, M.B. Kirkhan, D.M. Olsztk and C.E. Whitman (eds.). *Advances in Carbon Dioxide Effects Research. ASA Special Publication* No. 61, ASA-CSSA-SSSA, Madison, Wisconsin.

Arp, W.J. (1991). Effects of source-sink relations on photosynthetic acclimation to elevated CO_2. *Plant Cell and Environment* 14:869-875.

Awal, M.A. and T. Ikeda. (2002). Effects of changes in soil temperature on seedling emergence and phonological development in field-grown stands of peanut (*Arachis hypogaea*). *Environmental and Experimental Botany* 47: 101-113.

Baker, J.T. and L.H. Allen, Jr. (1993). Contrasting crop species responses to CO_2 and temperature-rice, soybean and citrus. *Vegetatio* 104:239-260.

Baker, J.T., L.H. Allen, Jr., K.J. Boote, P. Jones and J.W. Jones. (1989). Response of soybean to air temperature and carbon dioxide concentration. *Crop Science* 29: 98-105.

Bell, M.J., R. Shorter and R. Mayer. (1991). Cultivar and environmental effects on growth and development of peanut (*Arachis hypogaea* L.). 1. Emergence to flowering. *Field Crops Research* 27: 17-33.

Bhattacharya, N.C., S. Bhattacharya, P.K. Biswas and B.R. Strain. (1987). Partitioning of photoassimilates into carbohydrates and proteins of cowpea (*Vigna unguiculata* L.) to CO_2 enrichment environment during different stages of vegetative and reproductive growth. *Hortscience* 22:1113-1113.

Bhattacharya, S., N.C. Bhattacharya, P.K. Biswas and B.R. Strain. (1985). Response of cow pea (*Vigna unguiculata* L.) to CO_2 enrichment environment on growth, dry matter production and yield components at different stages of vegetative and reproductive growth. *Journal of Agricultural Sciences* (Cambridge) 105:527-534.

Biswas, P.K. and D.R. Hileman. (1985). *Response of Vegetation to Carbon Dioxide: Field Studies of Sweet Potato and Cowpea in Response to Elevated Carbon Diox-*

ide. Report 022, U.S. Department of Energy, Carbon Dioxide Research Division, Office of Energy Research, Washington, DC.

Boote, K.J., N.B. Pickering and L.H. Allen, Jr. (1997). Plant modeling: advances and gaps in our capability to predict future crop growth and yield. pp. 179-228. In: L.H. Allen, Jr., M.B. Kirkhan, D.M. Olsztk and C.E. Whitman (eds.). *Advances in Carbon Dioxide Effects Research*. ASA Special Publication No. 61, ASA-CSSA-SSSA, Madison, Wisconsin.

Bottomley, P.A. and H.H. Rogers. (1993). NMR imaging of root water distribution in intact *Vicia faba* L. plants in elevated atmospheric CO_2. *Plant Cell and Environment* 16: 335-338.

Bowes G. (1996) Photosynthetic responses to changing atmospheric carbon dioxide. pp. 387-407. In: N.R. Baker (ed.). *Photosynthesis and the Environment. Advances in Photosynthesis* Vol. 5, Kluwer, Dordrecht.

Bunce, J.A. (1998). The temperature dependence of the stimulation of photosynthesis by elevated carbon dioxide in wheat and barley. *Journal of Experimental Botany* 49:1555-1561.

Bunce, J.A. and F. Caulfield. (1991). Reduced respiratory carbon dioxide efflux during growth at elevated carbon dioxide in 3 herbaceous perennial species. *Annals of Botany* 67: 325-330.

Bunce, J.A. and L.H. Ziska. (1996). Responses of respiration to increases in carbon dioxide concentration and temperature in three soybean cultivars. *Annals of Botany* 77:507-514.

Campbell, W.J., L.H. Allen, Jr. and G. Bowes. (1988). Effects of CO_2 Concentration on rubisco activity, amount, and photosynthesis in soybean leaves. *Plant Physiology* 88:1310-1316.

Campbell, W.J., L.H. Allen, Jr. and G. Bowes. (1990). Response of soybean canopy photosynthesis to CO_2 concentration, light, and temperature. *Journal of Experimental Botany* 41:427-433.

Chen, J.J. and J.M. Sung. (1990). Gas exchange rate and yield responses of Virginia type peanut to carbon dioxide enrichment. *Crop Science* 30:1085-1089.

Chen, X.M., G.B. Begonia, D.M. Alm and J.D. Hesketh. (1995). Sooybean stomatal acclimation to long-term exposure to CO_2-enriched atmospheres. *Photosynthetica* 31: 51-57.

Clifford, S.C., C.R. Black, J.A. Roberts, I.M. Stronach, P.R. Singletonjones, A.D. Mohamed and S.N. Azamali. (1995). The effect of elevated atmospheric CO_2 and drought on stomatal frequency in groundnut (*Arachis hypogaea* L.). *Journal of Experimental Botany* 46:847-852.

Clifford, S.C., I.M. Stronach, A.D. Mohamed, S.N. Azamali and N.M.J. Crout. (1993). The effects of elevated atmospheric carbon dioxide and water stress on light interception, dry matter production and yield in stands of groundnut (*Arachis hypogaea* L.). *Journal of Experimental Botany* 44:1763-1770.

Clifford, S.C., I.M. Stronach, C.R. Black, P.R. Singleton-Jones, S.N. Azam-Ali and N.M.J. Crout. (2000). Effects of elevated CO_2, drought and temperature on the water relations and gas exchange of groundnut (*Arachis hypogaea*) stands grown in controlled environment glasshouses. *Physiologia Plantarum* 110:78-88.

Coleman, J.S., K.D.M. McConnaughay and F.A. Bazzaz. (1993). Elevated CO_2 and plant nitrogen use–is reduced tissue nitrogen concentration size dependent. *Oecologia* 93: 195-200.

Cowling, S.A. and R.F. Sage. (1998). Interactive effects of low atmospheric CO_2 and elevated temperature on growth, photosynthesis and respiration in *Phaseolus vulgaris*. *Plant Cell and Environment* 21: 427-435.

Craufurd, P.Q., E.H. Roberts, R.H. Ellis and R.J. Summerfield. (1996). A stability analysis of time to flowering as a screen for responsiveness to temperature and photoperiod in cowpea (*Vigna unguiculata*). *Euphytica* 88: 77-84.

Craufurd, P.Q., P.V.V. Prasad and R.J. Summerfield. (2002). Dry matter production and harvest index at high temperature in peanut. *Crop Science* 42: 146-151.

Craufurd, P.Q., P.V.V. Prasad, V.G. Kakani, T.R. Wheeler and S.N. Nigam. (2003). Heat tolerance in peanut. *Field Crops Research* 80: 63-77.

Cure, J.D. (1985). Carbon dioxide doubling response: a crop survey. pp 99-116. In: B.R. Strain and J.D. Cure (eds.). *Direct Effects of Increasing Carbon Dioxide on Vegetation*. DOE/ER-0238. U.S. Department of Energy, Carbon Dioxide Research Division, Washington, DC.

Cure, J.D. and B. Acock. (1986). Crop responses to carbon dioxide doubling: a literature survey. *Agricultural and Forest Meteorology* 38: 127-145.

Cure, J.D., T.W. Rufty and D.W. Israel. (1988). Phosphorus stress effects on growth and seed yield responses of nonnodulating soybean to elevated carbon dioxide. *Agronomy Journal* 80: 897-902.

Davis, W.J., T.A. Mansfield and A.M. Hetherington. (1990). Sensing of soil water status and the regulation of plant growth and development. *Plant Cell and Environment* 13: 709-719.

Dornbos, D.L. and R.E. Mullen. (1992). Soybean seed protein and oil content and fatty-acid composition adjustments by drought and temperature. *Journal of the American Oil Chemists Society* 69: 228-231.

Drake, B.G., J. Azcon-Bieto, J. Berry, J. Bunce, P. Dijkstra, J. Farrar, R.M. Gifford, M.A. Gonzalez-Meler, G. Koch, H. Lambers, J. Siedow and S. Wullschleger. (1999). Does elevated atmospheric CO_2 concentration inhibit mitochondrial respiration in green plants? *Plant Cell and Environment* 22: 649-657.

Ehlers, J.D. and A.E. Hall. (1996). Genotypic classification of cowpea based on responses to heat and photoperiod. *Crop Science* 36: 673-679.

Ehret, D.L. and P.A. Jolliffe. (1985). Photosynthetic carbon dioxide exchange of bean plants grown at elevated carbon dioxide concentrations. *Canadian Journal of Botany* 63:2026-2030.

Ellis, R.H., P.Q. Craufurd, R.J. Summerfield and E.H. Roberts. (1995). Linear relations between carbon dioxide concentration and rate of development towards flowering in sorghum, cowpea and soybean. *Annals of Botany* 75: 193-198.

FAO. (2001). *Food and Agricultural Organization: Statistical Data Base*. www.fao.org/.

Ferris, R., T.R. Wheeler, P. Hadley and R.H. Ellis. (1998). Recovery of photosynthesis after environmental stress in soybean grown under elevated CO_2. *Crop Science* 38:948-955.

Ferris, R., T.R. Wheeler, R.H. Ellis and P. Hadley. (1999). Seed yield after environmental stress in soybean grown under elevated CO_2. *Crop Science* 39: 710-718.
Finn, G.A. and W.A. Brun. (1982). Effects of atmospheric CO_2 enrichment on growth, nonstructural carbohydrate content and root nodule activity in soybean. *Plant Physiology* 69: 327-331.
Gibson, L.R. and R.E. Mullen. (1996a). Soybean seed composition under high day and night temperatures. *Journal of the American Oil Chemists Society* 73: 733-737.
Gibson, L.R. and R.E. Mullen. (1996b). Soybean seed quality reductions by high day and night temperature. *Crop Science* 36: 1615-1619.
Gifford, R.M., H. Lambers and J.I.L. Morrison. (1985). Respiration of crop species under CO2 enrichment. *Physiologia Plantarum* 63: 351-356.
Giller, K.E. and K.J. Wilson (1991). *Nitrogen Fixation in Tropical Cropping Systems.* CABI Publishing, Oxon, England, UK.
Gross, Y. and J. Kigel. (1994). Differential sensitivity to high temperature of stages in the reproductive development of common bean (*Phaseolus vulgaris* L.). *Field Crops Research* 36: 201-212.
Hall, A.E. (1992). Breeding for heat tolerance. *Plant Breeding Review* 10: 129-168.
Hall, A.E. and L.H. Allen. (1993). Designing cultivars for the climatic conditions of the next century. pp. 291-297. In: D.R. Buxton et al. (eds.). *International Crop Science Congress I*, CSSA, Madison, Wisconsin, USA.
Hammer, G.L., T.R. Sinclair, K.J. Boote, G.C. Wright, H. Meinke and M.J. Bell. (1995). A peanut simulation model: I. Model development and testing. *Agronomy Journal* 87: 1085-1093.
Hardman, L.L. and W.A. Brun. (1971). Effects of atmospheric carbon dioxide enrichment at different developmental stages on growth and yield components of soybeans. *Crop Science* 11: 886-888.
Hardy, R.W.F. and U.D. Havelka. (1973). Symbiotic N_2 fixation: multifold enhancement by CO_2 enrichment of field grown soybean. *Plant Physiology* 48: 35.
Hardy, R.W.F. and U.D. Havelka. (1975). Photosynthate as a major factor limiting nitrogen fixation by field-grown legumes with emphasis on soybean. pp. 421-439. In: P.S. Nutman (ed.). *Symbiotic Nitrogen Fixation in Plant.* Cambridge University Press, London, England.
Harlan, J.R. (1992). *Crops and Man*, 2nd edn. American Society of Agronomy and Crop Science Society of American, Madison, Wisconsin, USA.
Hesketh, J.D., J.T. Woolley and D.B. Peters. (1984). Leaf photosynthetic CO_2 exchange-rates in light and CO_2 enriched environments. *Photosynthetica* 18: 536-540.
Houghton J.T., Y. Ding, D.J. Griggs, M. Noguer, P.J. van der Linden and D. Xiaosu (Eds.). (2001). *Climate Change 2001: The Scientific Basis. Contribution of Working Group I to the Third Assessment Report of the Intergovernmental Panel on Climate Change (IPCC).* Cambridge University Press, UK. pp. 944.
Huber, S.C., H.H. Rogers and F.L. Mowry. (1984). Effects of water stress on photosynthesis and carbon partitioning in soybean (*Glycine max* [L.] Merr) plants grown in the field at different CO_2 levels. *Plant Physiology* 76:244-249.
Idso, S.B. and K.E. Idso. (2001). Effects of atmospheric CO_2 enrichment on plant constituents related to animal and human health. *Environmental and Experimental Botany* 45: 179-199.

Jablonski, L.M., X.Z. Wang and P.S. Curtis. (2002). Plant reproduction under elevated CO_2 conditions: a meta-analysis of reports on 79 crop and wild species. *New Phytologist* 156: 9-26.

Jahnke, S. (2001). Atmospheric CO_2 concentration does not directly affect leaf respiration in bean or poplar. *Plant Cell and Environment* 24: 1139-1151.

Jahnke, S. and M. Krewitt. (2002). Atmospheric CO_2 concentration may directly affect leaf respiration measurement in tobacco, but not respiration itself. *Plant Cell and Environment* 25: 641-651.

Jolliffe, P.A. and D.L. Ehret. (1985). Growth of bean plants at elevated carbon dioxide concentrations. *Canadian Journal of Botany* 63: 2021-2025.

Jones, P., J.W. Jones and L.H. Allen, Jr. (1985a). Seasonal canopy CO_2 exchange, water use, and yield components in soybean grown under different CO_2 and water stress conditions. *Transactions in American Society of Agricultural Engineers* 28: 2021-2028.

Jones, P., J.W. Jones and L.H. Allen, Jr. (1985c). Response of soybean canopy growth, photosynthesis, and transpiration to whole-day temperature changes in different CO2 environment. *Agronomy Journal* 77: 119-126.

Jones, P., L.H. Allen, Jr., J.W. Jones and R. Valle. (1985b). Photosynthesis and transpiration responses of soybean canopies to short term and long term CO_2 treatments. *Agronomy Journal* 77: 119-126.

Jones, P., L.H. Allen, Jr., J.W. Jones, K.J. Boote and W.J. Campbell. (1984). Soybean canopy growth, photosynthesis, and transpiration responses to whole season carbon dioxide enrichment. *Agronomy Journal* 76: 633-637.

Kakani, V.G., P.V.V. Prasad, P.Q. Craufurd and T.R. Wheeler. (2002). Response of *in-vitro* pollen germination and pollen tube growth of peanut to high temperature. *Plant Cell and Environment* 25: 1651-1661.

Kimball, B.A. (1983). Carbon dioxide and agricultural yield: an assemblage and analysis of 430 prior observations. *Agronomy Journal* 75: 779-788.

Kimball, B.A. (1985). Adaptation of vegetation and management practices to a higher carbon dioxide world. pp. 185-204. In: B.R. Strain and J.D. Cure (eds.). *Direct Effects of Increasing Carbon Dioxide on Vegetation.* U.S. Department of Energy, DOE/ER-0238, Washington, DC.

Kimball, B.A., J.R. Mauney, G. Guinn, F.S. Nakayama, P.J. Pinter, Jr., K.L. Clawson, S.B. Idso, G.D. Butler and J.R. Radin. (1984). *Response of Vegetation to Carbon Dioxide, Ser 023. Effect of Increasing Atmospheric CO_2 on the Yield and Water of crops.* U.S. Department of Energy and USDA. U.S. Water Conservation Laboratory and U.S. Western Cotton Research Laboratory, USDA-ARS, Phoenix, AZ.

Lawlor, D.W. and A.J. Keys. (1993). Understanding photosynthetic adaptation to changing climate. pp. 85-106. In: L. Fowden, T.A. Mansfield and J. Stoddart (eds.). *Plant Adaptation to Environmental Stress.* Chapman and Hill, London.

Leadley, P.W. and J.F. Reynolds. (1988). Effects of elevated carbon dioxide on estimation of leaf area and leaf dry weights of soybean. *American Journal of Botany* 75: 1771-1774.

Long, S.P. (1991). Modification of the response of photosynthetic productivity to rising temperature by atmospheric CO_2 concentration: has its importance been underestimated? *Plant Cell and Environment* 14: 729-739.

Long, S.P. and J.E. Hallgren. (1993). Measurement of CO_2 assimilation by plants in the field and the laboratory. pp. 129-167. In: D.O. Hall, J.M.O. Scurlock, H.R. Bolhar-Nordenkampf, R.C. Leegood and S.P. Long (eds.). *Photosynthesis and Production in Changing Environment. A Field and Laboratory Manual.* Chapman and Hall, London.

Lopez, F.B., T.L. Setter and C.R. McDavid. (1987). Carbon dioxide and light responses of photosynthesis in cowpea and pigeonpea during water deficit and recovery. *Plant Physiology* 85: 990-995.

Matsui, T., O.S. Namuco, L.H. Ziska and T. Horie. (1997). Effects of high temperature and CO_2 concentration on spikelet sterility in indica rice. *Field Crops Research* 51: 213-219.

Mjwara, J.M., C.E.J. Botha and S.E. Radloff. (1996). Photosynthesis, growth and nutrient changes in non-nodulated *Phaseolus vulgaris* grown under atmospheric and elevated carbon dioxide conditions. *Physiologia Plantarum* 97: 757-763.

Morison, J.I.L. and D.W. Lawlor. (1999). Interactions between increasing CO_2 concentration and temperature on plant growth. *Plant Cell and Environment* 22: 659-682.

Morison, J.I.L. and R.A. Gifford. (1984). Plant growth and water use with limited water supply high CO_2 concentrations. *Australian Journal of Plant Physiology* 11: 361-374.

Mortley, D.G., P.A. Loretan, J.H. Hill and J. Seminara. (1997). CO_2 enrichment influences yields of Florunner, Gerogia Red and New Mexico peanut cultivars. *Advances in Space Research* 20: 1905-1908.

Murray, D.R. (1995). Plant responses to carbon dioxide. *American Journal of Botany* 82: 690-697.

Murray, D.R. (1997). *Carbon Dioxide and Plant Responses.* Research Studies Press Ltd, Somerset, England.

Ntare, B.R., J.H. Williams and F. Dougbedji. (2001). Evaluation of groundnut genotypes for heat tolerance under field conditions in a Sahelian environment using a simple physiological model for yield. *Journal of Agricultural Science* 136: 81-88.

Ortega, J.L., F. Sanchez, M. Soberon and M.L. Flores. (1992). Regulation of nodule glutamine-synthetase by CO2 levels in bean (*Phaseolus vulgaris* L.). *Plant Physiology* 98: 584-587.

Overdieck, D., C. Reid and B.R. Strain. (1988). The effects of preindustrial and future CO_2 concentrations on growth, dry matter production and the C/N relationship in plants at low nutrient supply: *Vigna unguiculata* (Cowpea), *Abelmoschus esculentus* (Orka) and *Raphinus sativus* (Radish). *Angewandte Botanik* 62: 119-134.

Pan, D. (1996) *Soybean Responses to Elevated Temperature and Doubled* CO_2. PhD Dissertation. University of Florida, Gainesville, USA. Dissertation Abstracts International 57, no. 10B, (1996): 5987. Accession No. AAG709292.

Phillips, D.A., K.D. Newell, S.A. Hassell and C.E. Felling. (1976). The effect of CO_2 enrichment on root nodule development and symbiotic N_2 reduction in *Pisum sativum* L. *American Journal of Botany* 63: 356-362.

Pickering, N.B., L.H. Allen, Jr., S.L. Albrecht, P. Jones, J.W. Jones and J.T. Baker. (1994). Environmental plant chambers: controls and measurements using CR-10T data loggers. pp. 29-35. In: D.G. Watson, F.S. Zuzueta and T.V. Harrison. (eds.).

Computers in Agriculture: Proceedings of the 5th International Conference, Orlando, Florida 5-9 Feb. American Society of Agricultural Engineers, St. Joseph, Michigan, USA.
Piper, E.L. and K.J. Boote. (1999). Temperature and cultivar effects on soybean seed oil and protein concentrations. *Journal of American Oil Chemists Society* 76: 1233-1241.
Polley, H.W. (2002). Implications of atmospheric and climate change on crop yield and water use efficiency. *Crop Science* 42: 131-140.
Porter, M.A. and B. Grodzinski. (1984). Acclimation to high CO_2 in bean: carbonic anhydrase and ribulose bisphosphate carboxylase. *Plant Physiology* 74: 413-416.
Prasad, P.V.V., K.J. Boote, L.H. Allen, Jr. and J.M.G. Thomas. (2002). Effects of elevated temperature and carbon dioxide on seed-set and yield of kidney bean (*Phaseolus vulgaris* L.). *Global Change Biology* 8: 710-721.
Prasad, P.V.V., P.Q. Craufurd and R.J. Summerfield. (1999). Fruit number in relation to pollen production and viability in groundnut exposed to short episodes of heat stress. *Annals of Botany* 84: 381-386.
Prasad, P.V.V., P.Q. Craufurd and R.J. Summerfield. (2000). Effect of high air and soil temperature on dry matter production, pod yield and yield components of groundnut. *Plant and Soil* 222: 231-239.
Prasad, P.V.V., P.Q. Craufurd and R.J. Summerfield. (2001a). Response of groundnuts dependent on symbiotic and inorganic nitrogen to high air and soil temperatures.. *Journal of Plant Nutrition* 24: 623-637.
Prasad, P.V.V., P.Q. Craufurd, R.J. Summerfield and T.R. Wheeler. (2000). Effects of short episodes of heat stress on flower production and fruit-set of groundnut (*Arachis hypogaea* L.). *Journal of Experimental Botany* 51: 777-784.
Prasad, P.V.V., P.Q. Craufurd, V.G. Kakani, T.R. Wheeler and K.J. Boote. (2001b). Influence of high temperature during pre- and post-anthesis stages of floral development on fruit-set and pollen germination in peanut. *Australian Journal of Plant Physiology* 28: 233-240.
Prior, S.A., H.H. Rogers, G.B. Runion and J.R. Mauney (1994). Effects of free-air enrichment on cotton root growth. *Agricultural and Forest Meteorology* 70: 69-86.
Radoglou, K.M., P. Aphalo and P.G. Jarvis. (1992). Response of photosynthesis, stomatal conductance and water use efficiency to elevated CO_2 and nutrient supply in acclimated seedlings of *Phaseolus vulgaris* L. *Annals of Botany* 70: 257-264.
Reardon, J.C., J.R. Lambert and B. Acock. (1990). The influence of carbon dioxide enrichment on the seasonal patterns of nitrogen fixation in soybean. In: *Responses of Vegetation to Carbon Dioxide.* Series 016, US Department of Energy and USDA-ARS, Washington DC.
Reddy, K. R. and H.F. Hodges. (eds.) (2000). *Climate Change and Global Crop Productivity.* CABI Publishing, Oxon, UK.
Reddy, A.R., K.R. Reddy and H.F. Hodges. (1998). Interactive effects of elevated carbon dioxide and growth temperature on photosynthesis in cotton leaves. *Plant Growth Regulation* 26: 33-40.
Reddy, K.R., H.F. Hodges and J.M. McKinion. (1997). Crop modeling and applications: a cotton example. *Advances in Agronomy* 59: 225-290.

Reddy, V.R., B. Acock and M.C. Acock. (1989). Seasonal carbon and nitrogen accumulation in relation to net carbon dioxide exchange in a carbon dioxide enriched soybean. *Agronomy Journal* 81: 78-83.

Robbins, N.S. and D.M. Pharr. (1988). Effect of restricted root growth and carbohydrate metabolism and whole plant growth of *Cucumis sativus* L. *Plant Physiology* 87: 409-413.

Rogers, H.H., C.M. Peterson, J.N. McCrimmon and J.D. Cure. (1992). Response of plant roots to elevated atmospheric carbon dioxide. *Plant Cell and Environment* 15: 749-752.

Rogers, H.H., G.B. Runion and S.V. Krupa. (1994). Plant responses to atmospheric CO_2 enrichment with emphasis on roots and the rhizosphere. *Environmental Pollution* 83: 155-189.

Rogers, H.H., G.B. Runion, S.V. Krupa and S.A. Prior. (1997). Plant responses to atmospheric CO_2 enrichment: implications in root-soil-microbe interactions. pp. 1-34. In: L.H. Allen, M.B. Kirkhan, D.M. Olsztk and C.E. Whitman (eds.). *Advances in Carbon Dioxide Effects Research.* ASA Special Publication No. 61, ASA-CSSA-SSSA, Madison, Wisconsin.

Rogers, H.H., J.D. Cure and J.M. Smith. (1986). Soybean growth and yield response to elevated carbon dioxide. *Agriculture Ecosystems and Environment* 16: 113-128.

Rogers, H.H., J.D. Cure, J.F. Thomas and J.M. Smith. (1984). Influence of elevated CO_2 on growth of soybean plants. *Crop Science* 24: 361-366.

Rogers, H.H., S.A. Prior and J.M. Smith. (1986). Soybean growth and yield response to elevated carbon dioxide. *Agricultural Ecosystems and Environment* 16: 113-128.

Sage, R.F. (1994). Acclimation of photosynthesis to increasing atmospheric CO_2–the gas exchange perspective. *Photosynthesis Research* 39: 351-368.

Sage, R.F., T.D. Sharkey and J.R. Seemann. (1989). Acclimation of photosynthesis to elevated CO_2 in 5 C_3 species. *Plant Physiology* 89: 590-596.

Sapra, V.T. and A.O. Anaele. (1991). Screening soybean genotypes for drought and heat tolerance. *Journal of Agronomy and Crop Science* 167: 96-102.

Serraj, R., T.R. Sinclair and L.H. Allen, Jr. (1998). Soybean nodulation and N_2 fixation response to drought under carbon dioxide enrichment. *Plant Cell and Environment* 21: 491-500.

Sharkey, T.D., J.A. Berry and K. Raschke. (1985). Starch and sucrose synthesis in *Phaseolus vulgaris* as affected by light, CO_2, and abscisic acid. *Plant Physiology* 77: 617-620.

Sionit, N., B.R. Strain and E.P. Flint. (1987a). Interaction of temperature and CO_2 enrichment on soybean-growth and dry matter partitioning. *Canadian Journal of Plant Science* 67: 59-67.

Sionit, N., B.R. Strain and E.P. Flint. (1987b). Interaction of temperature and CO_2 enrichment on soybean-photosynthesis and seed yield. *Canadian Journal of Plant Science* 67: 629-636.

Sionit, N., H.H. Rogers, G.E. Bingham and B.R. Strain. (1984). Photosynthesis and stomatal conductance with CO_2-enrichment of container- and field-grown soybeans. *Agronomy Journal* 76: 447-451.

Socias, F.X., H. Medrano and T.D. Sharkey. (1993). Feedback limitation of photosynthesis of *Phaseolus vulgaris* L. grown in elevated CO_2. *Plant Cell and Environment* 16: 81-86.

Srinivasan, A., H. Takeda and T. Senboku. (1996). Heat tolerance in food legumes as evaluated by cell membrane thermostability and chlorophyll fluorescence techniques. *Euphytica* 88: 35-45.
Stanciel, K., D.G. Mortley, D.R. Hileman, P.A. Loretan, C.K. Bonsi and W.A. Hill. (2000). Growth, pod, and seed yield, and gas exchange of hydroponically grown peanut in response to CO_2 enrichment. *Hortscience* 35: 49-52.
Stitt, M. (1991). Rising CO_2 levels and their potential significance for carbon flow in photosynthetic cells. *Plant Cell and Environment* 14: 741-762.
Stronach, I.M., S.C. Clifford, A.D. Mohamed, P.R. Singletonjones, S.N. Azamali and N.M.J. Crout. (1994). The effects of elevated carbon dioxide, temperature and soil moisture on the water-use of stands of groundnut (*Arachis hypogaea* L.). *Journal of Experimental Botany* 45: 1633-1638.
Tardieu, F. and W.J. Davis. (1993). Integration of hydraulic and chemical signaling in the control of stomatal conductance and water status of droughted plants. *Plant Cell and Environment* 16: 341-349.
Thomas, J.F. and C.N. Harvey. (1983). Leaf anatomy of four species grown under continuous CO_2 enrichment. *Botanical Gazette* 144: 303-309.
Thomas, J.M.G. (2001). *Impact of Elevated Temperature and Carbon Dioxide on Development and Composition of Soybean Seed.* PhD Dissertation. University of Florida, Gainesville, USA. Dissertation Abstracts International 62, no. 09B, (2001): 3856. Accession No. AAI3025125.
Thomas, J.M.G., K.J. Boote, L.H. Allen, Jr., M. Gallo-Meagher and J.M. Davis. (2003). Elevated temperature and carbon dioxide effects of soybean seed composition and transcript abundance. *Crop Science* 43:1548-1557.
Thomas, R.B. and K.L. Griffin. (1994). Direct and indirect effects of atmospheric carbon dioxide enrichment on leaf respiration of *Glycine max* (L.) Merr. *Plant Physiology* 104: 355-361.
Valle, R., J.W. Mishoe, J.W. Jones and L.H. Allen, Jr. (1985b). Transpiration rate and water-use efficiency of soybean leaves adapted to different CO_2 environments. *Crop Science* 25: 477-482.
Valle, R., J.W. Mishoe, W.J. Campbell, J.W. Jones and L.H. Allen, Jr. (1985a). Photosynthetic responses of Bragg soybean leaves adapted to different CO_2 environments. *Crop Science* 25: 333-339.
Vu, J.C.V., L.H. Allen, Jr. and G. Bowes. (1989). Leaf ultra-structure, carbohydrates, and protein of soybeans grown under CO_2 enrichment. *Environmental and Experimental Botany* 29: 141-147.
Vu, J.C.V., L.H. Allen, Jr., K.J. Boote and G. Bowes. (1997). Effects of elevated CO_2 and temperature on photosynthesis and Rubisco in rice and soybean. *Plant Cell and Environment* 20: 68-76.
Vu, J.C.V., R.W. Gesch, A.H. Pennanen, L.H. Allen, Jr., K.J. Boote and G. Bowes. (2001). Soybean photosynthesis, Rubisco and carbohydrate enzymes function at supra-optimal temperatures in elevated CO_2. *Journal of Plant Physiology* 158: 295-307.
Wheeler, T.R., A. Chatzialioglou, P.Q. Craufurd, R.H. Ellis and R.J. Summerfield. (1997). Dry matter partitioning in groundnut exposed to high temperature stress. *Crop Science* 37: 1507-1513.

Wheeler, T.R., P.Q. Craufurd, R.H. Ellis, J.R. Porter and P.V.V. Prasad. (2000). Temperature variability and the yield of annual crops. *Agriculture Ecosystems and Environment* 82: 159-167.

Williams, L.E., T.M. DeJong and D.A. Phillips. (1981). Carbon and nitrogen limitations on soybean seedling development. *Plant Physiology* 68: 1206-1209.

Wolf, R.B., J.F. Cavins, R. Kleiman and L.T. Black. (1982). Effect of temperature on soybean seed constituents-oil, protein, moisture, fatty acids, amino acids and sugars. *Journal of the American Oil Chemists Society* 59: 230-232.

Wu, W.H., J.Y. Lu, D.G. Mortley, P.A. Loretan, C.K. Bonsi and W.A. Hill. (1997). Proximate composition, amino acid profile, fatty acid composition, and mineral content of peanut seeds hydroponically grown at elevated CO_2 levels. *Journal of Agricultural and Food Chemistry* 45: 3863-3866.

Zhang, J. and W.J. Davis. (1987). Increased synthesis of ABA in partially dehydrated root tips and ABA transport from roots to leaves. *Journal of Experimental Botany* 38: 2015-2023.

Ziska, L.H., J.A. Bunce and F.A. Caulfield. (2001). Rising atmospheric carbon dioxide and seed yield of soybean genotypes. *Crop Science* 41: 385-391.

Crop Responses to Elevated Carbon Dioxide and Interactions with Temperature: Cotton

K. Raja Reddy
P. V. Vara Prasad
V. Gopal Kakani

SUMMARY. Carbon dioxide concentration [CO_2] and temperature are major environmental factors that affect plant growth, development and yield. Increases in [CO_2] above present levels will improve productivity of cotton crop because of improved carbon exchange rates, vegetative growth and number of reproductive organs (bolls). In contrast, increase

---

K. Raja Reddy and V. Gopal Kakani are affiliated with the Department of Plant and Soil Sciences, Mississippi State University, Mississippi State, MS 39762 USA (E-mail: krreddy@ra.msstate.edu).

P. V. Vara Prasad is affiliated with the Department of Agronomy, University of Florida, Gainesville, FL 32611-0500 USA.

Address correspondence to: K. Raja Reddy at the above address.

The authors wish to thank David Brand, Gary Burrell, Kim Gourley, Wendell Ladner, Sam Turner and Dr. James McKinion for their valuable technical and programmatic support.

Part of this research was funded by the US DOE National Institute for Global Environment Change through the South Central Regional Center at Tulane University and the National Aeronautical and Space Administration-funded Remote Sensing Technology Center at Mississippi State University.

Contribution from the Department of Plant and Soil Sciences, Mississippi State University, Mississippi Agricultural and Forestry Experiment Station Paper No. J 10245.

[Haworth co-indexing entry note]: "Crop Responses to Elevated Carbon Dioxide and Interactions with Temperature: Cotton." Reddy, K. Raja, P. V. Vara Prasad, and V. Gopal Kakani. Co-published simultaneously in *Journal of Crop Improvement* (Food Products Press, an imprint of The Haworth Press, Inc.) Vol. 13, No. 1/2 (#25/26), 2005, pp. 157-191; and: *Ecological Responses and Adaptations of Crops to Rising Atmospheric Carbon Dioxide* (ed: Zoltán Tuba) Food Products Press, an imprint of The Haworth Press, Inc., 2005, pp. 157-191. Single or multiple copies of this article are available for a fee from The Haworth Document Delivery Service [1-800-HAWORTH, 9:00 a.m. - 5:00 p.m. (EST). E-mail address: docdelivery@haworthpress.com].

http://www.haworthpress.com/web/JCRIP

Digital Object Identifier: 10.1300/J411v13n01_08

in temperatures above optimum decrease cotton yield due to increased boll abscission and smaller boll size. Doubling of [CO_2] did not ameliorate the adverse effects of high temperature on reproductive growth (boll abscission or boll size) or fiber quality. Thus, it is concluded that in future climates, if increases in [CO_2] are associated with higher temperatures, cotton yield and quality will decrease particularly in regions where present temperature are near to optimum.

KEYWORDS. Carbon dioxide, cotton, temperature, climate change, carbon use, water use, gas exchange, growth, yield

INTRODUCTION

Chemical composition of the Earth's atmosphere has changed dramatically since the start of Industrial Revolution due to human activities (Mearns, 2000). The anthropogenic activity related to energy use and land use changes resulted in increased concentrations of many 'greenhouse' gases such as water vapor, atmospheric carbon dioxide [CO_2], methane, nitrous oxide, and sulphur dioxide. Changes in atmospheric concentrations of 'greenhouse' gases will impact natural and agricultural ecosystems by direct effects on vegetation and by indirect effects through changes in climate, particularly surface temperatures.

Burning of fossil fuel and rapid deforestation has increased atmospheric [CO_2] from 280 $\mu L\ CO_2\ L^{-1}$ (1800 A.D.) to 370 $\mu L\ CO_2\ L^{-1}$ (at present time) and such concentrations have not occurred in the past 40,000 years (Kump, 2002). At existing rates of emission, the most recent future scenarios of greenhouse gases in the atmosphere indicate that atmospheric [CO_2] could increase from current levels of 360 $\mu L\ CO_2\ L^{-1}$ to between 540 and 970 $\mu L\ CO_2\ L^{-1}$ by the end of the 21st century (Houghton et al., 2001). A projected consequence of this rise in atmospheric [CO_2] and other 'greenhouse' gases is warmer Earth's surface temperatures of 1.4 to 5.8°C as predicted by general circulation models (GCMs). Climatic patterns within any region however, will become more variable and are difficult to determine, but the rate of future warming over the northern hemisphere is expected to be faster than the mean global rate with more frequent occurrence of extremely hot years (Houghton et al., 2001).

Carbon dioxide is the most important substrate for photosynthesis, thus any change in atmospheric [CO_2] will influence growth and development of plants. On similar lines, most crop plants are also sensitive to supra-optimal temperatures. In future climates, increases in atmospheric [CO_2] and global warming will occur concurrently, and therefore it is important to understand and quantify the effects of atmospheric [CO_2] and temperature on crop carbon and water balance which in turn affects various facets of growth and development including quality.

Cotton is one of the major fiber-producing crops grown worldwide in both tropical and temperate regions. It is mostly cultivated as annual crop in regions between 47°N and 32°S latitudes. There are four domesticated species of cotton: *Gossypium hirsutum* L. (Upland cotton), *G. barbadense* L (Pima cotton), *G. herbaceum* L., and *G. arboreum* L. Upland cotton contributes to about 90% of the total world's cotton production, while Pima cotton contributes about 5% (Fageria et al., 1991). Cotton provides about 25% of the total world fiber market, while synthetic and other natural fibers provide the remaining 75% (Fageria et al., 1991). Apart from fiber, cottonseed is a good source of vegetable oil, and the oil cake residue is often used as protein-rich diet for animals. As population increases, the demand for cotton will continue to rise because of greater demand for fiber, oil and protein.

Cotton is indeterminate in growth habit; in that the mainstem apex continuously initiates leaves and axillary buds. The axillary buds on the lower nodes develop into vegetative or monopodial branches if conditions are favorable. The axillary buds in the upper nodes, normally above node five, develop into fruiting or sympodial branches. Vegetative branches behave much like the mainstem in that they produce both vegetative and fruiting branches. Fruiting branches, on the other hand, initiate one true leaf and then terminate as a flower. Branch elongation is accomplished by growth of axillary buds producing a sympodial zigzag structure (Mutsaers, 1983; Mauney, 1984). Quantification of physiological and phenological processes of this inherently plastic and indeterminate species to several environmental factors including atmospheric [CO_2] and temperature has and will provide a framework for developing a conceptual model that aid in understanding plant responses to global climate change (Doherty et al., 2003; K.R. Reddy et al., 1997a, 1997b, 2000, 2002a, 2002b).

Although cotton is a warm season tropical crop, it has relatively narrow temperature range compared with other tropical species. Cardinal temperature for cotton growth and development are in the range of 12 and 15°C (minimum temperature), 26 and 28°C (optimum temperature), and maximum temperature (ceiling) varies depending on the growth or developmental process and on the duration of exposure (K.R. Reddy et al., 1997a). Cotton is highly sensitive to temperatures above optimum, and the extent of damage depends upon the intensity and duration of high temperature stress. Furthermore, cotton being a C_3 plant is also highly responsive to changes in atmospheric [CO_2]. If the predictions of GCM's are true, then cotton crops grown in future environments will be certainly different from those of today. Therefore, it is important to quantify and understand response of cotton growth, development and physiology to predicted changes in environment to develop and design suitable cultivars and agronomic packages for future climates.

The objective of this chapter is to summarize the major findings of the studies conducted in the naturally lit controlled environment facility known as Soil-Plant-Atmosphere-Research (SPAR) at Mississippi State University over the past two decades on responses of cotton to climate change. However, wherever necessary, results will be supported by other published literature. Focus will be on the carbon and water balance, growth, development and yield responses of cotton plants to increasing [CO_2], and interactive effects of atmospheric [CO_2] and increase in air temperature.

CONTROLLED ENVIRONMENT FACILITIES AND EXPERIMENTAL DETAILS

Most of the research results presented in this chapter were obtained from experiments conducted in sunlit soil-plant-atmosphere research (SPAR) plant growth chambers at the Mississippi State University, Mississippi State, Mississippi, and USA. Details about structure, control capabilities, operation, data acquisition, quality of control of SPAR facilities are described elsewhere (Phene et al., 1978; K.R. Reddy et al., 2001). In brief, each SPAR consists of two main parts: bottom steel bin (2.0 m long, 0.5 m wide, and 1.0 m deep) for rooting medium and upper airtight Plexiglas chamber (2 m long, 1.5 m wide, and 2.5 m high) for aerial environment. The Plexiglas chambers transmit about 94% of the photosynthetically active radiation (PAR). Each unit is capable to control air temperature and chamber atmospheric [CO_2] at predetermined set points. Air temperature is measured using aspirated and shielded thermocouples and set points are controlled by opening and closing chilled ethylene glycol radiator and switching electrical heating system on and off, as necessary. To maintain [CO_2] in each chamber, pure CO_2 is injected through a system that includes a pressure regulator, solenoid and needle valve, and flowmeter. The flowmeters are calibrated with a gas displacement meter at the beginning and end of each experiment. Carbon dioxide flow rates are recorded three times a day and converted into mass quantity via gas law correction for temperature and pressure. A computer monitors the time intervals during which the solenoid valves are open, and thus the amount of gas injected is known. A leakage test is performed each night to derive the plant growth chamber leakage rate and to correct canopy gas measurements (Acock and Acock, 1989). Measurements on dew point temperature and canopy temperature, photosynthetic photon flux density (PPFD) are also measured every 10 s and averages stored every 900 s throughout the experimental period. List and details of all experiments conducted during the last two decades are given else where by K.R. Reddy et al. (2001). For details about experimental treatments and crop management practices see original references.

EFFECTS OF ATMOSPHERIC [CO_2] AND TEMPERATURE ON CARBON AND WATER USE

Photosynthesis

Leaf-level photosynthetic rates were 22 and 9% greater for plants grown at 700 and 450 μL CO_2 L^{-1} when compared to those grown at 350 μL CO_2 L^{-1} at 26°C (Figure 1). These responses are similar to those reported by Hendrey et al. (1993) and Hileman et al. (1994). Such increases in photosynthetic rates have typically been observed in most of the C_3 plant species including agronomic crops such as rice (Baker et al., 1990), wheat (Garcia et al., 1998), soybean (Allen et al., 1987; Campbell et al., 1988), vegetable crops (Peet and Wolf, 2000), and forest tree crops (Janssens et al., 2000).

At present day atmospheric [CO_2], carbon fixation rates of C_3 species including cotton are limited primarily by the amount of CO_2 available (Bowes,

FIGURE 1. Influence of carbon dioxide and temperature on leaf photosynthesis of recently-fully expanded mainstem leaves measured at 1500 photosynthetic photon flux density (adapted from A.R. Reddy et al., 1998).

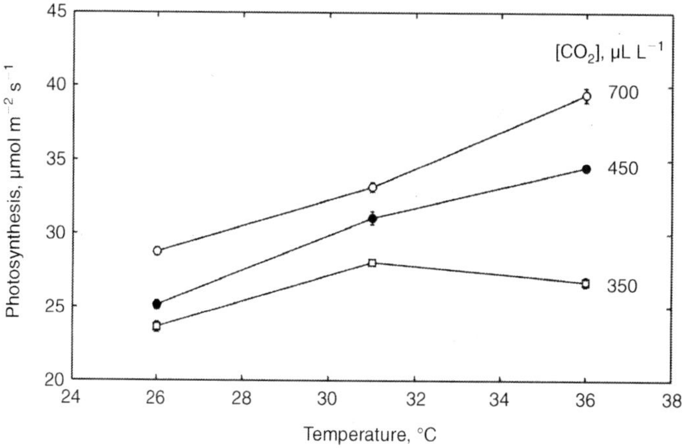

1996). In C_3 plants, CO_2 is taken up and reduced (i.e., gain of electrons or hydrogen) to carbohydrate (CH_2O) through photosynthetic carbon reduction cycle (PCR) that employs the enzyme ribulose 1,5-bisphosphate (Rubisco). Similarly, atmospheric O_2 also interacts with Rubisco to produce phosphoglycolate, which is partially metabolized to CO_2 in the photorespiratory carbon oxidation cycle (PCO). As Rubisco catalyses both carboxylation and oxygenase reactions, both CO_2 and O_2 compete for the same site on Rubisco. The balance between both reactions depends upon the of [CO_2] and O_2 concentration [O_2] in the mesophyll chloroplasts. It has been shown that through O_2 inhibition and photorespiration, the present ratio of [CO_2] and [O_2] causes about 35% decrease in photosynthesis of C_3 plants at 25°C, and higher temperatures can increase this inhibition (Bowes, 1996). Due to the competitive interaction, increase in [CO_2] will diminish the inhibitory effects of [O_2]. It has been shown that doubling of atmospheric [CO_2] can reduce photorespiration to extent of 50%, thus leading to greater photosynthesis rates in C_3 plants (Sharkey, 1988; Bowes, 1996).

Individual leaf photosynthesis response to atmospheric [CO_2] and temperature interactions were similar to those proposed by and reported by many in the literature (Long, 1991; Sage et al., 1995); where increases in leaf photosynthetic rates due to elevated [CO_2] were greater at higher temperature (Figure 1). Exposure to 700 μL CO_2 L^{-1} increased photosynthesis by 22% at 26°C, 22% at 31°C and 54% at 36°C. Similar linear increases in percentage enhancement of leaf photosynthesis with increasing growth temperature were observed in soybean (Vu et al., 1997). This phenomenon is not universal across all C_3 species (Long, 1991; Sage et al., 1995), for example, in bean the percentage increase in photosynthesis was greater at 28/18°C (maximum/minimum) compared to that at 40/30°C (Prasad et al., 2002), while in rice, percentage enhancement was not altered over a range of daytime temperature between 28 and 38°C (Vu et al., 1997). These species differences could partially be attributed to different temperature optima for photosynthesis and vegetative growth and differential effects of temperature and [CO_2] on Rubisco (Sage et al., 1989; Long, 1991; Vu et al., 1997; Bunce, 1998).

Furthermore, the fact that under ambient [CO_2] photosynthesis was maximum at 31°C and decreased at 36°C, whereas, at elevated [CO_2] photosynthesis linearly increased with temperature up to 36°C (Figure 1) suggests a greater optimum temperature for plants grown at elevated [CO_2]. This is in agreement with several studies (Sage et al., 1995) and mechanistic model predictions by Long (1991), where temperature optimum for light saturated photosynthesis increased by 5°C with increase of [CO_2] from 350 to 700 μL CO_2 L^{-1}.

A.R. Reddy et al. (1998) reported that there was up-regulation of Rubisco with increasing [CO_2] and temperature particularly at 700 μL CO_2 L^{-1} and 36/

28°C (day/night). This in contrast to several studies that generally show down regulation of Rubisco under elevated [CO_2] and temperatures (Vu et al., 1997; Sage et al., 1989). Similar up-regulation of Rubisco was observed in one study on soybean (Campbell et al., 1988). This suggests that CO_2 enrichment could ameliorate the adverse effects of high temperature on Rubisco, causing up-regulation of photosynthesis in cotton leaves (A.R. Reddy et al., 1998). However, further research is necessary to confirm this across other cotton cultivars and crop species.

Growth at elevated [CO_2] increased net photosynthetic rates of cotton canopies throughout the season (Figure 2). Canopy photosynthesis increased linearly between 30 and 80 days after emergence (DAE) in both ambient and elevated [CO_2], thereafter, photosynthetic rates of plants grown at ambient [CO_2] declined more rapidly than those grown at elevated [CO_2]. The response to elevated [CO_2] compared to ambient [CO_2] progressively increased with season and plant age with maximum absolute effects occurring during boll development and soon after boll opening. Comparisons of gross photosynthetic rates of cotton at different DAE showed that response of plants to elevated [CO_2] was greatest at 90 DAE, compared to earlier dates (V.R. Reddy

FIGURE 2. Temporal trends in net photosynthesis (P_n) of cotton canopies grown in 1995 ambient temperatures and in 360 and 720 μL CO_2 L^{-1} at Mississippi State, Mississippi. Photosynthesis was measured continuously, summarized at 900-s intervals, and normalized at 1200 photosynthetic photon flux density each day throughout the season from the light response curves similar to those presented in Figure 3. Time of first flower and first mature boll is indicated (adapted from K.R. Reddy et al., 1998b).

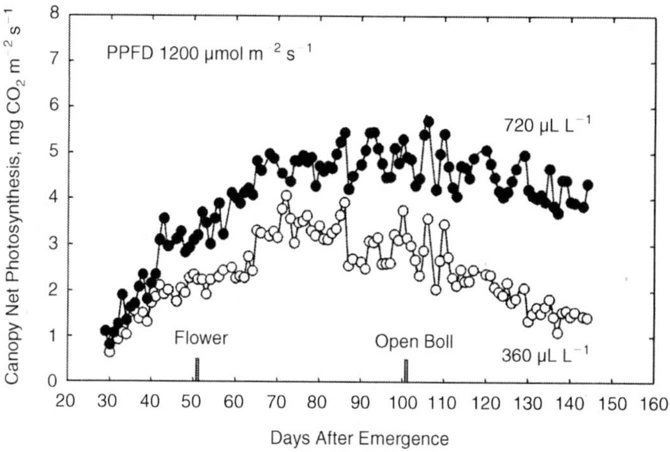

et al., 1991, 1995). Greater photosynthesis at elevated [CO_2] could be attributed to increased photosynthetic rates per unit leaf area, increased leaf area before flowering and continued vegetative growth and development during the boll-filling period through carbon-stress alleviation for plants grown in elevated [CO_2]. The effects of elevated [CO_2] on canopy CO_2 exchange rates have been reported for several crops, but the long-term responses are inconsistent. Both decline (Baker et al., 1990; Wolfenden and Diggle, 1995) and stimulation (Jones et al., 1985; Horie et al., 2000; Monje and Bugbee, 1998) have been reported in the literature. Contrary to the response in cotton, Sakai et al. (2001) reported that canopy photosynthetic enhancement to elevated [CO_2] (650 μL CO_2 L^{-1}) was 33% for three weeks before heading, but declined gradually and disappeared over time during the grain filling in rice, a determinate C_3 crop. The increase in rice biomass by elevated [CO_2] was attributed to more from the increase in carbon gain during early growth stages than at late stages. The decline in canopy carbon assimilation after heading was attributed to enhanced leaf senescence at elevated [CO_2] partially due to higher nitrogen re-translocation rates to meet the demands of greater number of ears produced in response to elevated [CO_2].

V.R. Reddy et al. (1995) exposed cotton canopies to five atmospheric [CO_2] varying from 350 to 900 for 22 days starting from first bloom and measured canopy photosynthesis for 14 days. It was observed that canopy photosynthesis of fruiting cotton plants gradually increased with increases in atmospheric [CO_2] above 350 μL CO_2 L^{-1} and reached maximum values at 700 μL CO_2 L^{-1} and there was no further increase at 900 μL CO_2 L^{-1}. Similar responses were reported for rice. Baker et al. (1990) grew rice plants in subambient (160, 250), ambient (330), and supra-ambient (500, 660, and 900 μL CO_2 L^{-1}) [CO_2] treatments. They reported that canopy photosynthetic rates increased with increasing [CO_2] from 160 to 500 μL CO_2 L^{-1}, but saturated beyond 500 μL CO_2 L^{-1}, about 200 μL CO_2 L^{-1} lower than that was observed in cotton.

Increase in solar radiation increased photosynthesis of cotton canopies at ambient and elevated CO_2 levels (Figure 3). However, the light response curve was higher for plants grown at elevated [CO_2] compared to ambient [CO_2]. Both the initial slopes of the light response curve and the asymptotes were greater in elevated [CO_2]. Similar responses were observed in our previous studies (V.R. Reddy et al., 1995). In contrast to acclimation of leaf-level photosynthesis to elevated [CO_2] (Moore et al., 1999), the canopy gross photosynthesis and canopy light response curves indicate that cotton canopy photosynthesis did not acclimate under elevated [CO_2] over the whole growing season provided other growth factors are not limiting (Figures 2 and 3).

Photosynthetic responses to elevated [CO_2] at the canopy are different from those at single-leaf level, because of variations in physiological and environ-

FIGURE 3. Influence of carbon dioxide and solar radiation (photosynthetic photon flux density–PPFD) on cotton canopy gross photosynthesis (P_g) at 60 days after emergence. P_g equals to net photosynthesis plus respiration (adapted from K.R. Reddy et al., 1995c).

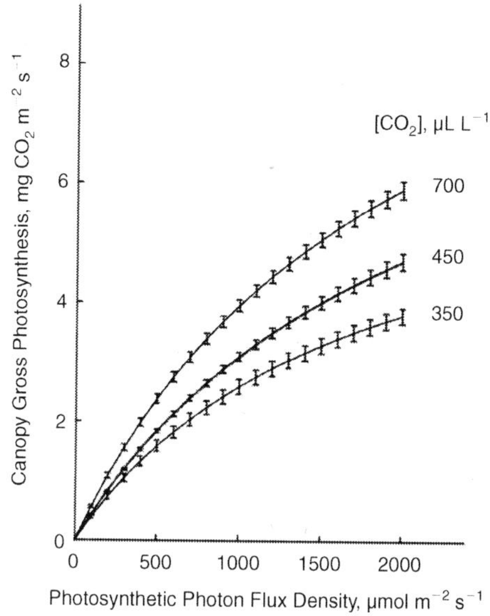

mental factors between plant organs and even between leaves in a canopy. Leaf-level photosynthesis rate, measured on recently, fully expanded leaves and expressed on per unit leaf area, does not always represent the most important determining or controlling factor for overall plant growth. Relative increases in canopy photosynthesis of plants grown in elevated $[CO_2]$ was greatest at 34°C, further decrease or increase in temperatures lead to decreased response with zero response at about 6 and 39°C (Figure 4). Similar interaction studies at elevated $[CO_2]$ and temperature showed that response of canopy photosynthetic rates to elevated $[CO_2]$ was greater at higher (35/27°C) temperature when compared to low (25/17°C) temperature (V.R. Reddy et al., 1995).

Thomas et al. (1993) showed that elevated $[CO_2]$ did not affect daytime respiration, but night time respiration of leaves grown at 650 μL CO_2 L^{-1} was greater than plants grown at 350 μL CO_2 L^{-1}. Responses of plant growth to both elevated $[CO_2]$ and temperature involve the balance between canopy photosynthesis and respiration. Furthermore, elevated $[CO_2]$ increased maintenance coefficient which were mainly related to increased starch accumula-

FIGURE 4. Influence of daytime growth temperature on relative photosynthesis of both Upland and Pima cotton. Relative photosynthesis was calculated as the ratio of rates at elevated (700/720 μL CO_2 L^{-1}) divided by rates at ambient (350/360 μL CO_2 L^{-1}), and each point is an average of at least 15 days.

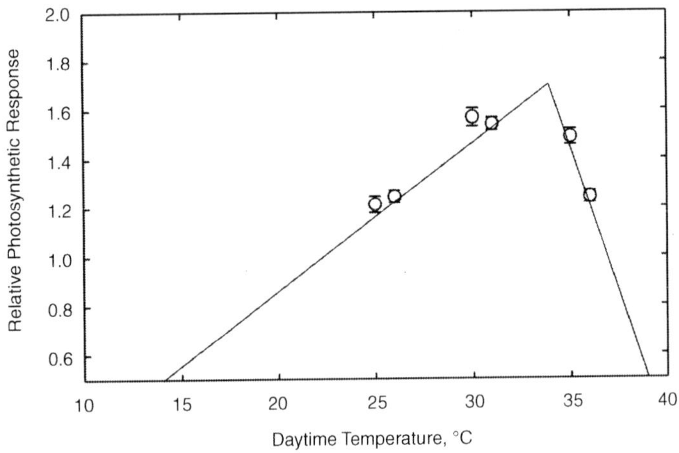

tion. However, responses of respiration to elevated [CO_2] are highly variable and not well understood (Boote et al., 1997). Few studies have been reported so far on simultaneous measurements on canopy photosynthesis and respiration to draw a general conclusion. Morison and Lawlor (1999) concluded that there is still considerable uncertainty about the balance of these two important plant processes; canopy photosynthesis and respiration and the extent and direction of acclimation of both processes to temperature and atmospheric [CO_2] alone and their interactions.

Stomatal Conductance, Transpiration, and Water Use

Exposure to double ambient [CO_2] decreased stomatal conductance of individual leaves (Figure 5) and canopy conductance to [CO_2] transfer (V.R. Reddy et al., 1995). Elevated [CO_2] decreased stomatal conductance of individual leaves by 43% at 26°C, 45% at 31°C, and 20% at 36°C. Similar responses have been observed in several C_3 species (Murray, 1997; Jarvis et al., 1999; Bunce, 2001) including cotton (Bhattacharya et al., 1994). Decreases in stomatal conductance were mainly due to partial closure of stomata but not to stomatal density (K.R. Reddy et al., 1998a). Although elevated [CO_2] decreases leaf stomatal conductance, it does not always decrease transpiration rate. This is because transpiration rate depends on the energy and vapor pres-

FIGURE 5. Influence of carbon dioxide and temperature on stomatal conductance recently-fully expanded mainstem leaves measured at 1500 photosynthetic photon flux density (adapted from A.R. Reddy et al., 1998).

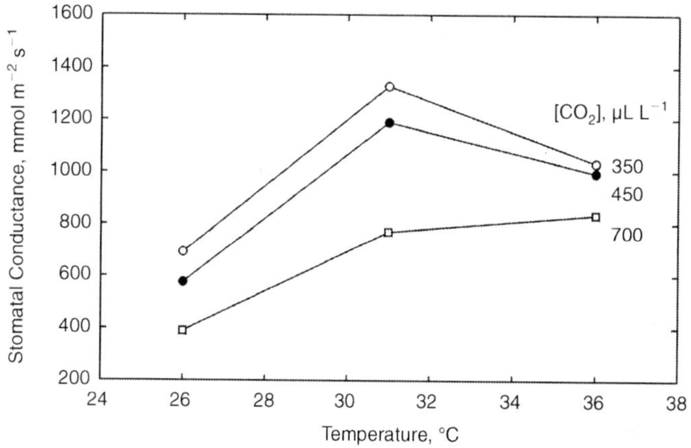

sure gradient between leaf and air and their temperature. Partial closure of stomata often increases leaf temperature, thus increasing the vapor pressure gradient between leaf and air, which leads to increase in transpiration (Boote et al., 1997; Allen, 1999). Furthermore, lower stomatal conductance under elevated [CO_2] improves leaf water potential and can accelerate leaf expansion (Morrison and Lawlor, 1999). Increased leaf area will lead greater surface available for transpiration, therefore transpiration rate expressed per plant or unit ground area may not decline owing to increased leaf area. This is evident form Figure 6, where there was very little effect (7% increase) of elevated [CO_2] on transpiration of cotton canopies. Similar responses have been shown in soybean (Allen, 1999).

Elevated [CO_2] increases leaf area and dry weight of individual plants, therefore the total water use of plants per plant and unit land area will remain the same or increased (Kimball, 1983; Allen, 1999). Increases in total water use by cotton plants were reported when grown at elevated [CO_2] (Samarakoon and Gifford, 1995; 1996). Water use efficiency expressed as a ratio of net photosynthesis to transpiration is increased with increasing [CO_2], reaching peak at 700 µL CO_2 L^{-1}, and reduced slightly thereafter at 800 and 900 µL CO_2 L^{-1} (Figure 7). These responses are similar to that observed for photosynthesis (V.R. Reddy et al., 1995). Therefore, increased water use efficiency under elevated [CO_2] is largely a result of increased photosynthesis rather than decrease in water use or transpiration. This confirms observations on other crops where similar results were reported (Allen et al., 1985; Baker and Allen, 1993).

FIGURE 6. Relation between daily total canopy transpiration of cotton plants grown at ambient (360 μL L^{-1}) and elevated (720 μL L^{-1}) [CO_2]. The data are from five temperature treatments of Upland cotton collected over several days during fruiting (adapted from K.R. Reddy et al., 2000).

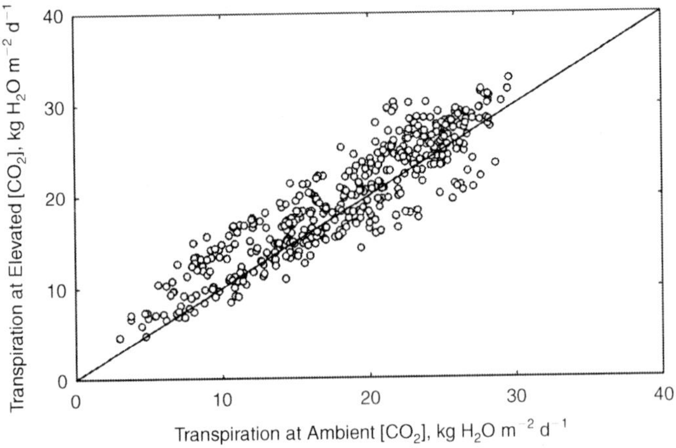

Interaction studies showed that transpiration rates of plants grown at elevated [CO_2] were significantly lower at all temperatures ranging between 25/17 and 35/27°C than plants grown in ambient [CO_2] (V.R. Reddy et al., 1995). Elevated [CO_2] decreased total daily transpiration across range of temperature between 25/17 and 35/27°C. These decreases in canopy transpiration occurred in spite of greater LAI in plants grown at elevated [CO_2]. In contrast, as temperature increased, there was gradual decrease in WUE due to increased water use at high temperatures. Similar responses of WUE and transpiration to elevated [CO_2] and temperature were observed in several other studies (Jones et al., 1985; Baker et al., 1990; Mauney et al., 1994; Allen, 1999; Hui et al., 2001).

EFFECT OF ATMOSPHERIC [CO_2] AND TEMPERATURE ON DEVELOPMENT, GROWTH, AND YIELD

With increased [CO_2], the rate of carbohydrate production is higher than in ambient [CO_2] due to reduction in photorespiration and faster carboxylation across a wide range of temperatures conducive to cotton growth. This increased rate of carbohydrate supply will have a myriad of secondary effects on cotton growth and development.

FIGURE 7. Influence of atmospheric carbon dioxide on canopy water-use efficiency of cotton during fruiting when the canopy was intercepting 90-95% of incoming solar radiation.

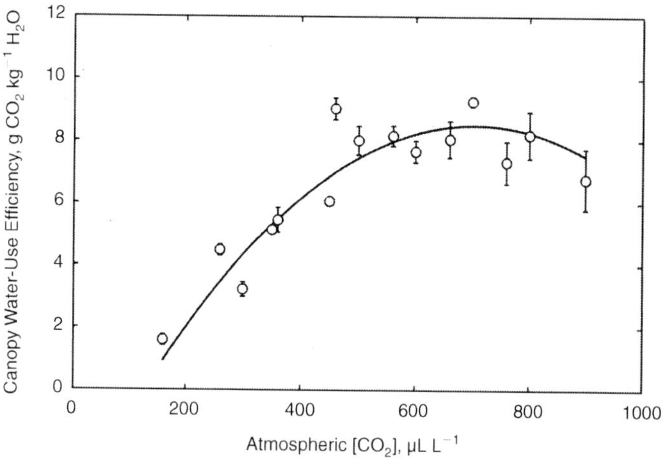

Development and Organ Growth

Temperature is the key environmental factor affecting ontogeny and morphogenesis, but often interacts with other environmental factors especially photoperiod. Most of the present day cotton cultivars are photoperiod insensitive, therefore temperature exerts dominant role in development. Plants require certain amount of heat units (thermal time) to reach each physiological development stage. At warmer temperatures, plants will achieve these requirements earlier, thus enhance physiological development in most plants including cotton (K.R. Reddy et al., 1993). Similarly, the initiation and expansion of leaves, shoots, and reproductive organs are mainly influenced by temperature and tissue water status (K.R. Reddy et al., 1997a, 1997c). However, the duration of each developmental phase declines as temperature increased. The triggering mechanisms for initiation or development of tillers and branches in determinate plants and vegetative and secondary branches in indeterminate plants such as cotton are however, less precisely known.

There was no effect of elevated [CO_2] on timing of physiological growth stages, i.e., days required from emergence to first square, square to flower, and from flower to open boll (Figure 8). In contrast, these developmental events were highly sensitive to temperature (K.R. Reddy et al., 1997a, 1997b, 1999) with different optimum temperatures for each of these developmental events (Figure 9). K.R. Reddy et al. (1997b) conducted experiments in naturally-lit

FIGURE 8. Relation between duration of major phonological events (emergence to square, Y = −0.5948 + 1.01162 * X, $r^2 = 0.99$; square to flower, Y = −1.1843 + 0.944, * X, $r^2 = 0.92$ and flower to open boll, Y = 1.6151 + 0.996 * X, $r^2 = 0.99$) of cotton plants grown at ambient (360 μL L^{-1}) and elevated (720 μL L^{-1}) [CO_2].

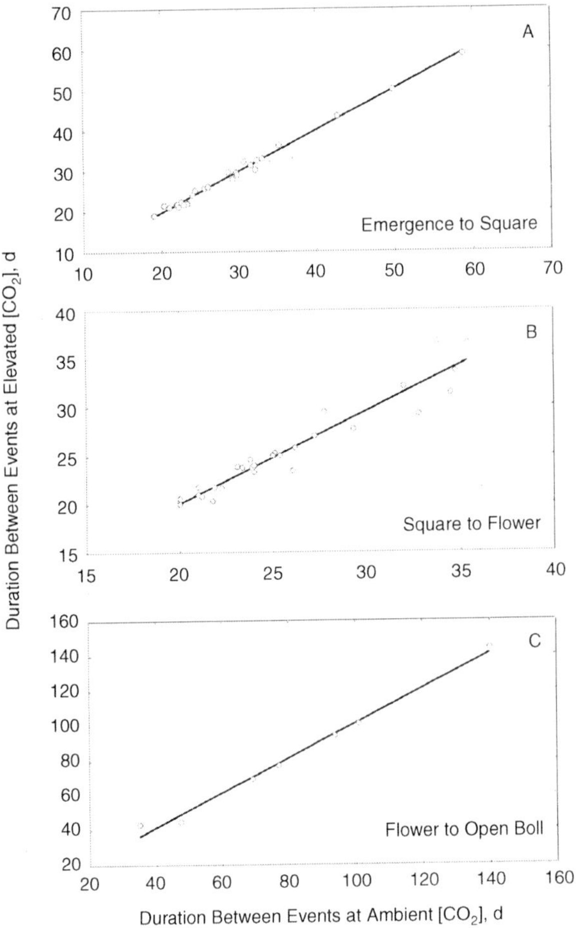

and under ambient average daily temperature conditions (23°C) and temperature regimes of ambient minus 2°C and ambient plus 2, 5, and 7°C at ambient and twice ambient [CO_2]. At all temperatures, there were no effects of CO_2 enrichment on major phenological events. But when compared to ambient temperature treatment, time to first square was increased by 7 d for ambient minus

FIGURE 9. Influence of temperature on daily progress of various phenological events in cotton: from plant emergence to first flower bud or square formation ($Y = -0.1264 + 0.01142 * X - 0.0001949 * X^2$, $r^2 = 0.95$), square to flower ($Y = -0.1212 + 0.0104 * X - 0.0001619 * X^2$, $r^2 = 0.88$) and flower to open boll ($Y = -0.006219 + 0.001 * X$, $r^2 = 0.96$).

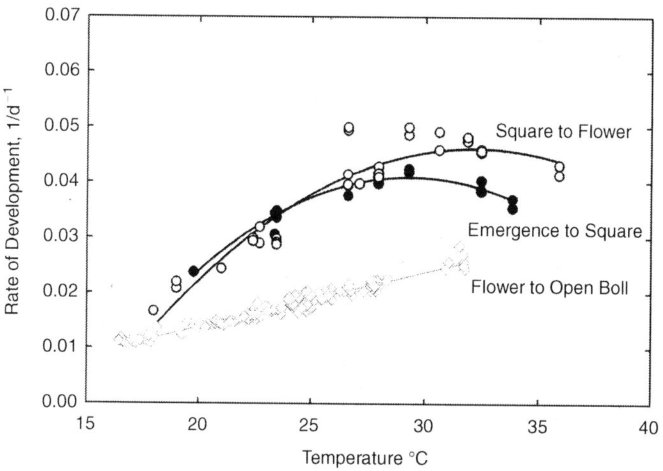

2°C, but was decreased by 2 d for ambient plus 2°C, and by 5 d for ambient plus 5°C and 7 d for ambient plus 7°C temperature treatments. Similar responses, although of different magnitudes, were observed for time to first open flower and open boll (K.R. Reddy et al., 1997b, 1999). The time required to produce squares, flowers and mature fruits was decreased by 1.6, 3.1, and 6.9 d per degree rise in temperature above 21°C (K.R. Reddy et al., 1997b). This clearly shows that among the changes projected in global climate change, the impact of elevated [CO_2] on plant development will be small in cotton and as in many other crops, but temperature increase expected in future climate will have considerable effect on crop development.

Main stem node addition rate, total number of mainstem nodes, leaf development rate on both main stem and fruiting branches, and rates of leaf expansion and internode elongation were not affected by elevated [CO_2], but increased with increasing temperature (Figures 10 and 11). Responses of developmental rates at these temperatures were best explained by quadratic functions.

Doubling of [CO_2] compared to ambient [CO_2] had little effect on plant height but branch length increased to a greater extent by about 17% (Figures 12a and 12b). It seems that elevated [CO_2] weakened apical dominance of the primary meristem resulting in more branches per plant and more number of

FIGURE 10. Influence of temperature on rates of mainstem ($Y = -0.6698 + 0.057 * X - 0.0006764 * X^2$, $r^2 = 0.94$) and fruiting branch node ($Y = -0.3645 + 0.03889 * X - 0.0005199 * X^2$, $r^2 = 0.84$) formation of cotton plants grown at ambient and elevated [CO_2]. The data are from several temperature treatments of Upland and Pima cotton grown under optimum water and nutrient conditions at both ambient and elevated carbon dioxide levels (adapted from K.R. Reddy et al., 1999).

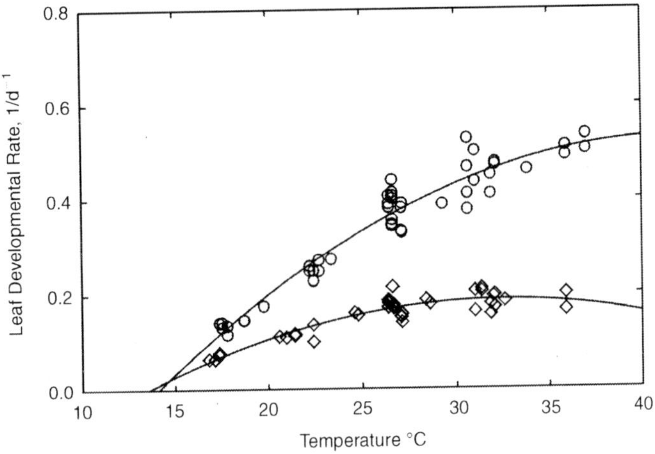

nodes/fruiting sites per branch. The result is a bushy plant with long branches in elevated [CO_2] conditions. Elevated [CO_2] increased individual leaf size by about 20% (K.R. Reddy et al., 1998a) and total plant leaf area by 27% across a range of environmental conditions (Figure 12c). Greater response on whole plant basis was caused by more number of branches and more leaves per branch.

A strong interaction between [CO_2] and temperature was recorded on number of vegetative and fruiting branches (K.R. Reddy et al., 1995a). Fewer vegetative branches, but more fruiting branches were produced at 30/22°C compared with either higher or lower temperature. Similar responses were observed in both Upland (V.R. Reddy et al., 1991; K.R. Reddy et al., 1992c; 1995a, 1995b) and Pima cotton (K.R. Reddy et al., 1992a). At optimum temperature, there was no effect of [CO_2] enrichment on the number of primary fruiting branches defined as those producing square on each node. However, CO_2 enrichment increased the number of secondary branches defined as those fruiting branches at any main or vegetative branch stem nodes. Similarly, time required for producing fruiting branch nodes at optimum temperature was slightly less at elevated [CO_2]. The combined effect of more secondary-

FIGURE 11. Influence of temperature on extension duration of leaves and internodes, and the equations describing these processes as functions of temperature are as follows: mainstem leaf expansion duration rate, $Y = -0.09365 + 0.01070 * X - 0.0001697 * X^2$; $r^2 = 0.95$ and mainstem internode elongation duration rate, $Y = -0.04312 + 0.007383 * X - 0.0001046 * X^2$; $r^2 = 0.96$ (adapted from K.R. Reddy et al., 1997c).

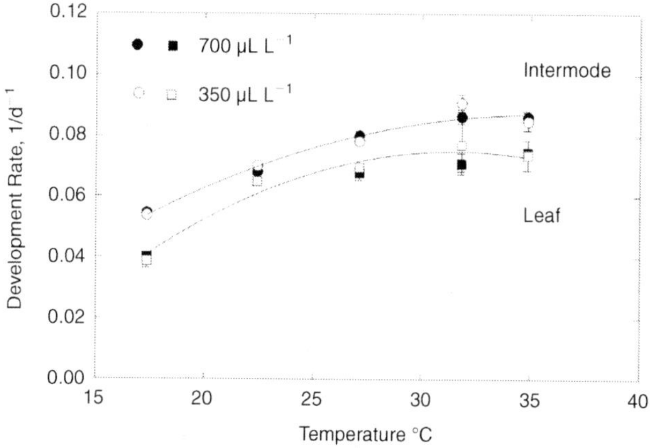

branches and more rapid node production lead to 40% increase in fruiting sites to be formed at 30°C for plants grown at elevated [CO_2] compared to those grown at ambient [CO_2] (K.R. Reddy et al., 1998a).

Carbon dioxide enrichment did not affect rate of root initiation or rate of root development, but increased the length of lateral roots (V.R. Reddy et al., 1994). This is in agreement with previous studies on cotton where exposure to elevated [CO_2] increased root mass, lengths and volume of tap roots, lateral roots and fine roots (Rogers et al., 1992; Prior et al., 1994). Although the number of lateral roots per unit length of taproot was not affected by elevated [CO_2], the overall taproot length under CO_2 enrichment resulted in greater number of lateral roots. Detailed studies on root architecture showed that root length and dry weight densities were altered due to CO_2 enrichment, with more growth in upper layers of the soil. Furthermore, as horizontal distance from row center increased, prolific spread of roots through soil profile was observed under elevated [CO_2] (Prior et al., 1994). Responses of rate of root initiation and growth were similar at both ambient and elevated [CO_2] and were mainly driven by temperature similar to phenological events of the aboveground events. As temperature increased from about 18°C to 23°C, rate of root initiation and growth increased, and further increase in temperature did not affect these parameters (Figure 13). Further research is necessary to clearly understand the effect of high temperature on root growth and development.

FIGURE 12. Relation between growth parameters of cotton plants grown at ambient (360 $\mu L\ L^{-1}$) and elevated (720 $\mu L\ L^{-1}$) [CO_2]. The equations describing these processes as growth processes at ambient [CO_2] are as follows: plant height ($Y = 2.4192 + 1.0263 * X$; $r^2 = 0.97$), branch length ($Y = 103.8843 + 1.1683 * X$; $r^2 = 0.96$) and whole plant leaf area ($Y = 0.02244 + 1.2485 * X$; $r^2 = 0.98$).

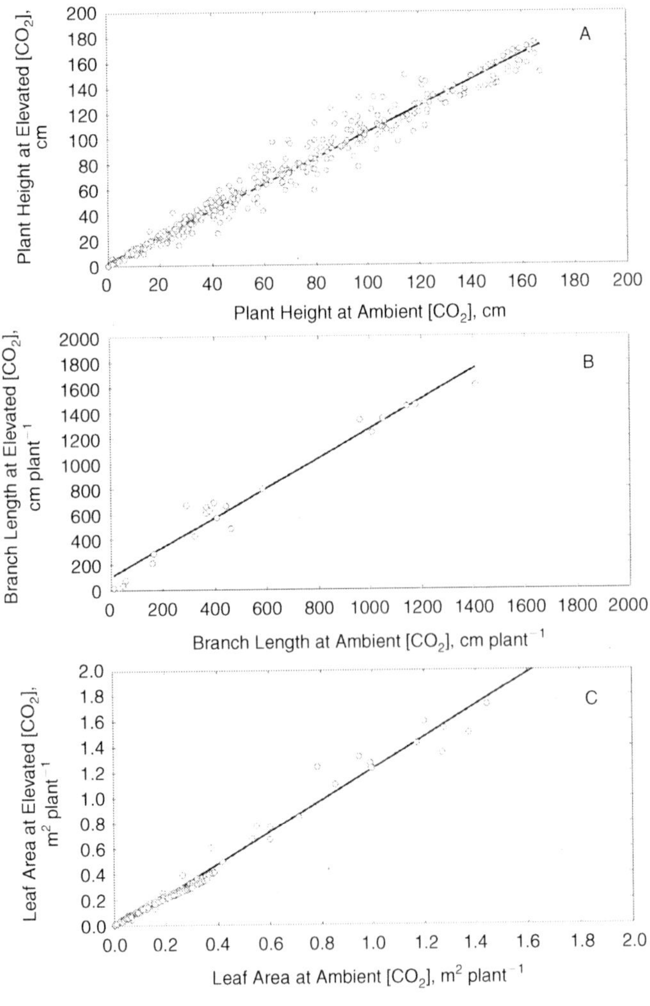

FIGURE 13. Influence of temperature on cotton root growth and development: number of roots ($Y = -115.6216 + 10.3364 * X - 0.1934 * X^2$, $r^2 = 0.75$), rate of root growth per unit surface area ($Y = -184.844 + 16.1199 * X - 0.2909 * X^2$, $r^2 = 0.88$) and rate of root growth per unit root length ($Y = -18.1514 + 1.6377 * X - 0.0279 * X^2$, $r^2 = 0.90$) (adapted from V.R. Reddy et al., 1994).

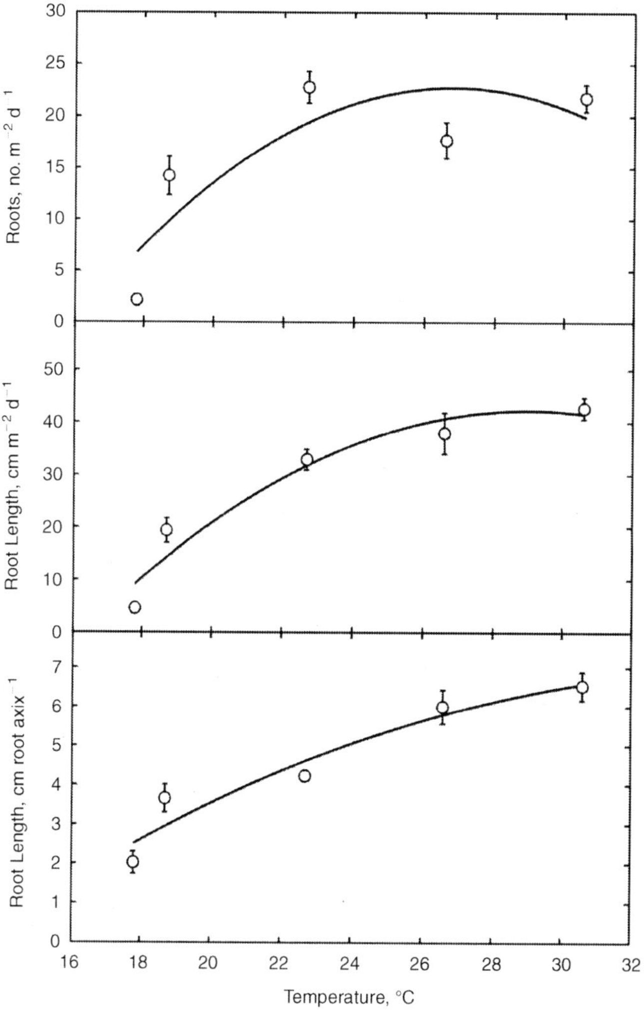

Accumulation of non-structural carbohydrates (TNC) in leaves was influenced by [CO_2] and temperature (K.R. Reddy et al., 1998a; A.R. Reddy et al., 1998). At temperatures ranging from 20/12 to 35/27°C, plants grown at elevated [CO_2] accumulated more than twice the concentration of TNC in leaves compared to those grown at ambient [CO_2] at those respective temperatures. However, there was no effect of [CO_2] at the highest temperature, 40/32°C. Highest concentration of TNC was observed in leaves grown at 20/12°C, and further increase in temperature to 25/17 through 40/32°C decreased TNC by > 75% under both ambient and elevated [CO_2]. Similarly, it was observed that as temperature increased from 26/18 to 36/28°C starch and sugar content of leaves were linearly decreased at 350, 450 and 700 μL CO_2 L^{-1} levels (A.R. Reddy et al., 1998). These responses reflect the growth and developmental rates of plants in different temperatures and [CO_2] (K.R. Reddy et al., 1997a, 1997b). This clearly suggests that photosynthesis of cotton was affected less by temperatures than by growth, resulting in accumulation of TNC. CO_2 enrichment favored production of more carbohydrates, but temperature limited growth and thus carbohydrate utilization (K.R. Reddy et al., 1997a).

Dry Matter Production and Yield

Growth at elevated [CO_2] compared to ambient [CO_2] increased dry matter production of cotton by 36% (Figure 14). K.R. Reddy et al. (1995b) reported that doubling [CO_2] caused plants to produce 40% more leaf, stem and root mass than when grown in ambient [CO_2]. Field studies on [CO_2] enrichment under ambient temperature conditions reported 37% increase in total biomass production (Mauney et al., 1994). Kimball and Mauney (1993) summarized the results of [CO_2] enrichment studies on cotton over 5 years, and reported that, on average, seed cotton yield (lint plus seed) and above ground biomass were increased 60 and 65%, respectively, by CO_2 enrichment to 650 μL CO_2 L^{-1}. Similar increases in dry matter production and growth under elevated [CO_2] have been observed in reports (Kimball et al., 2002 and reference within). The increase in biomass was mainly due to increase in early season leaf and stem growth in plants enriched with CO_2 and also to increase in photosynthetic rates. Whereas, the increased fruit yield was related to greater number of bolls under CO_2 enrichment due to increased branching resulting in more fruiting sites for each branch.

Studies on temperature effect on growth and partitioning showed that maximum dry matter was produced at 30/20°C, and decreased at lower (20/10 and 25/15°C) and higher temperature (35/25 and 40/30°C) regimes (V.R. Reddy et al., 1991). Similar responses were observed for boll weight. Plants grown at 30/20°C partitioned more than 40% of their total dry weight to reproductive structures compared to <15% at 20/10 and 25/15°C; and <4% at 35/25 and 40/

FIGURE 14. Relation between total plant weight of cotton plants grown at ambient (360 μL L^{-1}) and elevated (720 μL L^{-1}) [CO_2] ($Y = 2.3516 + 1.3237 * X$, $r^2 = 0.98$).

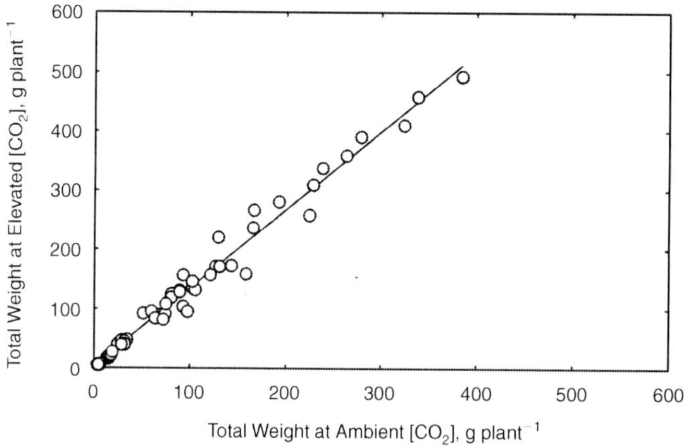

30°C. In contrast, partitioning to leaves, stems and roots was lower at 30/20°C compared to other temperatures. This is mainly due to increased demand for carbohydrates from stronger sinks (bolls) at optimum temperature conditions. As there were few sinks at high temperature most of the biomass was partitioned to leaves, stems and roots.

K.R. Reddy et al. (1992b, 1992c) studied the effects of temperature on formation of fruiting sites and retention of Upland cotton bolls and squares at five temperatures between 20/12 and 40/32°C. The total number of fruiting sites increased linearly with increasing temperature up to 40/32°C. In contrast, the number of bolls and squares retained reached maximum with similar numbers (40 plant^{-1}) at 30/22 and 35/27°C, with no fruiting structures retained at 40/32°C and few at 20/12 (2 plant^{-1}) and 25/17°C (18 plant^{-1}). Similar temperature effects on another species of cotton, Pima, were observed (K.R. Reddy et al., 1992a). The fewer numbers produced at temperatures below the optimum was due to slower development. In a recent study, the authors identified causes for fruit loss under high temperature. The length of the stigma was not affected and was about 80 mm irrespective of the temperature treatment. The length of style was 20 mm longer in the high temperature (36/28°C) treatment compared to that in optimum temperature (30/22°C). At optimum temperature, the length of the staminal column was 130 mm, while it was reduced to 100 mm at high temperature. Anthers failed to dehisce under high temperature that was attributed by the authors to reduction in pollen number per anther. Anthers formed at 30/22°C had about 205 pollen grains per anther, while those formed at 36/28°C had only 81 pollen grains per anther. Simultaneously, an *in-vitro* study was carried out by the authors to study the viability of pollen produced under different temperatures on an artificial pollen growth media. It was found that pollen germination percentage was about 60 percent from flowers produced at 30/22°C, in contrast, the pollen produced under 36/28°C failed to germinate. In summary, the causes for boll loss are five fold: (1) elongation of style before the anthers dehisce, (2) decrease in length of staminal column, (3) failure of anthers to dehisce, (4) decrease in pollen number per anther, and (5) failure of pollen produced to germinate.

Lower fruit-set at high temperatures were therefore related to decreased pollen production, poor pollen viability and pollen germination in several other species (peanut–Prasad et al., 1999, 2001; common bean–Gross and Kigel, 1994; cowpea–Hall, 1992). Although studies have shown that male gametes are generally more sensitive than female gametes, the possible role of stigma receptivity cannot be ruled out.

There were strong interactions between temperatures and atmospheric [CO_2] on plant biomass accumulation (K.R. Reddy et al., 1995b). Carbon dioxide enrichment increased dry matter accumulation in leaves, stems, and roots and total biomass of cotton plants in high [CO_2] environments at all temperatures ranging from 19 to 36°C, but maximum response occurred at near optimum temperature (31°C) than at warmer or cooler temperatures. For fruiting structures, bolls and squares, maximum response occurred at 27°C with 45% increase in dry matter with [CO_2] enrichment. While at 31°C, there was 78% reduction in fruiting structure mass produced at elevated [CO_2] when compared to plants grown in ambient [CO_2] and 31.3° average temperature condition. At 35°C, no fruiting structures were formed at ambient or elevated [CO_2].

K.R. Reddy et al. (1997b) found vegetative matter production at maturity in cotton grown at ambient (23°C) and ambient plus 2, 5, and 7°C were similar within each [CO_2] treatment. Boll weights produced at ambient and ambient plus or minus 2°C were similar, but no bolls were retained till maturity at ambient plus 5 or 7°C. They measured the number of bolls produced and retained for a 28 d period during early fruiting period and related to the temperature during the corresponding period. There were no effects of temperature on the total number of bolls produced with approximately 200-250 boll produced per plant in 28 d. At temperatures = 29.6°C, most of the bolls produced were retained, but at temperature 32.6 and 34.6°C, the numbers of bolls retained were negligible. Similar responses were obtained at both ambient and elevated [CO_2] treatment, suggesting that there were no beneficial effect of elevated [CO_2] on reproductive growth at supra-optimum temperatures.

Fruit production efficiency, defined as the dry weight of bolls per total dry weight, increased as temperature increased from about 26 to 29°C, then declined rapidly at temperature > 29°C under both ambient and elevated [CO_2] (K.R. Reddy et al., 1992c, 1997c; Figure 15) due to decreased fruit retention. Temperature optima for vegetative growth measured early in the season (66 DAE) was around 22 to 28°C, and growth decreased either at lower or higher temperatures (Figure 16). Elevated [CO_2] did not change either the optima or the rate. The fact that lower fruit number and fruit dry weight occurred at elevated [CO_2] and temperature levels in spite of increase in photosynthetic rates clearly suggest that cause was not related to carbohydrate supply. Studies on other crops have showed that this response is due mainly to failure of fertilization processes or early abortion of fruits (Matsui et al., 1997; Prasad et al., 2002). Studies in cotton showed reduction in pollen production and pollen sterility at daytime temperatures of 38°C (Meyer, 1969; Powell, 1969).

Growth rates of both squares and boll were influenced by temperature (Figure 17). As temperature increased from 17 to 36°C, there was curvilinear increase in the square growth rates with maximum rates occurring at 31°C. Where as boll growth rate increased from 17 to 26°C, but further increase in temperature decreased boll growth rates. Vegetative growth, stem elongation and leaf area expansion rates, on the other hand, increased linearly with increase in temperature up to 30°C, and declined at higher temperatures.

Elevated [CO_2] increased fruit dry weight to the extent of 45% at optimum temperatures (K.R. Reddy et al., 1995b). As there was no effect of elevated [CO_2] on individual boll weight (Figure 18), greater fruit dry weight were due

FIGURE 15. Fruit production efficiency for cotton (boll mass/total dry weight produced) as affected by temperature and atmospheric [CO_2] ($Y = -6232.0859 + 471.3171 * X - 8.53456 * X^2$, $r^2 = 0.92$) (adapted from K.R. Reddy et al., 1997b).

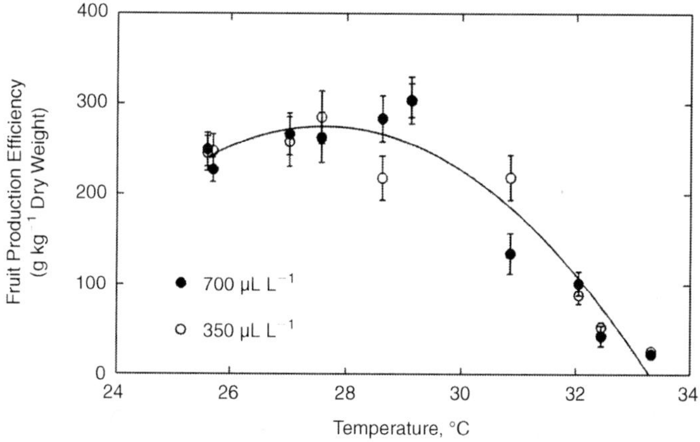

FIGURE 16. Influence of temperature and atmospheric [CO_2] on vegetative dry matter production of cotton harvested 66 days after emergence.

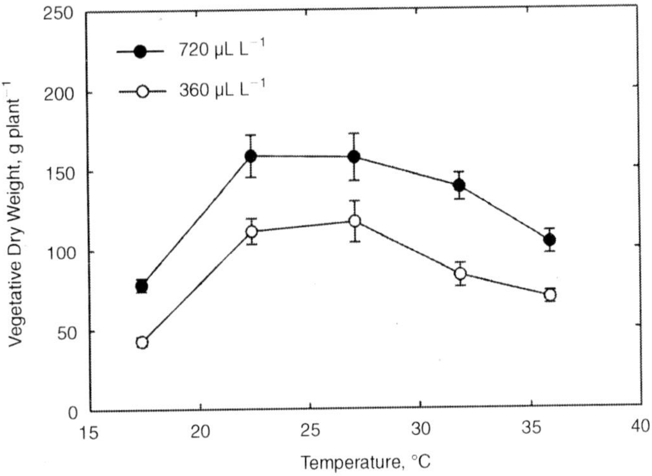

mainly to increased number of fruits. In cotton, exposure to double ambient [CO_2] increased number of fruiting sites by about 32% (Figure 19). This was due mainly to greater number of secondary branches and more fruiting sites per branch produced in [CO_2] enriched plants (Figure 20). Increase in temperature decreased boll size to same the extent in plants grown at ambient and elevated [CO_2]. This clearly shows that, doubling of [CO_2] did not ameliorate the adverse effect of high temperature on cotton boll retention and boll size. These results are in agreement with those of rice (Horie et al., 2000), wheat (Wheeler et al., 1996), soybean (Baker et al., 1989), and kidney bean (Prasad et al., 2002).

Fiber Properties

Although fiber is the main economic product of cotton crop, little research has been conducted on the effects of elevated [CO_2] and/or temperature on fiber properties and quality. Fiber properties of cotton were not affected by elevated [CO_2] and no interaction between temperatures and [CO_2] were observed (Figure 21). However, there were significant effects of temperature on fiber properties of plants grown at ambient and elevated [CO_2]. As temperature during the boll-filling period increased from 20 to 30°C, the percentage of short-fiber motes increased in both ambient and elevated [CO_2]. Small motes are formed on days when daytime temperature was near 38°C (Pearson, 1949).

FIGURE 17. Influence of temperature on vegetative (leaf and stem elongation) and reproductive growth (boll and square growth) rates of cotton grown under optimum water and nutrient conditions. The equations describing various growth processes are as follows: leaf area expansion rate ($Y = -337.6094 + 19.2017 * X$, $r^2 = 0.93$, temperature < 29°C), stem elongation rate ($Y = -46.6789 + 2.4683 * X$, $r^2 = 0.97$, temperature > 29°C), square growth rate ($Y = -0.01017 + 0.001253 * X - 0.00001948 * X^2$, $r^2 = 0.96$) and boll growth ($Y = -0.23303 + 0.02797 * X - 0.0005852 * X^2$, $r^2 = 0.93$).

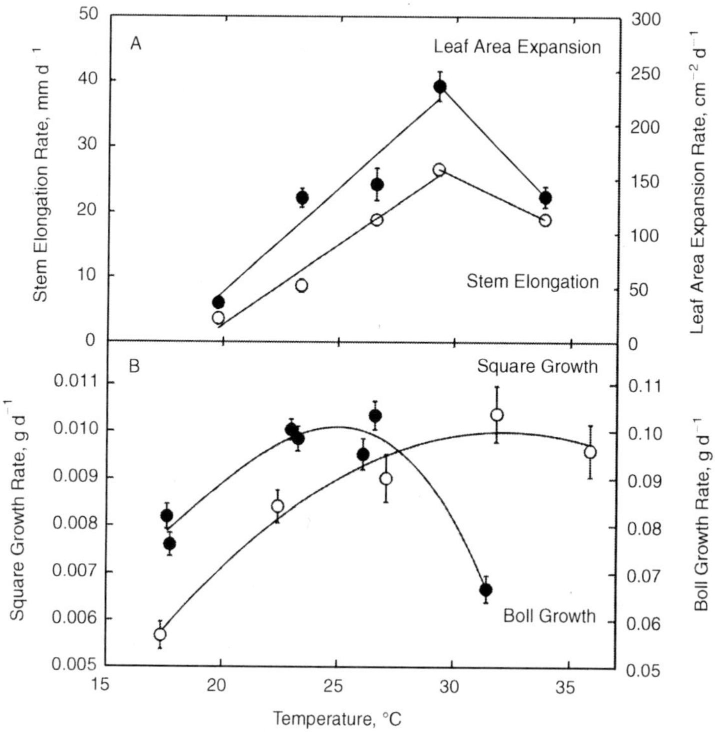

K.R. Reddy et al. (1999) suggested that since high temperatures can alter pollen and ovule development, leading to inefficient fertilization, small fiber motes are derived from unfertilized ovules. Small motes can also be formed from fertilized ovules that abort shortly. Under elevated [CO_2], more ovules are produced because of greater number bolls produced and retained per plant and the potential for ovule and seed abortion is greater.

Some other fiber quality measurements include fiber length, fiber maturity, fiber fineness, and percentage of short fiber content. It is generally conceived that greater the fiber length, optimum maturity and fineness and smaller the

FIGURE 18. Influence of temperature on mature boll weight of cotton grown at ambient and elevated atmospheric $[CO_2]$ ($Y = 2.0095 + 0.05818 * X - 0.01788 * X^2$, $r^2 = 0.97$).

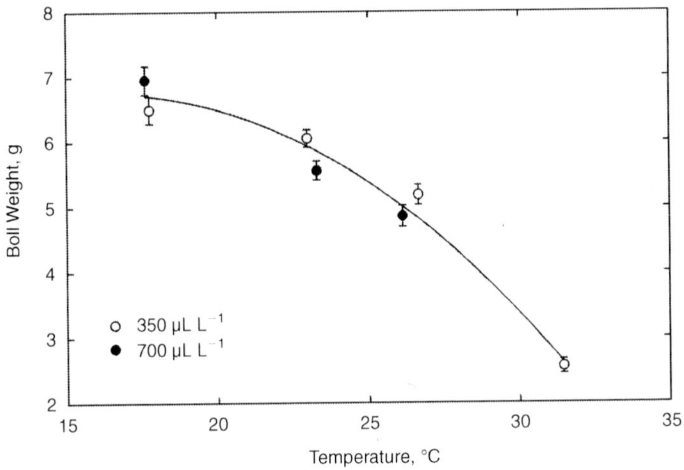

FIGURE 19. Relation between total fruiting sites produced on cotton plants grown at ambient (360 µL L^{-1}) and elevated (720 µL L^{-1}) $[CO_2]$ ($Y = -0.9997 + 1.3844 * X$, $r^2 = 0.95$).

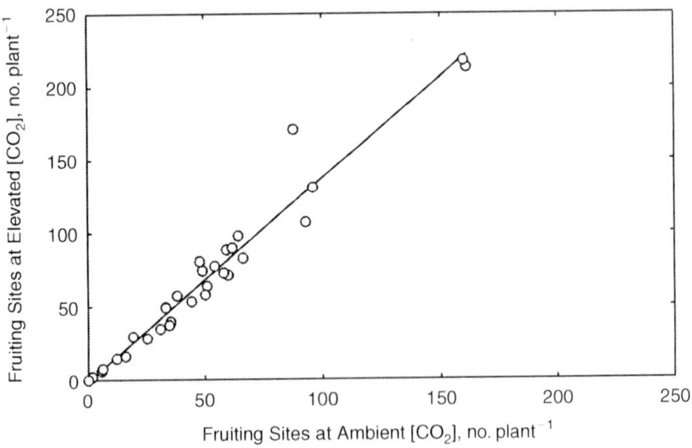

FIGURE 20. Influence of atmospheric [CO_2] on primary and secondary branches of cotton plants grown in optimum water and nutrient conditions.

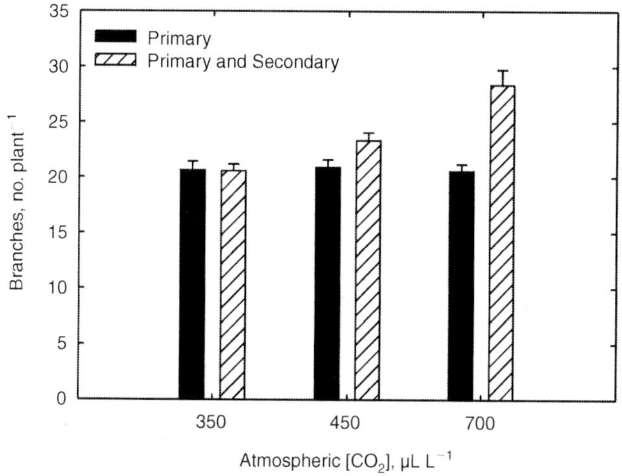

percentage of short fibers better is the fiber quality of cotton. Response of mean fiber length was best described by quadratic response, and mean fiber length decreased with increasing temperatures from 17 to 31°C during the period from flowering to open boll (Figure 22). The lowest percentage of short-fiber was observed at 26°C, and increased at lower or higher temperatures under both ambient and elevated [CO_2]. In contrast, fiber fineness and maturity (as described degree of circularity, cross sectional area and micronafis) increased linearly with an increase in temperature up to 26°C, but decreased at 32°C at both ambient and elevated [CO_2] (Figure 22).

CONCLUSION

Even though, the responses of cotton, an indeterminate C_3 crop, growth and development to changes in atmospheric [CO_2] and temperature are complex, several generalizations are evident from our results. In general, increased [CO_2] stimulated growth of cotton at most temperatures suitable for growth, but greater response occurred at near optimum temperature. Although there is a larger stimulation of leaf-level net CO_2 assimilation rates by increased [CO_2] at higher temperature, season-long canopy-level photosynthesis, which integrates all facets of growth and development of the crop, exhibited a bilinear response to temperature. Relative increases in canopy photosynthesis of plants

FIGURE 21. Influence of temperature on (A) fiber length by weight ($Y = 25.7846 + 0.5154 * X - 0.0158 * X^2$, $r^2 = 0.90$) and (B) short fiber content by weight ($Y = 5.3149 - 0.0773 * X$, $r^2 = 0.85$, temperature < 26°C) (adapted from K.R. Reddy et al., 1999).

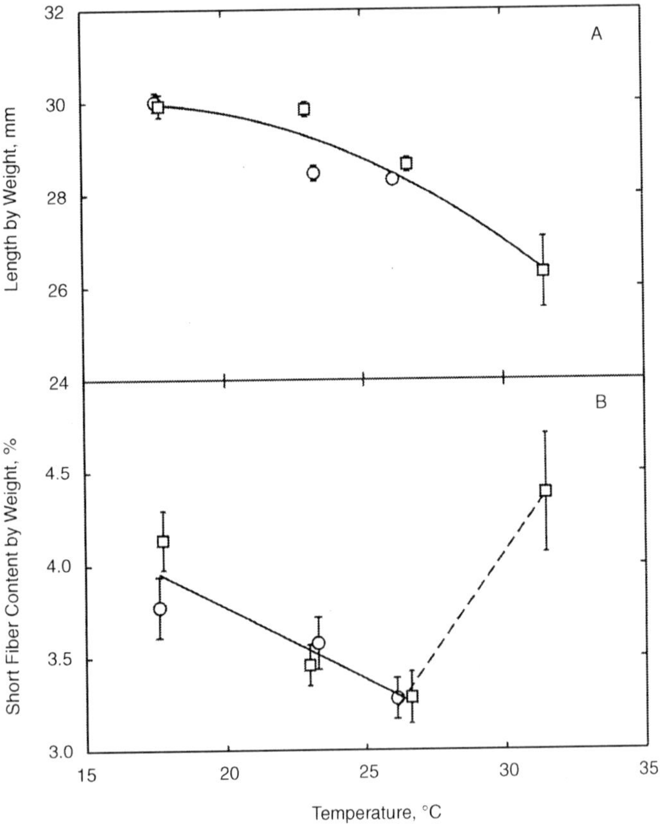

grown in elevated [CO_2] was greatest at 34°C, further decrease or increase in temperatures lead to decreased response with zero response at about 6 and 39°C. Long-term rates of canopy photosynthesis were always higher than that were observed in the current atmospheric [CO_2] resulting in higher accumulation of carbohydrates. As cotton growth is plastic to changes in environment, many of secondary growth and developmental responses were stimulated to increase in carbohydrates. As such, increased [CO_2] resulted in increased number of several organs and organ sizes (to a lesser extent except fruit size). Therefore, decreased stomatal conductivity and transpiration at the leaf-level

FIGURE 22. Influence of temperature on fiber properties of cotton; (A) degree of circularity, theta (Y = 52568 + 0.01378 * X, $r^2 = 0.94$, temperature < 26°C), (B) cross-sectional area, A (n) (Y = 59.4578 + 2.6903 * X, $r^2 = 0.91$, temperature < 26°C, and (C) micronafis (Y = 0.4265 + 0.2091 * X, $r^2 = 0.94$, temperature < 26°C), which is similar to micronaire for plants grown in ambient (circles) and elevated (squares) atmospheric [CO_2] (adapted from K.R. Reddy et al., 1999).

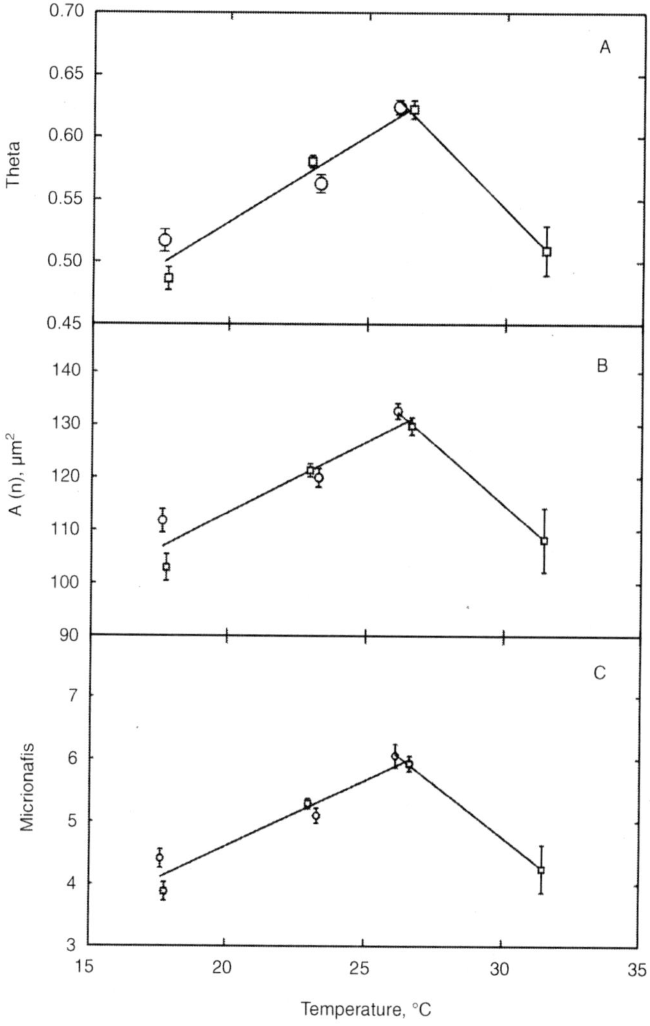

caused by elevated [CO_2] was compensated by increased leaf area, and the net result being that there is no difference in total water use. As for respiration is concerned, there is still considerable uncertainty even though it is considered that warmer temperatures normally increase the ratio of respiration to assimilate supply and elevated [CO_2] decreases the ratio.

Cotton major phenostages, such as emergence to square, square to flower, and flower to open boll and thus crop maturity, and mainstem phyllochrons are unaffected by elevated atmospheric [CO_2]. These phenological rates are mainly driven by temperature with different optimum for each of those developmental events. Warmer temperatures however, cause faster rates of development resulting less time for assimilate to fill the additional sinks that are produced in response to elevated carbohydrates. Therefore, elevated [CO_2] increased total biomass production mainly due to increased and sustained stimulation of photosynthesis over the season resulting in larger leaf area, greater number of branches and number of leaves per branch. Boll weights were increased under high [CO_2] when grown under optimum temperatures owing to greater number of fruiting sites and bolls caused by increased branching and more fruiting sites per branch. Temperatures above 30°C decreased the number of bolls retained, boll-filling period, rate of boll filling and boll size under both ambient and elevated CO_2 conditions. Doubling of [CO_2] did not offset the adverse effects of supra-optimal temperature on reproductive growth and development. Therefore, in future, if the increases in atmospheric [CO_2] are associated with higher temperature, productivity of cotton will reduce particularly in regions where present temperatures are close to optimum. However, in the regions where cool temperatures presently suppress yields would be benefited by predicted changes in climate. Future research should be aimed to explore species and cultivars differences for high temperature tolerance and response to CO_2 enrichment.

REFERENCES

Acock, B. and M.C. Acock. (1989). Calculating air leakages rates in controlled environment chambers containing plants. *Agronomy Journal* 81: 619-623.

Allen, L.H., Jr. (1999). Evapotranspiration responses of plants and crops to carbon dioxide and temperature. *Journal of Crop Production* 2: 37-70.

Allen, L.H., Jr., E.C. Bisbal, W.J. Campbell and K.J. Boote. (1987). Responses of vegetation to rising CO_2: photosynthesis, biomass and seed yield of soybean. *Global Biochemical Cycles* 1: 1-14.

Allen, L.H., Jr., P. Jones and J.W. Jones. (1985). Rising atmospheric CO_2 and evapotranspiration. pp. 13-27. In: *Advances in Evapotranspiration*. ASAE Publication 14-85. American Society of Agricultural Engineers, St. Joseph, Michigan, USA.

Baker, J.T., and L.H. Allen, Jr. (1993). Contrasting crop species responses to CO_2 and temperature: rice, soybean and citrus. *Vegetatio* 104/105: 239-260.

Baker, J.T., L.H. Allen, Jr., and K.J. Boote. (1990). Growth and yield response of rice to carbon dioxide concentration. *Journal of Agricultural Sciences* 115: 313-320.

Baker, J.T., L.H. Allen, Jr., K.J. Boote, P. Jones and P.W. Jones. (1989). Response of soybean to air temperature and carbon dioxide concentration. *Crop Science* 29: 98-105.

Bhattacharya, N.C., J.W. Radin, B.A. Kimball, J.R. Mauney, G.R. Hendrey, J. Nagy, K.F. Lewin and D.C. Ponce. (1994). Leaf water relations of cotton in a free-air-CO_2-enriched environment. *Agricultural and Forest Meteorology* 70: 171-182.

Boote, K.J., N.B. Pickering and L.H. Allen, Jr. (1997). Plant modeling: advances and gaps in our capability to predict future crop growth and yield. pp. 179-228. In: L.H. Allen, Jr., M.B. Kirkhan, D.M. Olsztk, C.E. Whitman (eds.). *Advances in Carbon Dioxide Effects Research.* ASA Special Publication No. 61, ASA-CSSA-SSSA, Madison, Wisconsin, USA.

Bowes G. (1996) Photosynthetic responses to changing atmospheric carbon dioxide. pp. 387-407. In: N.R. Baker (ed.). *Photosynthesis and the Environment.* Advances in Photosynthesis Vol. 5, Kluwer, Dordrecht, The Netherlands.

Bunce, J.A. (1998). The temperature dependence of the stimulation of photosynthesis by elevated carbon dioxide in wheat and barley. *Journal of Experimental Botany* 49: 1555-1561.

Bunce, J.A. (2001). Direct and acclamatory response of stomatal conductance to elevated carbon dioxide in four herbaceous crop species in the field. *Global Change Biology* 7: 323-331.

Campbell, W.J., L.H. Allen, Jr. and G. Bowes. (1988). Effects of CO_2 Concentration on rubisco activity, amount, and photosynthesis in soybean leaves. *Plant Physiology* 88:1310-1316.

Doherty, R.M., L.O. Mearns, K.R. Reddy, M. Downton and L.M. Daniel. (2003). Impacts of spatial scale of climate change scenarios on simulated cotton production in the Southeastern USA. *Climate Change*, 60:99-129.

Fageria, N.K., V.C. Baligar and C.A. Jones. (1991). *Growth and Mineral Nutrition of Field Crops.* Marcel Dekker, New York, USA.

Garcia, R.L., S.P. Long, G.W. Wall, C.P. Osborne, B.A. Kimball, G.Y. Nie, P.J. Pinter, R.L. Lamorte and F. Wechsung. (1998). Photosynthesis and conductance of spring wheat leaves: field response to continuous free-air atmospheric CO_2 enrichment. *Plant Cell and Environment* 21: 659-669.

Gross, Y. and J. Kigel. (1994). Differential sensitivity to high temperature of stages in the reproductive development of common bean *Phaseolus vulgaris* L. *Field Crops Research* 36: 201-212.

Hall, A.E. (1992). Breeding for heat tolerance. *Plant Breeding Reviews* 10: 129-167.

Hendrey, G.R., K.F. Lewin and J. Nagy. (1993). Free air carbon dioxide enrichment: development, progress, results. *Vegetatio* 104: 17-31.

Hileman, D.R., G. Hukula, P.K. Kenjige, N. Sinha, N.C. Bhattacharya, P.K. Biswas, K.F. Lewin, J. Nagy and G.R. Hendrey. (1994). Canopy photosynthesis and transpiration of field-grown cotton exposed to free air CO_2 enrichment (FACE) and differential irrigation. *Agricultural and Forest Meteorology* 70: 189-207.

Horie, T., J.T. Baker, H. Nakagawa, T. Matsui and H.Y. Kim. (2000). Crop ecosystems responses to climate change: rice. pp. 81-106. In: K.R. Reddy and H.F. Hodges (eds.), *Climate Change and Global Crop Productivity*. CABI Publishing, Oxon, UK.

Houghton, J.T., Y. Ding, D.J. Griggs, M. Noguer, P.J. van der Linden and D. Xiaosu (eds.). (2001). *Climate Change 2001: The Scientific Basis. Contribution of Working Group I to the Third Assessment Report of the Intergovernmental Panel on Climate Change (IPCC)*. 944 pages, Cambridge University Press, Cambridge, UK.

Hui, D.F., Y.Q. Luo, W.X. Cheng, J.S. Coleman, D.W. Johnson and D.A. Sims. (2001). Canopy radiation and water use efficiency as affected by elevated [CO_2]. *Global Change Biology* 7: 75-91.

Janssens, I.A., M. Mousseau and R. Ceulemans. (2000). Crop ecosystems to climate change: tree crops. pp. 245-270. In: K.R. Reddy and H.F. Hodges (eds.), *Climate Change and Global Crop Productivity*. CABI Publishing, Oxon, UK.

Jarvis, A.J., T.A. Mansfield and W.J. Davies. (1999). Stomatal behavior, photosynthesis and transpiration under rising CO_2. *Plant Cell and Environment* 22: 639-648.

Jones, P., L.H. Allen, Jr. and J.W. Jones. (1985). Responses of soybean canopy photosynthesis and transpiration to whole-day temperature change in different CO_2 environments. *Agronomy Journal* 77: 242-249.

Kimball, B.A. (1983). Carbon dioxide and agricultural yield: An assemblage of 430 prior observations. *Agronomy Journal* 75: 779-788.

Kimball, B.A., K. Kobayashi and M. Bindi. (2002). Responses of agricultural crops to free-air CO_2 enrichment. *Advances in Agronomy* 77: 293-368.

Kimball, B.A. and J.R. Mauney. (1993). Response of cotton to varying CO_2, irrigation, and nitrogen: yield and growth. *Agronomy Journal* 85: 706-712.

Kump, L.R. (2002). Reducing uncertainty about carbon dioxide as a climate driver. *Nature* 419: 188-190.

Long, S.P. (1991). Modification of the response of photosynthetic productivity to rising temperature by atmospheric CO_2 concentration: has its importance been underestimated? *Plant Cell and Environment* 14: 729-739.

Matsui, T., O.S. Namuco, L.H. Ziska and T. Horie. (1997). Effects of high temperature and CO_2 concentration on spikelet sterility in indica rice. *Field Crops Research* 51: 213-219.

Mauney, J.R. (1984). Cotton square shedding. Why they fall; what it means to yields. *Crops and Soil Magazine* 37: 20-22.

Mauney, J.R., B.A. Kimball, P.J. Pinter Jr., R.L. LaMorte, K.F. Lewin, J. Nagay and G.R. Hendrey. (1994). Growth and yield of cotton in response to free-air carbon dioxide enrichment (FACE) environment. *Agricultural and Forest Meteorology* 70: 49-67.

Mearns, L.O. (2000). Climate change and variability, pp. 7-35. In: K.R. Reddy and H.F. Hodges (eds.), *Climate Change and Global Crop Productivity*. CABI Publishing, Oxon, UK.

Meyer, V.G. (1969). Some effects of genes, cytoplasm and environment on male sterility in cotton. *Crop Science* 9: 237-242.

Monje, O. and B. Bugbee. (1998). Adaptation to high CO_2 concentration in an optimal environment: radiation capture, canopy quantum yield and carbon use. *Plant Cell and Environment* 21: 315-324.

Moore, B.D., S.H. Cheng, D. Sims and J.R. Seemann. (1999). The biochemical and molecular basis for photosynthesis acclimation to elevated atmospheric CO_2. *Plant Cell and Environment* 22: 567-582.

Morrison, J.I.L. and D.W. Lawlor. (1999). Interactions between increasing CO_2 concentration and temperature on plant growth. *Plant Cell and Environment* 22: 659-682.

Murray, D.R. (1997). *Carbon Dioxide and Plant Responses.* Research Studies Press Ltd., Somerset, England, UK.

Mutsaers, H.J.W. (1983). Leaf growth in cotton. II. The influence of temperature, light, water stress and root restrictions on growth and initiation of leaves. *Annals of Botany* 51: 521-529.

Pearson, N.L. (1949). Mote types in cotton and their occurrence as related to variety, environment, position in lock, lock size and number of locks per boll. USDA Technical Bulletin. U.S. Government Print Office, Washington, DC, USA.

Peet, M. and D.W. Wolf. (2000). Crop ecosystems to climate change: vegetable crops. pp. 213-243. In: K.R. Reddy and H.F. Hodges (eds.), *Climate Change and Global Crop Productivity.* CABI Publishing, Oxon, UK.

Phene, C.J., D.N. Baker, J.R. Lambert, J.E. Parsons and J.M. McKinion. (1978). SPAR–a soil-plant-atmospheric-research system. *Transactions of ASAE* 21: 924-930.

Powell, R.D. (1969). Effect of temperature on boll set and development of *Gossypium hirsutum. Cotton Growers Review* 46: 29-36.

Prasad, P.V.V., K.J. Boote, L.H. Allen, Jr. and J.M.G. Thomas. (2002). Effects of elevated temperature and carbon dioxide on seed-set and yield of kidney bean (*Phaseolus vulgaris* L.). *Global Change Biology* 8: 710-721.

Prasad, P.V.V., P.Q. Craufurd and R.J. Summerfield. (1999). Fruit number in relation to pollen production and viability in groundnut exposed to short episodes of heat stress. *Annals of Botany* 84: 381-386.

Prasad, P.V.V., P.Q. Craufurd, V.G. Kakani, T.R. Wheeler and K.J. Boote. (2001). Influence of high temperature during pre- and post-anthesis stages of floral development on fruit-set and pollen germination in peanut. *Australian Journal of Plant Physiology* 28: 233-240.

Prior, S.A., H.H. Rogers, G.B. Runion and J.R. Mauney (1994). Effects of free-air enrichment on cotton root growth. *Agricultural and Forest Meteorology* 70: 69-86.

Reddy, A.R., K.R. Reddy and H.F. Hodges, (1998). Interactive effects of elevated carbon dioxide and growth temperature on photosynthesis in cotton leaves. *Plant Growth Regulation* 26: 33-40.

Reddy, K.R., G.H. Davidonis, A.S. Johnson and B.T. Vinyard. (1999). Temperature regime and carbon dioxide enrichment alter cotton boll development and fiber properties. *Agronomy Journal* 91: 851-858.

Reddy, K.R., P.R. Doma, L.O. Mearns, H.F. Hodges A.G. Richardson, M.Y.L. Boone and V.G. Kakani. (2002a). Simulating the impacts of climate change on cotton production in the Mississippi delta. *Climate Research* 22: 271-281.

Reddy, K.R., H.F. Hodges and B.A. Kimball. (2000). Crop ecosystems to climate change: cotton. pp. 161-187. In: K.R. Reddy and H.F. Hodges (eds.), *Climate Change and Global Crop Productivity.* CABI Publishing, Oxon, UK.

Reddy, K.R., H.F. Hodges and J.M. McKinion. (1993). A temperature model for cotton phenology. *Biotronics* 22: 47-59.

Reddy, K.R., H.F. Hodges and J.M. McKinion. (1995a). Carbon dioxide and temperature effects on Pima cotton development. *Agronomy Journal* 87: 820-826.

Reddy, K.R., H.F. Hodges and J.M. McKinion. (1995b). Carbon dioxide and temperature effects on Pima cotton growth. *Agriculture, Ecosystems & Environment* 54: 17-29.

Reddy, K.R., H.F. Hodges and J.M. McKinion. (1995c). Cotton crop responses to a changing environment. pp. 3-30. In: Cynthia Rosenzweig et al. (eds.). *Climate Change and Agriculture: Analysis of Potential International Impacts.* American Society of Agronomy Special Publication no. 59, Madison, Wisconsin, USA.

Reddy, K.R., H.F. Hodges and J.M. McKinion. (1997a). Crop modeling and applications: a cotton example. *Advance in Agronomy* 59: 225-290.

Reddy, K.R., H.F. Hodges and J.M. McKinion (1997b). A comparison of scenarios for the effect of global climate change on cotton growth and yield. *Australian Journal of Plant Physiology* 24: 707-713.

Reddy, K.R., H.F. Hodges and J.M. McKinion. (1997c). Modeling temperature effects on cotton internode and leaf growth. *Crop Science* 37: 503-509.

Reddy, K.R., H.F. Hodges, J.M. McKinion and G.W. Wall. (1992a). Temperature effects on Pima cotton growth and development. *Agronomy Journal* 84: 237-243.

Reddy, K.R., H.F. Hodges, J.J. Read, J.M. McKinion, J.T. Baker, L. Tarpley and V.R. Reddy. (2001). Soil plant atmosphere research (SPAR) facility: a tool for plant research and modeling. *Biotronics* 30: 27-50.

Reddy, K.R., H.F. Hodges and V.R. Reddy. (1992b). Temperature effects on cotton fruit retention. *Agronomy Journal* 84:26-30.

Reddy, K.R., V.G. Kakani, J.M. McKinion and D.N. Baker. (2002b). Applications of a cotton simulation model, GOSSYM, for crop management, economic and policy decisions. pp 33-73. In: L.R. Ahuja, Liwang Ma and T.A. Howell (eds.) *Agricultural System Models in Field Research and Technology Transfer*, CRC Press, LLC, Boca Raton, Florida, USA.

Reddy, K.R., V.R. Reddy and H.F. Hodges. (1992c). Temperature effects on early season cotton growth and development. *Agronomy Journal* 84: 237-243.

Reddy, K.R., R.R. Robana, H.F. Hodges, X.J. Liu and J.M. McKinion (1998a). Influence of atmospheric CO_2 and temperature on cotton growth and leaf characteristics. *Environmental and Experimental Botany* 39: 117-129.

Reddy, K.R., H.F. Hodges and J.M. McKinion (1998b). Photosynthesis and environmental factors. pp. 1443-1450. In: P. Dugger and D. Richter (eds.), *Proceedings Beltwide Cotton Conferences*, San Diego, California, USA, 5-9 January 1998. National Cotton Council, Memphis, USA.

Reddy, V.R., K.R. Reddy, M.C. Acock and A. Trent. (1994). Carbon dioxide enrichment and temperature effects on root growth in cotton. *Biotronics* 23: 47-57.

Reddy, V.R., K.R. Reddy and D.N. Baker. (1991). Temperature effects on growth and development of cotton during the fruiting period. *Agronomy Journal* 83: 211-217.

Reddy, V.R., K.R. Reddy and H.F. Hodges. (1995). Carbon dioxide enrichment and temperature effects on cotton canopy photosynthesis, transpiration, and water use efficiency. *Field Crops Research* 41: 13-23.

Rogers, H.H., C.M. Peterson, J.N. McCrimmon and J.D. Cure. (1992). Response of plant roots to elevated atmospheric carbon dioxide. *Plant Cell and Environment* 15: 749-752.

Sage, R.F., J. Santrucek and D.J. Grise. (1995). Temperature effects on the photosynthetic response of C_3 plants to long-term CO_2 enrichment. *Vegetatio* 121: 67-77.

Sage, R.F., T.D. Sharkey and J.R. Seemann (1989). Acclimation of photosynthesis to increasing atmospheric CO_2: the gas exchange perspective. *Photosynthesis Research* 39: 351-368.

Sakai, H., K. Yagi, K. Kobayashi and S. Kawashima. (2001). Rice carbon balance under elevated CO_2. *New Phytologist* 150: 241-249.

Samarakoon, A.B. and R.M. Gifford. (1995). Soil water content under plants at high CO_2 concentration and interactions with direct CO_2 effects: a species comparison. *Journal of Biogeography* 22: 193-202.

Samarakoon, A.B. and R.M. Gifford. (1996). Water use and growth of cotton in response to elevated CO_2 in wet and drying soil. *Australian Journal of Plant Physiology* 23: 63-74.

Sharkey, T.D. (1988). Estimating the rate of photorespiration in leaves. *Physiologia Plantarum* 73: 147-152.

Thomas, R.B., C.D. Reid, R. Ybema and B.R. Strain. (1993). Growth and maintenance components of leaf respiration of cotton grown in elevated carbon dioxide partial pressure. *Plant Cell and Environment* 16: 539-546.

Vu, J.C.V., L.H. Allen, Jr., K.J. Boote and G. Bowes. (1997). Effects of elevated CO_2 and temperature on photosynthesis and Rubisco in rice and soybean. *Plant Cell and Environment* 20: 68-76.

Wheeler, T.R., G.R. Batts, R.H. Ellis, P. Hadley and J.I.L. Morison. (1996). Growth and yield of winter wheat (*Triticum aestivum*) crops in response to CO_2 and temperature. *Journal of Agricultural Science* 127: 37-48.

Wolfenden, J. and P.J. Diggle. (1995). Canopy gas exchange and growth of upland pasture swards in elevated CO_2. *New Phytologist* 130: 369-380.

The Effects of Natural CO_2 Enrichment on the Growth of Maize

D. Vodnik
H. Šircelj
D. Kastelec
I. Maček
H. Pfanz
F. Batič

SUMMARY. Maize plants (*Zea mays*) were grown in a field influenced by a geothermal source of CO_2 (CO_2 spring Stavešinci, Slovenia). Yield parameters and the content of photosynthetic pigments, mineral nutrients, antioxidants and soluble sugars were measured and the photosynthetic performance was followed in plants growing at different CO_2 conditions. Growth parameters were negatively correlated with the soil CO_2 concentration as measured at 20 cm depth. Analysis of *A-Ci* curves revealed lower maximum photosynthesis (CO_2-saturated photosynthesis) and lower carboxylation efficiency in high CO_2 plants when compared to low CO_2 plants. Several biochemical parameters, as the decrease

---

D. Vodnik, H. Šircelj, D. Kastelec, I. Maček and F. Batič are affiliated with the Agronomy Department, Biotechnical Faculty, University of Ljubljana, Jamnikarjeva 101, SI-1001 Ljubljana, Slovenia (E-mail: dominic.vodnik@uni-lj.si).

H. Pfanz is affiliated with the Institute of Applied Botany, University of Essen, D-45117, Essen, FRG.

[Haworth co-indexing entry note]: "The Effects of Natural CO_2 Enrichment on the Growth of Maize." Vodnik, D. et al. Co-published simultaneously in *Journal of Crop Improvement* (Food Products Press, an imprint of The Haworth Press, Inc.) Vol. 13, No. 1/2 (#25/26), 2005, pp. 193-212; and: *Ecological Responses and Adaptations of Crops to Rising Atmospheric Carbon Dioxide* (ed: Zoltán Tuba) Food Products Press, an imprint of The Haworth Press, Inc., 2005, pp. 193-212. Single or multiple copies of this article are available for a fee from The Haworth Document Delivery Service [1-800-HAWORTH, 9:00 a.m. - 5:00 p.m. (EST). E-mail address: docdelivery@haworthpress.com].

http://www.haworthpress.com/web/JCRIP

Digital Object Identifier: 10.1300/J411v13n01_09

in chlorophyll content and the increase in antioxidants indicate stress in plants exposed to a high CO_2 environment.

KEYWORDS. Natural CO_2 springs, *Zea mays*, photosynthesis, soil CO_2, pigments, antioxidants

INTRODUCTION

On the basis of early experiments and due to the presence of carbon dioxide concentrating mechanisms of C4 plants, it was assumed that responsiveness to CO_2 should be much lower in C4 than in C3 plants (Bowes, 1993, Drake et al. 1997). There is, however, mounting evidence that growth of many C4 plants responds positively to an elevated CO_2 concentration (e.g., Ziska and Bunce 1997, Wand et al. 1999, Ghannoum et al. 2000). Yet, underlying mechanisms of this response are still largely unexplained.

Beside forage and range grasses and weeds, which contribute substantially to the global primary productivity, C4 plants include some of the world's most important crops. It is therefore understandable that many studies of the CO_2 effects have focused on the two main C4 crops, sorghum and maize (an NADP-ME-type monocot). Under double ambient CO_2 the growth response of maize varied from no stimulation of growth (Hunt et al. 1991) to 3-6% (Morison and Gifford 1984, Ziska and Bunce 1997), 20% (Wong 1979), 24% (Carlson and Bazzaz 1980), 36% (Morison and Gifford 1984), and up to 50% stimulation (Rogers and Dahlman 1993). It is reported that the average growth stimulation of C4 plants in response to a doubled CO_2 concentration is about 22-33%, which is lower than the average for C3 plants (40-44%) (Poorter 1993, Wand et al. 1999).

In the study by Maroco et al. (1999) it was observed that maize plants grown under ca. three times atmospheric CO_2 concentration (1100 μmol mol^{-1}) accumulate 20% more biomass than plants grown under ambient CO_2 (350 μmol mol^{-1}). This increase in biomass was found despite lower carboxylation efficiency and a reduction of the activity of several enzymes of the C4 pathway. It was concluded that growth of maize might benefit from an elevated CO_2 concentration through regulation of content and activity of certain photosynthetic enzymes (acclimation) and by improved water-use efficiency. On the contrary, a similar photosynthetic behaviour has been rarely found in other C4 plants. This can be easily explained by low concentrations of Rubisco in C4 plants relative to C3 plants and the fact that Rubisco operates under high CO_2 concentration in bundle sheath cells (Ghannoum et al. 2000).

How C4 plants respond in an enriched CO_2 environment will depend not only on the effectiveness of the CO_2 concentrating mechanism of photosynthesis and possible acclimation strategies but also on the interaction of the prevailing environmental conditions. It has been shown that a CO_2-induced growth stimulation of C4 plants increases with decreasing soil water availability and increasing leaf-to-air water vapour pressure deficit, air temperature, light intensity and nitrogen supply (Ghannoum et al. 2000, Kang et al. 2002). For a proper evaluation of C4 crop plant response to elevated CO_2 studies have to be performed under conditions that are optimal for C4 photosynthesis (high light and high temperature; Maroco et al. 1999) and close related to standard agronomic practises. For this purpose FACE systems can be efficiently used.

Natural carbon dioxide sources (CO_2 springs) provide an alternative method to study both short- and long-term response of plants to elevated CO_2 (Raschi et al. 1997). Studies at these sites have been mainly focused on natural vegetation in order to evaluate long-term effects of CO_2 enrichment. It has been shown that a natural high-CO_2 environment provided by a geothermal source, can be exploited for studies on crop species too (e.g., Badiani et al. 1997). Beside the research linked to the problems of a global atmospheric CO_2 increase, CO_2 springs offer an opportunity to study the plant's physiological response to an extreme and fluctuating gaseous regime.

A few years ago the yield of maize had been dramatically reduced at the part of the Stavešinci field (NE Slovenia) influenced by geothermal source of CO_2. Therefore, we were interested in the physiological response of maize to natural CO_2 enrichment. The objectives of the study were to compare performance of plants differently exposed to CO_2 by measuring yield, gas exchange and different biochemical parameters. Results are discussed with regard to what has been found for several C3 and C4 plant species investigated at Stavešinci mofette during last three years (Pfanz and Vodnik, unpublished; Vodnik et al. 2002).

MATERIALS AND METHODS

Site Description

The study was conducted in 2002 at the geothermal mofette field near Stavešinci, NE Slovenia (see Batiè et al. 1999, Kaligariè 2001, Turk et al. 2002). At this site in a flat area of ca. 3000 m^2 very pure CO_2 is released into the atmosphere by several vents. Air CO_2 concentration is very much depended on weather (wind) conditions and it can fluctuate from 0.036% to at least 1%. At the part of the mofette field agriculture was stopped in 1998 due to low yields presumably caused by gas emissions. Present vegetation of the abandoned area consists of several C3 and C4 grasses and several (pioneer) herbs. Growing plants of different species can be found at locations with soil CO_2 ranging from 0.4 to 26% (v/v). Individual plant height is highly variable, the smallest plants can be found in the close neighbourhood to the vents.

The soil at the studied area is a reduced gley on quaternary alluvium, consisting of silty clay material derived from different parent rocks. The following soil properties were determined: pH ($CaCl_2$) 4.4; organic matter (Walkley-Black method) 5.3%, total N (Kjeldahl digestion) 0.25%; C/N = 12.3, available P_2O_5 293 mg kg^{-1}, available K_2O 232 mg kg^{-1}, respectively, CEC = 204 $mmol_C$ kg^{-1}.

Experiment

Before the sowing N:P:K (7:20:30) fertilizer was applied to the soil in the concentration of 600 kg per hectare. In the first week of May, 2002 maize (*Zea mays* DK 312, Dekalb, USA) was sown at the part of the field, which is partly influenced by CO_2 releasing vents. Altogether 400 m^2 were sown, later on, precise growth and soil gas measurements were made in an area of 153 m^2 (17 m $\times$ 9 m) (see bellow). Maize was grown according to standard agronomic practices. Nitrogen fertilizer KAN (27% N) was applied in the second week of June in the concentration of 500 kg per hectare. No irrigation treatment was used. Due to relatively wet summer 2002 plants experienced no severe drought during the season.

In the frame of experimental field area two subplots of 15 m^2 were selected. Before the season preliminary, pilot soil CO_2 measurements (see Methods) revealed differences in soil CO_2 concentration. Higher soil CO_2 concentrations were measured on the first plot (plot H, 1-10% CO_2) when compared to the second (plot L, 0.1-0.3% CO_2). Differences in soil [CO_2] were later, after the harvest, confirmed by measurements using regular sampling grid (see bellow), although variability within the single plot could also be observed. Since previous studies of some plant species revealed good negative correlation between plant height and high soil [CO_2] (Pfanz et al., submitted; Vodnik et al. 2002), the plant height in the first half of the growing season was an additional criterion for selection of subplots. Plants from the plot H were substantially lower when compared to the plants from plot L. As the plots were relatively close to each other (Figure 1) we presume that the basic soil characteristics were similar.

From these two subplots plants were used for more detailed measurements and analyses. Plants from randomly selected grids within the subplot were measured or sampled. In the first week of August (90 days after sowing) gas exchange measurements were performed and leaves were sampled for biochemical analyses. In the middle of September, before physiological maturity,

FIGURE 1. Spatial distribution of soil CO_2 concentration. Maize field, Stavešinci, September 2002. Measurements were taken at the depth of 20 cm by using portable gas analyser GA2000 (Ansyco). Subplots with low- (Plot L) and high CO_2 exposure (Plot H) are shown.

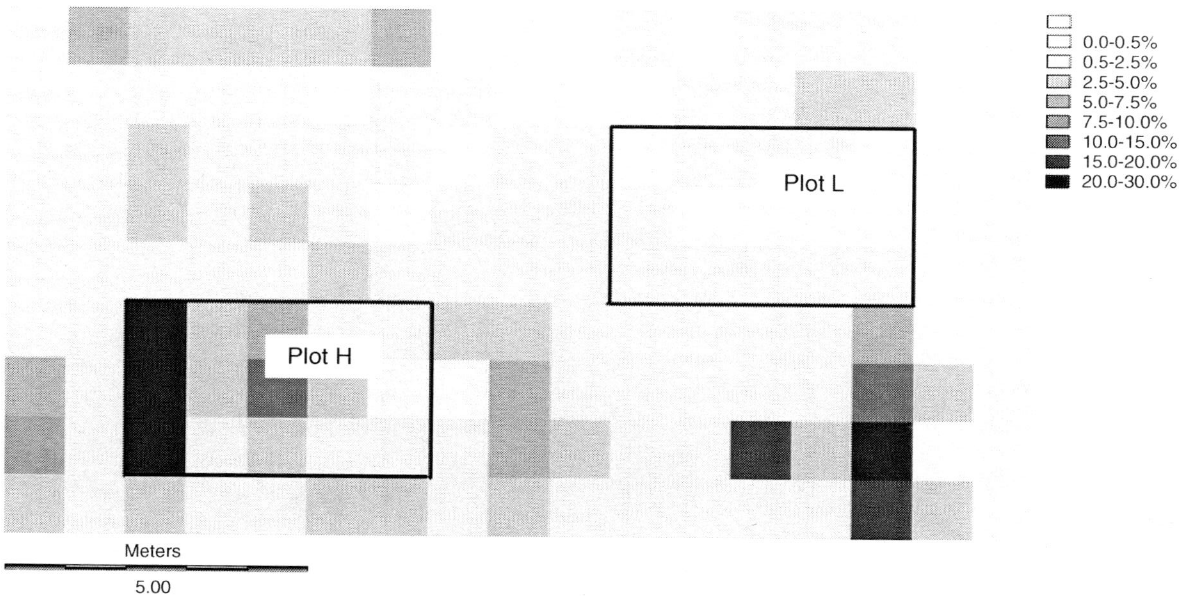

plants were harvested separately for every square meter in a 17 m × 9 m grid. All plants per the area of 153 m^2 were harvested (801 plants). Fresh and dry mass of corn stalk (including both leaf blades and stem) and cobs were measured for every single plant. Since the number of plants differed between grids, growth parameters were calculated per plant (for every square meter) and are not given per area.

Soil CO_2 Measurement

Our previous studies at natural CO_2 spring Stavešinci have shown that the soil CO_2 concentration can be used to describe CO_2 regime during growth.

One week after the final harvest soil CO_2 concentrations were measured by using portable landfill gas analyser GA2000 (Ansyco, Germany). Holes of 20 cm depth with 2 cm diameter were drilled at the centre of squares following the same grid as at harvest. Gas piping tube of the analyser was then inserted into the hole and soil air was sucked and analysed by infrared analyser. Values were recorded in percentages (v/v) and map of soil [CO_2] was created. Short-term fluctuations of air CO_2 concentration caused by wind and air turbulence prevented efficient similar spatial analysis of air [CO_2] at canopy height (see also comments in Badiani et al. 2000).

Gas Exchange Measurements

Gas exchange measurements were made with a portable photosynthetic system LI-6400 (LI-COR Inc., Lincoln, USA) for 5 plants from each subplot (plot L, plot H). The response of net CO_2-exchange (A, μmol m^{-2} s^{-1}) to changing intercellular CO_2 concentrations (C_i, μmol mol^{-1}) was conducted at 27°C, 50% RH and at light-saturating conditions (1000 μmol photons m^{-2} s^{-1}). Net CO_2-exchange was measured 10-15 min after CO_2-supply when stable leaf chamber conditions were reached.

Biochemical Analyses

Leaf samples for measurements of the chosen biochemical stress parameters were collected at midday (4th fully developed leaf). One leaf per plant, and three plants per plot were analysed. Sampled leaves were immediately frozen in liquid nitrogen, lyophilised and stored at −20°C until analysis.

Chloroplast Pigments

Chloroplast pigments (neoxanthin, violaxanthin, antheraxanthin, zeaxanthin, lutein, α-carotene, β-carotene, chlorophyll *a*, chlorophyll *b*) were determined using the method described by Pfeifhofer (1989). Pigments were extracted from the dry leaf powder with ice-cold acetone. All extraction procedures were performed in dim light. Acetone extracts were subjected to an HPLC gradient analysis (column Speherisorb S5 ODS-2 250 × 4.6 mm with precolumn S5 ODS-2 50 × 4.6 mm), using following solvents: solvent A: acetonitrile: methanol:water (100:10:5, v/v/v), solvent B: acetone:ethylacetate (2:1, v/v), at a flow rate of 1 mL min^{-1}, linear gradient from 10% solvent B to 70% solvent B in 18 min was applied. Run time was 30 min, detection was photometric at 440 nm.

Ascorbic Acid

Concentration of ascorbic acid was measured according to Tausz et al. (1996). Ascorbic acid was extracted from the lyophilised leaf powder with 1.5% (w/v) metaphosphoric acid containing 1 mM EDTA. Extracts were subjected to an isocratic HPLC analysis (column Lichrosorb RP-8 250 × 4.6 mm with precolumn Lichrosorb RP-8 50 × 4.6 mm) using methanol:water (1:3, v/v) containing 1 mM hexadecylammoniumbromide and 0.05% (w/v) sodium dihydrogen phosphate monohydrate (pH 3.6) as solvent, at a flow rate of 1 ml min^{-1}. Run time was 20 min, detection was photometric at 248 nm.

Soluble Carbohydrates

Soluble carbohydrates were extracted from the lyophilised leaf powder with 80% ethanol. After evaporating the supernatant to dryness, carbohydrates were redissolved in HPLC grade water and subjected to an isocratic HPLC analysis (column Aminex HPX-87C 300 × 7.8 mm) using bidistilled water as solvent, at a flow rate of 0.6 ml min^{-1}, run time 60 min and detection with RI detector (extraction method by Pelah et al. 1997, chromatographic conditions reported by Dolenc and Stampar 1997).

CNS Analysis

Dry leaf material was ball-milled (Retsch). The amount of total carbon, nitrogen and sulphur in plant samples was measured by CNS-analyser CNS-2000 (LECO). Infrared detection method (C and S) and conductivity detection method (N) were used (Manual 1994).

Statistical Methods

The correlation between corn yield parameters and soil CO_2 concentration was analysed using Pearson correlation coefficient.

The non-linear regression models were fitted to the net CO_2-exchange (A) as response variable, and intercellular CO_2 concentrations (C_i) as explanatory variable. The best fits for A-C_i curves were obtained using exponential function with three parameters: $A = \beta_0 \cdot (1 - \exp(-\beta_1(C_i - \Gamma)))$. β_0 describes the asymptotic value of the A-C_i curve and it can be interpreted as CO_2 saturated photosynthesis ($\mu mol\ m^{-2}\ s^{-1}$). β_1 describes the initial slope of the A-C_i curve and can be interpreted as carboxylation efficiency ($mol\ m^{-2}\ s^{-1}$). Γ presents the value of intercellular CO_2 at the point where the A-C_i curve crosses the abscise and the net CO_2 exchange is zero. So, Γ is equivalent to the CO_2 compensation point ($\mu mol\ mol^{-1}$) in our exponential model.

The t-test was used to compare the parameters of A-Ci curves and all other parameters measured for two subplots (plot L and plot H). Statistical analysis was made by Statgraphics Plus (Manugistics, USA) software.

RESULTS AND DISCUSSION

Soil CO_2 Concentration and Maize Growth

Measurements at the non-arable part of the Stavešinci mofette area revealed that concentrations up to 80% can be frequently measured and that pure CO_2 (100%) can be detected in gas transducing cracks. At sites not influenced by geothermal sources concentrations between 0.1-0.5% have been measured (Pfanz et al. submitted, Vodnik et al. 2002). In our maize field soil CO_2 concentrations up to 26.7% were measured (Figure 1). These concentrations are substantially lower than at the non-ploughed part of the mofette area. Measurements revealed also a high spatial variability of the soil CO_2 concentration (Figure 1) that probably results from high heterogeneity in compaction and porosity of arable soil. Less variable and more spatially correlated values have been measured at other parts of the moffete field (Vodnik and Pfanz, unpublished). Fluctuations of air [CO_2] prevented efficient gas analysis in the bulk air above the plots and differentiation of plots on the basis of air [CO_2]. This problem could be overcome by using a dense net of fast response CO_2 analysers and time integration of the concentrations.

Despite the high variability of the soil CO_2 and the fact that soil gas measurements reflect instantaneous gas conditions, we were able to prove a significant linear correlation between plant growth parameters (Figures 2 and 3) and transformed soil CO_2 concentration (logarithmic transformation). When the data from the whole studied area (17 m × 9 m) were taken into account the Pearson correlation coefficient for average dry mass of corn stalk per plant and ln(soil CO_2) was $r = -0.36$ ($p = 0.0001$) and for average dry mass of corn cobs per plant and ln(soil CO_2) was $r = -0.35$ ($p = 0.0001$) (Figure 3). More signifi-

FIGURE 2. Spatial distribution of dry mass of corn stalk (*Zea mays*) at field in Stavešinci, September 2002.

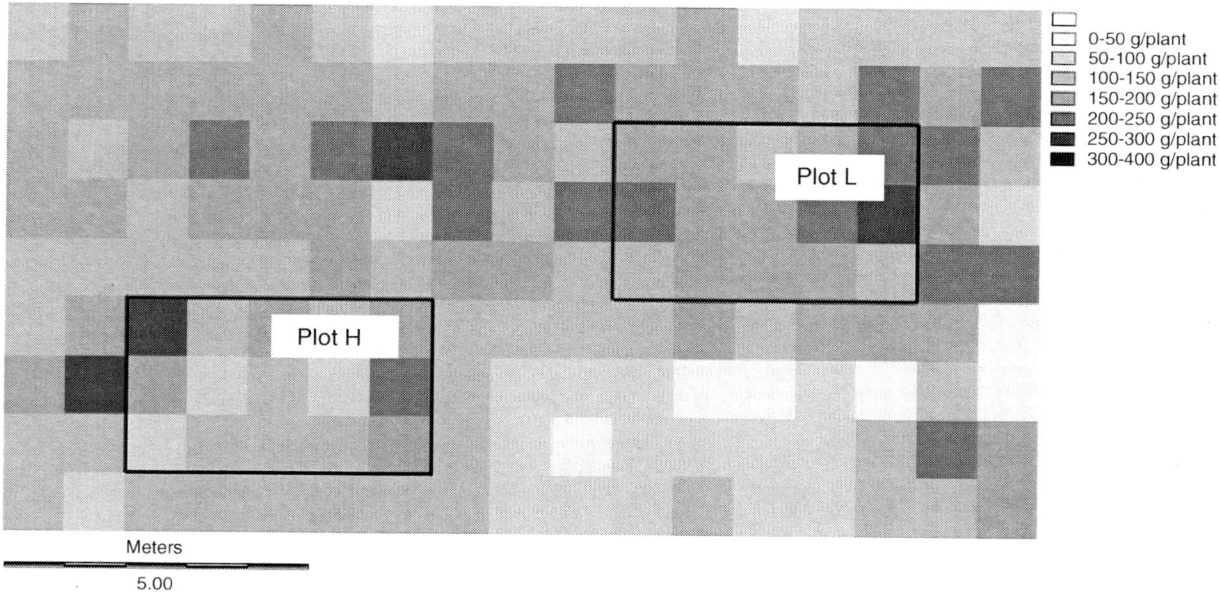

FIGURE 3. Correlation of maize yield (corn stalk dry mass (a, c), corn cob dry mass (b, d)) and soil CO_2. Stavešinci, September 2002. Data are shown for whole studied area of 17 m × 9 m (a, b) and for selected subplots, plots L and H (c, d). The Pearson correlation coefficient for average dry mass of corn stalk per plant and ln(soil CO_2) was $r = -0.36$ ($p = 0.0001$) and for average dry mass of corn cobs per plant and ln(soil CO_2) was $r = -0.35$ ($p = 0.0001$) (for whole area). More significant correlation ($r = -0.45$, $p = 0.013$ and -0.46, $p = 0.010$) was obtained when correlation between the same variables was estimated for the data from the two subplots, only.

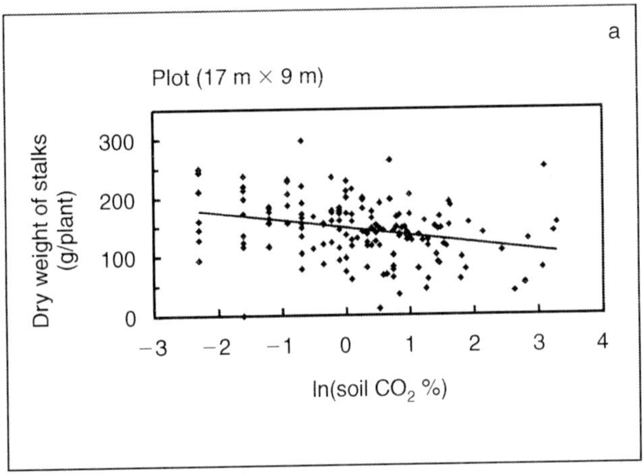

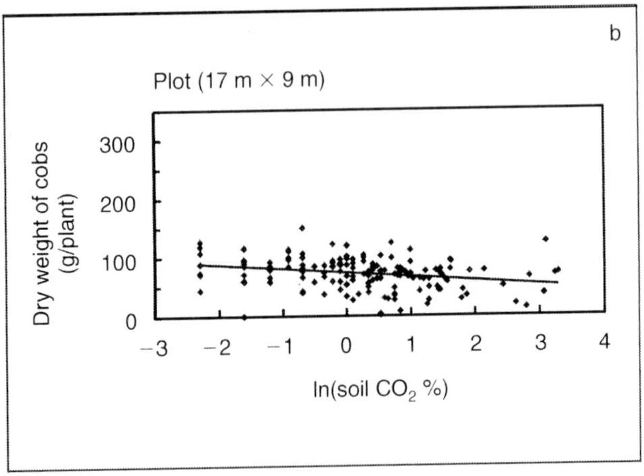

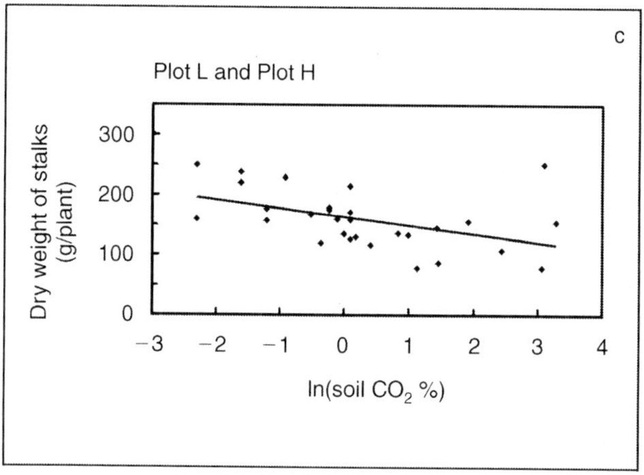

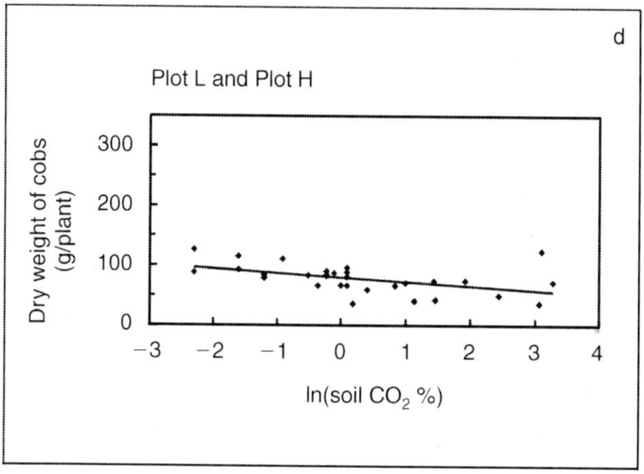

cant correlation ($r = -0.45$, $p = 0.013$ and -0.46, $p = 0.010$) was obtained when correlation between the same variables was estimated for the data from the two subplots, only.

Average growth parameters of maize from two subplots were compared by t-test. Average mass of corn stalk was significantly higher at plot L (0.533 kg ± 0.030) when compared to plot H (0.426 kg ± 0.038) ($p = 0.036$) (Table 1). Similar differences between average mass of cobs were obtained ($p = 0.043$) (Table 1).

TABLE 1. Growth parameters of maize (*Zea mays*) sampled at two different sites (Plot L = low CO_2, Plot H = high CO_2) at the geothermal mofette field Stavešinci, NE Slovenia. Means ± standard errors are presented, N = 68-74 for height and N = 15 for mass , t-test. Different letters indicate statistically significant difference ($p < 0.05$).

	Low CO_2	High CO_2	p value
Plant height [cm]	197 ± 3 a	174 ± 4 b	0.000
Dry mass of corn stalks [g plant^{-1}]	533 ± 30 a	426 ± 38 b	0.036
Dry mass of corn cobs [g plant^{-1}]	261 ± 17 a	210 ± 18 b	0.043

Previous studies at the Stavešinci mofette showed that the growth of several (semi-) natural species may be strongly inhibited in response to high [CO_2] in the air and–probably more important–in the rooting horizon. So far the growth has never been shown to be stimulated (Vodnik et al. 2002). The result of our study confirmed that similar inhibition of vegetative and reproductive growth can be expected also in crops in agriculturally fertilised sites, which are principally not nutrient limited. This indicates that elevated CO_2 at the mofette area is a strong stressor.

Photosynthesis

To clarify whether maize growing under different CO_2 regimes differ in their photosynthetic response, gas exchange was measured in plants from these two subplots. Statistical inference on the parameters of *A-Ci* curves showed that the carboxylation efficiency is significantly decreased in the case when maize plants were grown in the conditions of high soil CO_2 concentrations. The results indicate that the maximum photosynthesis (CO_2-saturated photosynthesis) is also lower for maize plants growing in higher soil CO_2 concentrations. Due to the higher variability, significant difference for this parameter was obtained only at $p < 0.10$ ($p = 0.086$) (Figure 4, Table 2).

Studies at other natural CO_2 springs showed reduction in maximum photosynthesis in *Phragmites australis* (Miglietta et al. 1995) and in different grassland species (Miglietta et al. 1998, Cook et al. 1998) but not in *Scirpus lacustris* (Bettarini 1997). In general, published data on photosynthetic parameters of mofette plants are inconsistent (Raschi et al. 1997, Miglietta et al. 1995, Badiani et al. 2000). According to our experience reductions of maximum photosynthesis and reductions of carboxylation efficiency are present in

FIGURE 4. A-C_i curves of maize (*Zea mays*) differently exposed to naturally elevated CO_2 concentration. Five individuals were measured for low-CO_2 (A) and high-CO_2 (B) exposure. Models were calculated by using equation $A = a^*(1 - \exp(-b^*(Ci - \Gamma)))$.

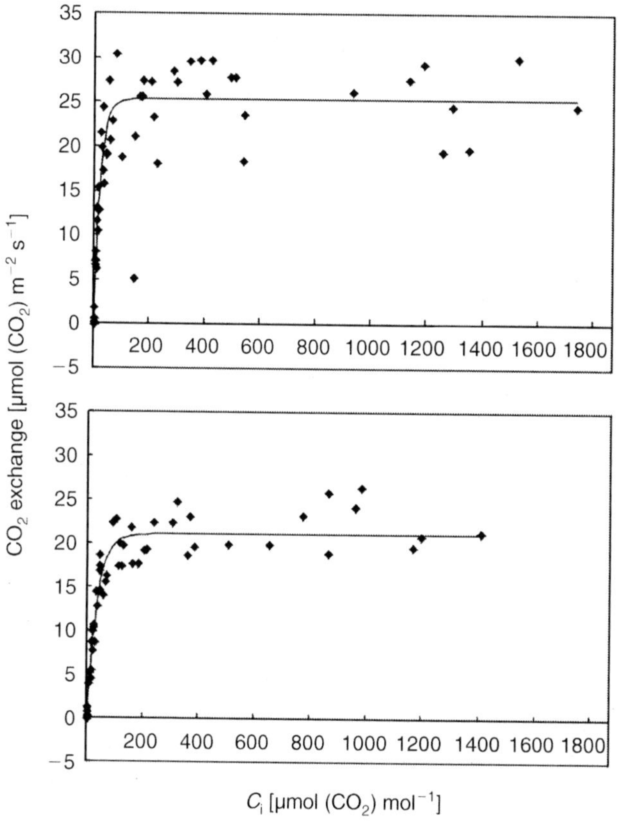

several C3 and C4 plant species growing at the Stavešinci mofette. A similar response as in maize was found in the C4 grass *Setaria pumila* (Vodnik et al., 2002), but not in cockspur (*Echinochloa crus-galli*) (Vodnik et al. 2002), another C4 grass which is a common weed in maize fields.

Yet, the photosynthetic response of plant growing under naturally elevated carbon dioxide can hardly be compared to the response of plants exposed to double atmospheric CO_2 concentration. Prolonged fumigation often results in down-regulation of photosynthesis (Stitt 1991, Bowes 1993, Webber et al. 1994, Drake et al. 1997). In the experiment by Maroco et al. (1999), a decrease

TABLE 2. Photosynthetic parameters of maize plants (*Zea mays*) from two different sites (Plot L = low CO_2, Plot H = high CO_2) at the mofette Stavešinci, NE Slovenia. Parameters were derived from models of *A*-*Ci* curves; low-CO_2: $A = 25.29*(1-EXP(-0.048*(Ci - 2.69)))$, high-$CO_2$: $A = 21.08*(1-EXP(-0.030*(Ci - 3.36)))$. Means ± standard errors are presented, N = 5, t-test. Different letters indicate statistically significant difference ($p < 0.05$).

	Low CO_2	High CO_2	p value
Maximum photosynthesis [$\mu mol\ m^{-2}\ s^{-1}$]	25.29 ± 0.61 a	21.08 ± 0.43 a	0.086
CO_2 compensation point [$\mu mol\ mol^{-1}$]	2.69 ± 1.04 a	3.36 ± 1.18 a	0.45
Carboxylation efficency [$mol\ m^{-2}\ s^{-1}$]	0.048 ± 0.006 a	0.030 ± 0.003 b	0.036

in the carboxylation efficiency was found in maize fumigated by 1100 μmol CO_2 mol^{-1}. This corresponds to a 40% reduction of the carboxylation efficiency observed in our study. Since the maximum carboxylation efficiency of C4 plants depends on PEPCase (von Caemmerer and Furbank 1999), it is to conclude that the observed decrease is a result of a lowered C4 cycle capacity. Significantly decreased levels of total leaf nitrogen (Table 3, p = 0.00005), which could lead to a decrease in total soluble protein per leaf area, support this assumption. In a number of C4 species it was shown that the point of inflection of the *A*-*Ci* curve (i.e., the *Ci* at which saturation occurs) decreases with nitrogen deprivation (Ghannoum et al. 2000, Leegood and von Caemmerer 1989).

Mineral Nutrition and Biochemical Parameters

It is well known that mineral nutrition is of crucial importance for plant performance at elevated CO_2. This issue is of particular importance at natural CO_2 springs where extreme soil conditions decrease the availability of mineral nutrients in the soil. In addition, it is to presume that nutrient uptake could be drastically affected due to the effects of soil hypoxia on the roots. Preliminary results (Maèek et al. 2002) indicate that the root respiratory potential of young maize plants is reduced when plants are growing at high CO_2 conditions similar to those in our study.

Decreased concentrations of leaf nutrients other than nitrogen have also been detected in plants from the Stavešinci mofette. In *Phleum pratense* growing at Stavešinci macro-nutrients decreased the closer the plants grew to the

TABLE 3. The content of nutrients, soluble sugars, pigments and antioxidants in the leaves of maize (*Zea mays*) sampled at two different sites (Plot L = low CO_2, Plot H = high CO_2) at the geothermal mofette field Stavešinci, NE Slovenia. Means ± standard errors are presented, N = 3-5, t-test. Different letters indicate statistically significant difference ($p < 0.05$), ns = non-significant.

	Low CO_2	**High CO_2**	**p value**
Carbon [%]	45.70 ± 0.26 a	46.42 ± 0.08 b	0.032
Nitrogen [%]	3.69 ± 0.14 a	1.98 ± 0.17 b	0.000
Sulphur [%]	0.310 ± 0.018 a	0.150 ± 0.014 b	0.000
Soluble sugar [mg/gDW]	102.7 ± 112.7 a	86.8 ± 8.3 a	ns
Violaxanthin [Mg/gDw]	0.283 ± 0.048 a	0.146 ± 0.042 b	0.021
Antheraxanthin [mg/gDW]	0.071 ± 0.016 a	0.083 ± 0.011 a	ns
Zeaxanthin [mg/gDW]	0.092 ± 0.029 a	0.122 ± 0.043 a	ns
Neoxanthin [mg/gDW]	0.236 ± 0.004 a	0.120 ± 0.025 b	0.001
Lutein [mg/gDW]	1.843 ± 0.102 a	1.044 ± 0.145 b	0.002
β-carotene [mg/gDW]	2.039 ± 0.089 a	0.988 ± 0.193 b	0.001
Carotenoids [mg/gDW]	4.563 ± 0.271 a	2.503 ± 0.350 b	0.005
Chlorophyll *a* [mg/gDW]	10.42 ± 2.67 a	5.64 ± 1.11 b	0.046
Chlorophyll *b* [mg/gDW]	4.04 ± 0.609 a	2.03 ± 0.398 b	0.009
Ascorbate [Mg/gDw]	2.74 ± 0.02 a	3.50 ± 0.16 b	0.001

DW–dry weight

CO_2 emitting vents. The decrease was around 40% in N and P, and 20% in K and S (Pfanz et al., submitted), yet, no differences in total carbon concentration were found. In our study, total carbon concentration is even slightly increased in plants exposed to higher CO_2 concentrations. It is known that non-structural carbohydrates can accumulate under high CO_2 and this can contribute to photosynthetic down-regulation (Rogers et al. 1994, Barnes et al. 1995). In our case, there were no significant differences in the leaf soluble sugars content (glucose, fructose, and sucrose) between high- and low-CO_2 grown plants. On the other hand, a significantly higher ratio of soluble sugars/chlorophylls was measured in high CO_2-plants (11.3) than in low-CO_2 plants

(7.1). Hence, a negative regulation of photosynthesis by a reduced sink activity cannot be ruled out.

In our study maize growing under high CO_2 conditions had lower leaf chlorophyll content than plants growing under a low CO_2 environment. This is in agreement with the results reported by Maroco et al. (1999) for the second leaves of maize. A decrease in foliar chlorophyll content was also reported for other crop species (e.g., Tuba et al. 1994 for wheat) and is documented for several plant species growing at the Stavešinci mofette (Pfanz and Vodnik, unpublished). Lower chlorophyll content could be a limiting factor for optimal photosynthesis and lower ratio of chlorophyll/xanthophyll cycle pigments in high CO_2 plants could contribute to a lower maximal photosynthesis. On the other hand, the photosynthetic down-regulation (inhibition) found in *Phleum pratense* was not accompanied by a decrease in chlorophyll (Pfanz, personal communication).

We also found significantly lower foliar contents of neoxanthin, lutein, violaxanthin, β-caroten and total carotenoids in high CO_2 plants compared to low CO_2 grown plants. Chlorophyll a/b and chlorophyll/total carotenoids ratios were not significantly different in high CO_2 plants as compared to low CO_2 plants, neither there were any differences for the levels of antheraxanthin, zeaxanthin, total xanthophyll cycle pigments and the de-epoxidation state. The ratio of chlorophyll/xanthophyll cycle pigments however, was significantly lower in high-CO_2 plants when compared to low-CO_2. This effect can also be related to a nitrogen deprivation as is reported for maize by Toth et al. (2002). High values of xanthophylls indicated the need for the dissipation of the excess light energy in the chloroplast of maize leaves.

Ascorbate plays an important role in protecting the photosynthetic apparatus against the destructive effect of reactive oxygen species. In our research foliar ascorbate content was significantly higher in high-CO_2 plants compared to low-CO_2 plants. Studies of antioxidants in connection of elevated CO_2 are scarce and have yielded variable results (Badiani et al. 2000). Also in our case, further studies are needed to find out if high foliar content of ascorbate in maize growing under high CO_2 conditions is the result of depressive effects of elevated CO_2 on the activity of ascorbate peroxidase (Thibaud et al. 1995, Badiani et al. 1993) or an increased biosynthesis of ascorbate as an adaptation to oxidative stress (Foyer 1993; Smirnoff and Pallanca 1996). According to the analyses of foliar photosynthetic pigments and antioxidants we suggest that the survival strategy of maize under high CO_2 conditions is to attenuate the oxidative risk by decreasing the chlorophyll content, and by increasing of the antioxidative agents (zeaxanthin and ascorbate per chlorophyll molecule).

Changes in biochemical parameters (photosynthetic pigments, antioxidants and soluble sugars), disturbances of mineral nutrition (decreased nitrogen content) and changes of photosynthetic parameters (decreased maximum photosynthesis and a lower carboxylation efficiency) are interrelated. Since plants are exposed to high soil and high atmospheric CO_2 concentration, both effects on roots and on shoots can be expected.

CONCLUSIONS

We were able to show that a reduction in the yield of maize can directly be negatively correlated with high soil CO_2 concentrations. High [CO_2] negatively affected plant photosynthetic performance (decreased carboxylation efficiency, lower CO_2-saturated photosynthesis) and has resulted in a lower content of leaf chlorophylls and nitrogen. Increased levels of antioxidants indicate that maize growing at elevated CO_2 was severely stressed. Due to the direct and indirect action of elevated CO_2 on below- and above-ground plant organs the causal analysis of the effects is not simple. Direct and indirect effects of high CO_2 (and concomitant low partial pressures of soil O_2) on roots are very likely. Therefore, future research should focus on root functioning under naturally elevated CO_2.

REFERENCES

Badiani, M., D'Annibale, A., Paolacci, A.R., Miglietta, F., Raschi, A. (1993). The antioxidants status of soybean (*Glycine max* Merill.) leaves grown under natural CO_2 enrichment in the field, *Aust. J. Plant Physiol.* 20: 275-284.

Badiani, M., Paolacci, A.R., D'Annibale, A., Miglietta, F., Raschi, A. (1997). Can rising CO_2 alleviate oxidative risk for the plant cell? Testing the hypothesis under natural CO_2 enrichment. In *Plant responses to elevated CO_2; Evidence from natural springs*, ed. A. Raschi, F. Miglietta, R. Tognetti, P.R. van Gardingen, Cambridge: Cambridge University Press, pp. 221-241.

Badiani, M., Raschi, A., Paolacci, A.R., Miglietta, F. (2000). Plants responses to elevated CO_2; a perspective from natural CO_2 springs. In *Environmental pollution and plant response*, ed. S.B. Agrawal, M. Agrawal: Lewis Pub. Boca Raton, pp. 45-81.

Barnes, J.D., Ollerenshaw, J.H., Whitfield, C.P. (1995). Effects of elevated CO_2 and/or O_3 on growth, development and physiology of wheat (*Triticum aestivum* L.), *Global Change Biol.* 1:129-142.

Batiè, F., Bernik, R., Pfanz, H., Wittmann, C., Šircelj, H., Schmidt, J., Turk, B. (1999). Effects of elevated CO_2 concentrations from natural sources on anatomy and physiology of selected plant species. In *SNACE, FACE and OTC's CO_2 enrichment at the leaf/air interface and/or the root/soil interface; results in growth and development of plants*, ed. M. Kaligariè, S. Škornik, Maribor: University of Maribor, pp. 10-11.

Bettarini, I., Miglietta, F., Raschi, A. (1997) Studying morpho-physiological response of *Scirpus lacustris* from naturally CO_2-enriched environments. In Raschi, A., Miglietta, F., Tognetti, R., van Gardingen, P.R. (Eds.). *Plant responses to elevated CO_2. Evidence from natural springs*, Cambridge: Cambridge University Press, 272 p.

Bettarini, I., Vaccari, F.P., Miglietta, F. (1998). Elevated CO_2 concentration and stomatal density: observations from 17 plant species growing in a CO_2 spring in central Italy, *Glob. Chang. Biol.* 4: 17-22.

Bowes, G. (1993). Facing the inevitable: Plants and increasing atmospheric CO_2, *Annu. Rev. Plant. Physiol. Plant. Mol. Biol.* 44: 309-332.

Carlson, R., Bazzaz, F. (1980). The effects of elevated CO_2 concentrations on growth, photosynthesis, transpiration and water use efficiency of plants. In *Environmental and climatic impact of coal utilization*, ed. J. Singh, A. Deepak, New York: Academic Press, pp. 609-623.

Cook, A.C., Tissue, D.T., Roberts, S.W., Oechel, W.C. (1998). Effects of long-term elevated [CO_2] from natural CO_2 springs on *Nardus stricta*: photosynthesis, biochemistry, growth and phenology, *Plant, Cell Environ.* 21: 417-425.

Dolenc, K., Štampar, F. (1997). An investigation of the application and conditions of analyses of HPLC methods for determining sugars and organic acids in fruits, *Research Reports, Biotechnical Faculty, University of Ljubljana* 69: 99-106.

Drake, B.G., Gonzàlez-Meler, M.A., Long, S.P. (1997). More efficient plants: a consequence of rising atmospheric CO_2, *Annu. Rev. Plant Physiol. Plant Mol. Biol.* 48: 609-639.

Foyer, C.H. (1993). Ascorbic acid. In *Antioxidants in higher plants*, ed. R.G. Alscher, J.L. Hess, Boca Raton, CRC Press, Inc., pp. 32-58.

Ghannoum, O., von Caemmerer, S., Ziska, L.H., Conroy, J.P. (2000). The growth response of C4 plants to rising atmospheric CO_2 partial pressure: a reassessment, *Plant, Cell and Environ.* 23: 931-942.

Hunt, R., Hand, D., Hannah, M., Neal, A. (1991). Response to CO_2 enrichment in 27 herbaceous species. *Funct. Ecol.* 5: 410-421.

Kaligariè, M. (2001). Vegetation patterns and responses to elevated CO_2 from natural CO_2 springs at Strmec (Radenci, Slovenia), *Acta Biol. Slov.* 44: 31-38.

Kang, S.Z., Zhang, F.C., Hu, X.T., Zhang, J.H. (2002). Benefits of CO_2 enrichment on crop plants are modified by soil water status, *Plant Soil*, 238 (1): 69-77.

Leegood, R.C., von Caemmerer, S. (1989). Some relationships between contents of photosynthetic intermediates and the rate of photosynthetic carbon assimilation in leaves of *Zea mays* L., *Planta* 178: 258-266.

Maček, I., Pfanz, H., Batič, F., Vodnik, D. (2002). Root respiration and root respiratory potential of *Echinochloa crus-galli* under CO_2 enrichment. In *3rd Slovenian Symposium on Plant Physiology, September 25-27, 2002, Ljubljana, Book of Abstracts*, ed. J. Dolenc-Koce, D. Vodnik, M. Dermastia, Ljubljana: Slovenian Society of Plant Physiology, p. 74.

Maroco, J.P., Edwards, G.E., Ku, M.S.B. (1999). Photosyntehtic acclimation of maize to growth under elevated levels of carbon dioxide. *Planta* 210: 115-125.

Miglietta, F., Badiani, M., Bettarini, I., van Gadingen, P.R., Selvi, F., Raschi, A. (1995). Preliminary studies of a long-term CO_2 response of Mediterranean vegetation around natural CO_2 vents. In *Global changes and Mediterranean-type ecosystems*, ed. J.M. Moreno, E.C. Dechela, Heidelberg, New York: Springer Verlag, pp. 102-120.

Miglietta, F., Bettarini, I., Raschi, A., Körner, C., Vaccari, F.P. (1998). Isotope discrimination and photosynthesis of vegetation growing in the Il Bossoleto CO_2 spring, *Chemosphere* 36: 771-776.

Morison, J.I.L., Gifford, R. (1984). Plant growth and water use with limited water supply in high CO_2 concentrations. II. Plant dry weight, partitioning and water use efficiency, *Aust. J. Plant Physiol.* 11: 375-384.

Pelah, D., Wang, W., Altman, A., Shoseyov, O., Bartles, D. (1997). Differential accumulation of water stress-related proteins, sucrose synthase and soluble sugars in *Populus* species that differ in their water response, *Physiologia Plantarum* 99: 153-159.

Pfanz, H., Wittmann, C., Vodnik, D., Aschan, G., Batiè, F., Turk, B., Maèek, I. Photosynthetic performance of timothy grass (*Phleum pratense* L.) is strongly affected by elevated carbon dioxide in mofettes, *Photosynth Res* (unpublished).

Pfeifhofer, W. (1989). Evidence of chlorophyll b and lack of lutein in *Neottia nidus-avis* plastids, *Biochem. Physiol. Pflanzen*, 184: 66-51.

Poorter, H. (1993). Interspecific variation in the growth response of plants to an elevated ambient CO_2 concentration, *Vegetatio* 104/105: 77-97.

Raschi, A., Miglietta, F., Tognetti, R., van Gardingen, P.R. (1997). *Plant responses to elevated CO_2. Evidence from natural springs*, Cambridge: Cambridge University Press, 272 p.

Rogers, H.H., Brett Runion, G., Krupa, S.V. (1994). Plant responses to atmospheric CO_2 enrichment with emphasis on roots and the rhizosphere, *Environ. Pollut.* 83: 155-189.

Rogers, H.H., Dahlman, R.C. (1993). Crop responses to CO_2 enrichment. *Vegetatio* 104/105: 117-131.

Smirnoff, N., Pallanca, J.E. (1996). Ascorbate metabolism in relation to oxidative stress, *Biochem. Soc. Transactions* 24: 472-478.

Stitt, M. (1991). Rising CO_2 levels and their potential significance for carbon flow in photosynthetic cells, *Plant Cell & Environment* 14: 741-762.

Tausz, M., Kranner, I., Grill, D. (1995). Simultaneous determination of ascorbic acid and dehydroascorbic acid in plant materials by high performance liquid chromatography. *Phytochemical Analysis* 17: 136-141.

Thibaud, M.C., Cortez, N., Rivere, H., Betsche, T. (1995). Photorespiration and related enzymes in pea (*Pisum sativum*) grown in high CO_2, *J. Plant Physiol.* 146: 596-603.

Tóth, V.R., Mészaros, I., Veres, S., Nagy, J. (2002). Effects of the available nitrogen on the photosynthetic activity and xanthophylls cycle pool of maize in field, *J. Plant Physiol.* 159 (6): 627-634.

Tuba, Z., Szente, K., Koch, J. (1994). Response of photosynthesis, stomatal conductance, water use efficiency and production to long-term elevated CO_2 in winter wheat, *J. Plant Physiol.* 144: 661-668.

Turk, B., Pfanz, H., Vodnik, D., Batiè, F., Sinkoviè, T. (2002). The effects of elevated CO_2 in natural CO_2 springs on bog rush (*Juncus effusus* L.) plants. I. Effects on shoot anatomy, *Phyton* 42:13-23.

UN ECE-ICP Programme Coordinating Centers. (1994). Manual on methods and criteria for harmonized sampling, assessment, monitoring and analysis of the effects of air pollution on forests. Hamburg, Praga, 177 p.

Vodnik, D., Pfanz, H., Wittmann, C., Maèek, I., Kastelec, D., Turk, B., Batiè, F. (2002). Photosynthetic acclimation in plants growing near carbon dioxide spring, *Phyton. Ann. Rei. Bot. "Responses of Plant Metabolism to Air Pollution and Global Change"* 42: 239-244.

von Caemmerer, S., Furbank, R.T. (1999). Modelling C4 photosynthesis. In *The biology of C4 photosynthesis*, ed. R.F. Sage, R.K. Monson, New York: Academic Press, pp. 173-211.

Webber, A.N., Nie, G.Y., Long, S.P. (1994). Acclimation of photosynthetic proteins to rising atmospheric CO_2, *Photosynthesis Research*, 39: 413-425.

Wand, S.J.E., Midgley, G.F., Jones, M.H., Curtis, P.S. (1999). Responses of wild C4 and C3 grasss (*Poaceae*) species to elevated atmospheric CO_2 concentration: a test of current theories and perceptions, *Global Change Biol.* 5: 723-741.

Wong, S.C. (1979). Elevated atmospheric partial pressure of CO_2 and plant growth. Interactions of nitrogen nutrition and photosynthetic capacity in C3 and C4 plants, *Oecologia* 44: 68-74.

Ziska, L.H., Bunce, J.A. (1997). Influence of increasing carbon dioxide concentration on the photosynthetic and growth stimulation of selected C4 crops and weeds, *Photosynth. Res.* 54: 199-208.

The Effects of Elevated CO_2 and Its Interaction with Water Stress on a C_4 Pasture Species (*Blepharis linariifolia*) Native to Sudan

S. Zaki-Eldeen
M. B. Jones

SUMMARY. A doubling of the pre-industrial atmospheric CO_2 concentration (~280 ppmv) is expected before the end of this century, if the current rate of growth in the annual emission of fossil carbon is maintained. The effects of elevated CO_2 concentrations on native tropical species has received little attention to date. In this study the effects of elevated CO_2, under both well-watered and water-stressed condition, on a C_4 pasture species (*Blepharis linariifolia* (dicot) native to the western part of Sudan was investigated. Elevated CO_2reduced whole plant tran-

---

S. Zaki-Eldeen is affiliated with the Institute of Environmental Studies, Khartoum University, Sudan.

M. B. Jones is affiliated with the Botany Department, Trinity College at the University of Dublin, Dublin 2, Ireland.

Address correspondence to: Dr. S. Zaki-Eldeen, Institute of Environmental Studies, Khartoum University, Sudan (E-mail: zakields@hotmail.com).

Thanks are due to the staff and colleagues of the Botany Department, Trinity College, Dublin for their continuous support during the period of this research.

The Irish Departments of Education and Foreign Affairs financially supported this work.

[Haworth co-indexing entry note]: "The Effects of Elevated CO_2 and its Interaction with Water Stress on a C_4 Pasture Species (*Blepharis linariifolia*) Native to Sudan." Zaki-Eldeen, S., and M. B. Jones. Co-published simultaneously in *Journal of Crop Improvement* (Food Products Press, an imprint of The Haworth Press, Inc.) Vol. 13, No. 1/2 (#25/26), 2005, pp. 213-237; and: *Ecological Responses and Adaptations of Crops to Rising Atmospheric Carbon Dioxide* (ed: Zoltán Tuba) Food Products Press, an imprint of The Haworth Press, Inc., 2005, pp. 213-237. Single or multiple copies of this article are available for a fee from The Haworth Document Delivery Service [1-800-HAWORTH, 9:00 a.m. - 5:00 p.m. (EST). E-mail address: docdelivery@haworthpress.com].

http://www.haworthpress.com/web/JCRIP

Digital Object Identifier: 10.1300/J411v13n01_10

spiration, particularly for juvenile plants. This resulted in wetter soils and higher leaf water potentials under elevated CO_2 in both well-watered and water-stressed conditions. Late in the growth cycle, elevated CO_2 had less effect on whole plant transpiration as there was a significant increase of both leaf area and leaf temperature. Assimilation rates and dry weight were enhanced by CO_2 enrichment under both water regimes. The results support the hypothesis that CO_2-induced growth enhancements in C_4 species occurs as a result of direct stimulation of photosynthesis as well as improved water relations. Carbon dioxide enrichment also hastened flowering and increased the inflorescence dry weight of *B. linariifolia* particularly under well-watered conditions. The results suggest a significant increase in yield and persistence of *B. linariifolia* in it's natural environment as atmospheric CO_2 concentrations increase.

KEYWORDS. C_4 photosynthesis, elevated CO_2, pastures, water stress, Sudan

ABBREVIATIONS. A: CO_2 assimilation rate (μmol m^{-2} s^{-1}); Aws: plants grown under ambient CO_2 and water-stressed conditions; Aww: plants grown under ambient CO_2 and well-watered conditions; C_4 plant: plant with C_4 pathway of carbon assimilation; C_a: ambient CO_2 concentration (ppm); C_i: leaf intercellular CO_2 concentration (ppm); CO_2: carbon dioxide; DAT: days after the start of treatment; Ews: plants grown under elevated CO_2 and water-stressed conditions; Eww: plants grown under elevated CO_2 and well-watered conditions; FC: field capacity; g_s: stomatal conductance (mmol m^{-2} s^{-1}); RWC: relative water content (%); WAT: weeks after the start of treatment;WUE: water-use efficiency

INTRODUCTION

The well-documented increase in atmospheric CO_2 concentration (IPCC, 1996) acts to increase plant biomass production of many plant species (Knapp et al., 1996). Carbon dioxide has a direct effect on photosynthesis through its role as a substrate, and indirect effects on both photosynthesis and transpiration through its effect on stomatal conductance (Beadle et al., 1993; Woodward, 1987). The reduced water loss at lower g_s under elevated CO_2 has the potential to forestall water-stress, thereby maintaining optimal carbon fixation, particularly in natural grasslands ecosystems which are often dominated by C_4 species and where water-stress is frequent (Ghannoum et al., 1997; Owensby et al., 1997).

Despite the fact that approximately half of the world's grass species possess the C_4 photosynthetic pathway, few studies have tested the responses of tropical and temperate C_4 species to elevated CO_2 (Wand et al., 1999). Results obtained from early studies appeared to confirm the hypothesis that C_4 plants do not show significant growth responses to elevated CO_2, due to the presence of a CO_2 concentrating mechanism in the bundle sheath cells (Lemon, 1984). However, there have been several subsequent reports of significant growth responses to elevated CO_2 in a number of C_4 species (Ghannoum et al., 1997; Hunt et al., 1996; LeCain & Morgan, 1998; Morgan et al., 1998; Read & Morgan, 1996). In a review paper, Poorter (1993) found the average growth stimulation (to a doubling of present CO_2 concentration) in C_4 species to be less than in C_3 but far from insignificant (22 vs. 41%). Furthermore, in a later review Wand et al. (1999) reported an average growth enhancement of 33% for C_4 species. Some reports of CO_2-enhanced growth in C_4 species can be attributed to the beneficial effect of elevated CO_2 on plant water relations, due to reduced stomatal conductance and improved water use efficiency at elevated CO_2 (Owensby et al., 1997). However, significant enhancement of growth of C_4 species under elevated CO_2 in an environment where water is non-limiting has also been reported (Ghannoum et al., 1997; Ghannoum & Conroy, 1998; LeCain & Morgan, 1998; Read & Morgan, 1996).

The response of native Sudanese plant species to elevated atmospheric carbon dioxide has received little or no attention to date. The natural vegetation cover of the Sudan comprises a high proportion of C_4 dominated grasslands, which are the main source of forage for livestock and wildlife (Elamin & Mustafa, 1992). *Blepharis linariifolia* Pers, dicot of the Acanthaceae family found throughout Sudan, was selected for the present study. It has been established that this species has the C_4 pathway of carbon assimilation, which is indicated by a zero CO_2 compensation point, a distinct Kranz anatomy and typical C_4 carbon stable isotope values (Zaki-Eldeen, 2001).

B. linariifolia grows in environments where it naturally experiences a wide variety of drying patterns. In the following experiment plants were subjected to regular cycles of short-term water deficits, under both ambient and elevated CO_2. Because the species is C_4, it was hypothesized that the most positive growth responses (biomass production) to elevated CO_2 will occur under water stressed compared to well-watered conditions.

MATERIALS AND METHODS

Controlled environment experiment was made with *Blepharis linariifolia*, which was grown in high clay content soil. The soil was first sieved (2 mm) and then sterilized in a drying oven at 80°C for 48 hr, following which it was mixed with sand in a ratio of 25:75. Plastic pots (0.8 L) were filled with the soil mixture, which was then watered to field capacity.

Seeds of *B. linariifolia* were germinated in a growth cabinet at a day/night temperature of 25/18°C. One week after germination they were transplanted into small pots (0.25 L) to ensure they were well established. Three weeks after germination (3 WAG) seedlings were transplanted to larger pots (0.8 L). The surface of the growing medium was covered with 2 cm depth of plastic beads to reduce soil evaporation to minimum. The pots were randomly allocated to two identical growth cabinets (Fitotron, Sanyo-Gallenkamp plc, UK), at either ambient CO_2 (360 ± 10 ppmv) or twice-ambient CO_2 (700 ± 10 ppmv) concentration. An infrared gas analyser (ADC, UK) was used to control the CO_2 concentration by injecting pure carbon dioxide from a cylinder into the elevated cabinet to maintain the required concentration. The day/night temperatures were 30/22°C, relative humidities were 50/60%, and PAR was 700 $\mu mol\ m^{-2} s^{-1}$ at plant height over a 12-h photoperiod. To reduce effects due to differences between the growth cabinet (the growth cabinet factor), the plants were rotated between the two growth cabinets twice a week and CO_2 concentration were changed at the same time. Thirty-two pots were allocated to each CO_2 treatment, and those plants were divided equally into well-watered plants [ambient CO_2 well-watered (Aww) and elevated CO_2 well-watered plants (Eww)] that were watered daily to their field capacity (FC), and water-stressed plants [ambient CO_2 water-stressed (Aws) and elevated CO_2 water-stressed plants (Ews)] that were exposed to repeated drought cycles. The plants were fed with 1/2 strength Long Ashton solution (complete nutrient solution based on macronutrient of nitrate-type and mixed with micronutrients), every week. Blank pots, filled with the same growing medium used in the experiment, and covered with plastic beads, were used to measure direct evaporation from the soil surface. Three harvests were made to coincide with the end of a drought cycle, the first two harvests were made after the water-stressed plants had been exposed to two drought cycles (of one week in each cycle), while the third harvest was made after a further three drought cycles.

Soil Water Content and Transpiration

Soil water content was expressed as percent (%) FC. After transplanting the seedlings the soil was covered with the plastic beads and the pots were saturated with water and left overnight, before percent (%) FC was determined. At the same time on each day of the experiments (18:00 hrs), the potted plants were removed from the growth cabinets and weighed on an electronic balance (Sartorius Type 1801). The change in percent (%) FC was determined and the well-watered plants were then watered to 100% FC. The water-stressed plants, were only re-watered to 100% FC at the end of the drought cycles. During drought cycles, percent (%) FC was calculated from the initial water content less the cumulative water loss per pot and related to water content at FC.

Daily transpiration E (g day^{-1}) was calculated as:

$$E = (\Delta W - S) \text{ (Morrison \& Gifford, 1984a)}$$

Where ΔW is the change in weight (g) since the previous weighing, S is the water loss from the soil (estimated from pots without plants covered with plastic beads).

Growth Measurements

Destructive harvests of *B. linariifolia* plants were made 2, 4, and 7 weeks after the start of treatments (WAT). Each harvest coincided with the end of a stress cycle and plants were re-watered on the day of the harvest. At each harvest 7 plants per treatment were selected randomly, removed from the chambers and separated into root, leaf blade and stems. Roots were carefully washed by hand and fine root material recovered by sieving. Fresh weights were determined, and then plants parts were dried to constant weight in a forced air oven at 80°C. Leaf area was estimated non-destructively, at intervals of approximately one week, by measurements of leaf length and width and the use of the equation that related width × length to area (Zaki-Eldeen, 2001).

Water Use Efficiency

Water use efficiency (WUE, mg DW/g H_2O) was calculated from the equation (Morrison & Gifford, 1984a):

$$WUE = \Delta DW/(\Delta W - S)$$

Where ΔDW is the total (aerial and root) plant dry weight increase, ΔW is the change in pot weight over the period, S is the loss of water direct from the soil surface. In the current experiment the values of WUE were 'normalized' by multiplying by the vapour pressure deficit (Beale et al., 1999). The VPD for the current experiments was 2.1 kPa for the day (Temp. 30°C and RH 50%)

and 1.06 kPa for the night (Temp. 22°C and RH 50%). The day-night VPD average was 1.6, which was used for normalizing the WUE.

Plant Water Relations

Leaf water potential (Ψ_L) of mature leaves of similar ages was measured at 08:00 h on the day of each harvest, using pressure chamber (Type SKPM 1400, Sky, UK), following the procedure of Boyer (1995).

Relative water content was determined six weeks after the start of the treatment, and 3 days after water had been withheld in the water stress treatment. For each treatment 4 leaves of similar age were excised in the afternoon (13:00 h). Leaf fresh weight (W_f) was determined using an electronic balance, and the leaves were then floated on distilled water overnight in petri dishes, at a very low light level. Leaves were then weighed to determine the turgid weights (W_t), and dried at 80°C to determine the dry weight (W_d).

The percent (%) relative water content (RWC) was determined as:

$$RWC\% = \left(\frac{Wf - Wd}{Wt - Wd} \right) \times 100$$

Stomatal Conductance Measurements

Measurements of stomatal conductance (g_s) were made *in situ* in each growth cabinet using a transit-time diffusion porometer (AP4, Delta-T Devices, Burwell, Cambridge, UK). On the day of measurements conductance was determined in the morning (10:00 a.m.), the afternoon (2:00 p.m.) and the evening (5:00 p.m.). Measurements were made on five plants for each treatment, on the upper (adaxial) and the lower (abaxial) surfaces on mature leaves of similar age. The total conductance was calculated from the sum of the abaxial + adaxial values for each leaf (Knapp et al., 1996; Morrison, 1998; Mott, 1988). The measurements were made frequently during the treatment periods, both when the plants were water stressed and when re-watered.

PHOTOSYNTHESIS MEASUREMENTS

The light saturated rate of photosynthesis was determined *in vivo* on mature leaves of similar age. An Infrared gas analysis (IRGA) based photosynthesis system (LCA4, Analytical Development Company, UK) was used to measure the gas exchange parameters. The measurements were made both when the water-limited plants were water-stressed as well as when they were re-watered. A leaf microclimate control system was used in conjunction with the

LCA4. The leaf microclimate control unit consisted of a control unit (LCA4-LMC-002/B) with an integral unit for the supply of enriched CO_2 and a temperature controlled leaf chamber (PLC4N/TC/LMC-002). Light (1400 $\mu mol\ m^{-2}\ s^{-1}$) was supplied by a light source attached to the leaf chamber. The temperature of the leaf chamber was maintained at 30°C and the VPD was controlled by setting the chamber humidity at 50% RH (growth condition in the GC). Carbon dioxide assimilation rates were measured at 50, 90, 100, 150, 200, 300, 500, 700, 800, and 900 ppm CO_2 concentration. The assimilation rates (A, $\mu mol\ m^{-2}\ s^{-1}$) at each of the CO_2 concentrations, the values of stomatal conductance, transpiration rates and the internal CO_2 concentration (C_i, ppm) were calculated by the instrument software of the LCA4. A/C_i curves were analysed using the GraphPad Prism Programme (Version 3, GraphPad Software, Inc. USA) that compares equations with F-test and determine the best fit at 95% confident intervals.

Leaf Temperature

Leaf temperature measurements were made at 14:00 h at different stages of the drought cycles, using an electronic thermometer (1604 Cr/Ai, Comark, England).

Statistical Analysis

Data were analysed using the statistical package Data desk (Version 6.0) and $P \leq 0.05$ was used as a critical level of significance. Data were subjected to analysis of variance (ANOVA) using a factorial design for each species separately. Two-ways interactions (2 CO_2 treatments × 2 water treatments) were used to examine results of each harvest. Three-ways interactions (2 CO_2 treatments × 2 water treatments × Time) were also used for the analysis of some of the parameters (e.g., temperature, leaf area). Data was log transformed as needed, and normal probability plots and residuals plots were inspected to ensure that the data met the assumptions of ANOVA. *Scheffe´ Post Hoc* tests were used to compare means of the different treatments when interactions were significant.

RESULTS

Soil Water Content and Transpiration

Carbon dioxide enrichment of *B. linariifolia* plants reduced the rate of soil drying under both well-watered and water-stressed conditions (Figure 1), but the size of the effect varied over the treatment period. Between the first and the

FIGURE 1. Soil water content expressed as percent (%) field capacity (FC) of pots containing *Blepharis linariifolia* plants at ambient (A) or elevated (E) CO_2 concentrations under either well watering (ww) or water stress (ws) conditions. Values represent means of 15, 10 or 5 replicates (plants) over 7 drying cycles and time of re-watering, respectively. Arrows show the time of each of harvest, at an end of a drought cycle.

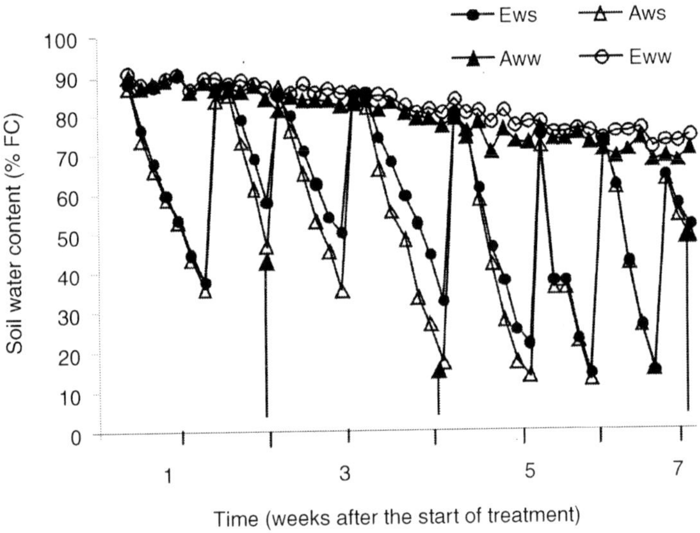

second harvests (2-4 WAT) the percent (%) FC was higher ($P < 0.05$) under all plants grown at elevated CO_2, but by the end of the treatment period (4-7 WAT) elevated CO_2 concentration had positive ($P < 0.05$) effects only under well-watered conditions.

The mean daily transpiration from the start of treatments until the first harvest (2 WAT) was significantly affected by both elevated CO_2 concentration and water-stress ($P < 0.05$), although, over time the effect of water-stress in reducing transpiration became greater. The reduction in transpiration caused by the water-stress treatment increased from 37% to 53% by the end of the treatment period, while the reduction caused by elevated CO_2 concentration decreased over time from 17% to an insignificant amount by the end of the treatment period. Under well-watered conditions elevated CO_2 concentration reduced transpiration by 13% ($P < 0.05$) both early (0-2 WAT) and late in the treatment period (4-7 WAT) although, no significant difference was found in the middle of the treatment period (2-4 WAT). Under drought conditions elevated CO_2 concentration also reduced transpiration, but the reduction was found to decrease from 20% ($P < 0.05$) at the first harvest to zero between the second and the third harvests (Figure 2).

Because elevated CO_2 concentration had a significant effect on the leaf area development of *B. linariifolia* (see growth measurement results), when transpiration was expressed on leaf area basis (mg/cm^2), elevated CO_2 concentration was found to cause a reduction over the whole treatment period. Under both well-watered and water-stressed conditions, elevated CO_2 concentration caused average reductions of 34% and 35%, respectively.

Leaf Water Potential

The morning values of $Ø_L$ were generally higher at elevated CO_2 concentration (Figure 3). Under water stressed conditions $Ø_L$ was higher at elevated CO_2 concentration during the whole course of the experiment ($P < 0.001$), but under well-watered conditions $Ø_L$ it varied from significantly higher (2 WAT) to no difference (7 WAT). This was probably caused by the increased total plant transpiration found later in the experiment.

FIGURE 2. Cumulative mean daily transpiration (g plant^{-1}) of *Blepharis linariifolia* plants grown at ambient (A) or elevated CO_2 (E) concentrations under either well-watered (ww) or water-stressed (ws) conditions. Values represent means of 15, 10 or 5 plants for 0-2, 2-4, 4-7 weeks after the start of treatments, respectively.

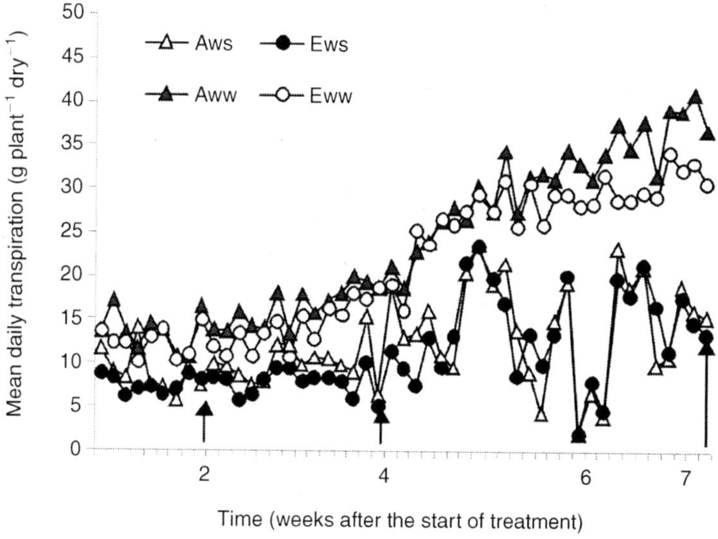

FIGURE 3. Leaf water potential of *Blepharis linariifolia* plants grown at either ambient (A) or elevated CO_2 (E) under well-watered (ww) or water-stressed (ws) conditions. Means ± (SE), n = 4. Time was weeks after the start of treatment.

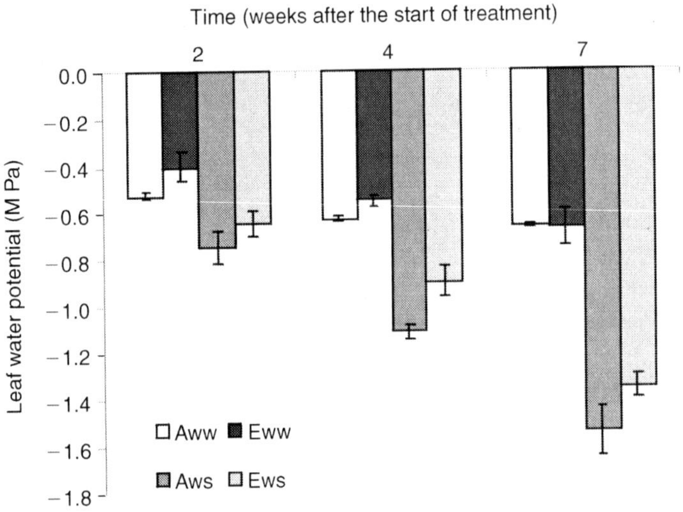

Relative Water Content

Carbon dioxide enrichment had no effect on the midday relative water contents, of well-watered plants of *B. linariifolia* but under water-stressed conditions doubling of CO_2 concentration increased the relative water content from 35 to 52% (P < 0.05) (Table 1). Towards the end of the drying cycles the stressed plants wilted, particularly under ambient CO_2 conditions, however they recovered quickly when they were re-watered. The relative water content was found to decrease by 50% in water-stressed compared to well-watered plants.

Water Use Efficiency

Elevated CO_2 concentration caused a significant increase ($P < 0.001$) in WUE which was also found to increase ($P < 0.001$) over time (Figure 4). Between the first and the second harvests (2-4 WAT) there was no difference in WUE between well-watered and water-stressed plants at ambient CO_2 concentration, but between the second and the third harvests (4-7 WAT) it was higher in the water-stressed plants. The interaction between water supply and

TABLE 1. Leaf relative water content (RWC) of *Blepharis linariifoli* plants grown at either ambient (A) or elevated CO_2 (E) under well-watered (ww) or water-stressed conditions (ws). Means ± (SE), n = 4.

Treatments	RWC
Aww	83 ± 1[a]
Aws	35 ± 2[b]
Eww	83 ± 1[a]
Ews	52 ± 1[c]

In the row means with the same letters are not significantly different at $P < 0.05$.

FIGURE 4. Water use efficiency (WUE, mg dry wt/g water) of *Blepharis linariifolia* plants grown at either ambient or (A) or elevated (E) CO_2 under well-watered (ww) or water-stressed (ws) conditions. Values represent mean ± (SE), n = 5. Time interval was weeks after the start of treatments.

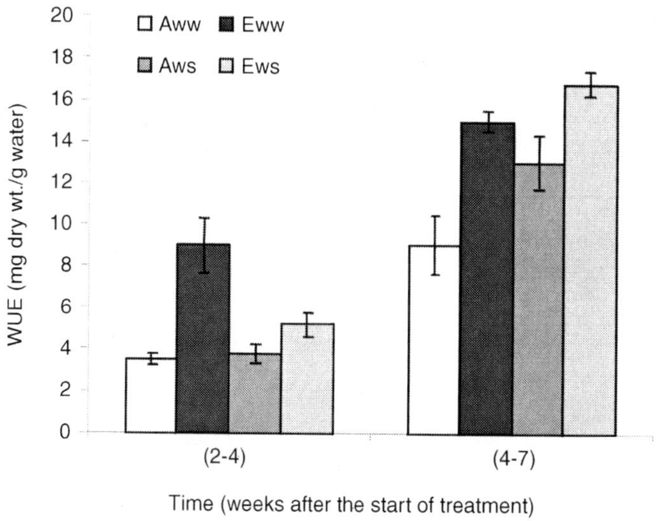

CO_2 was significant, but there was no significant difference in WUE between water-stressed and non-stressed plants at elevated CO_2 concentration. Under well-watered conditions, elevated CO_2 concentration increased WUE by 91% while under water-limited conditions the increase was 25% compared to plants grown at ambient CO_2 concentration.

Stomatal Conductance

Stomatal conductance was reduced by both elevated CO_2 concentration and water-stress treatment. Under well-watered conditions elevated CO_2 concentration consistently caused significant (36%) reduction in g_s ($P < 0.01$). Under water stress condition elevated CO_2 concentration had a smaller but significant effect on g_s (Figure 5). The conductance of water-stressed plants that were grown under ambient CO_2 was more sensitive to drought than plants under the elevated CO_2 concentration. However, immediately after re-watering, g_s of water-stressed plants at ambient CO_2 increased considerably. At ambient CO_2, g_s of water-stressed plants was found to recover fully to the level of the well-watered plants only at the early stages of the experiment (until 20 DAT). However, at elevated CO_2 concentration, g_s of water stressed plants was more stable and was found to respond to re-watering by a moderate increase in g_s (Figure 5). By the end of drought cycles, elevated CO_2 concentration had caused an average reduction of 20% in g_s in water-stressed plants compared to their counterparts at ambient CO_2 concentration.

FIGURE 5. Stomatal conductance (g_s, mmol $m^{-2}s^{-1}$) of *Blepharis linariifolia* plants that were grown under well-watered (ww) and water-stressed (ws) conditions under either ambient (A) or elevated CO_2 (E). Each point represents mean of 15 measurements. Time was days after the start of treatments. Error bar represent ± (SE). Arrows at the bottom of the figure indicate the time of re-watering after drought cycles.

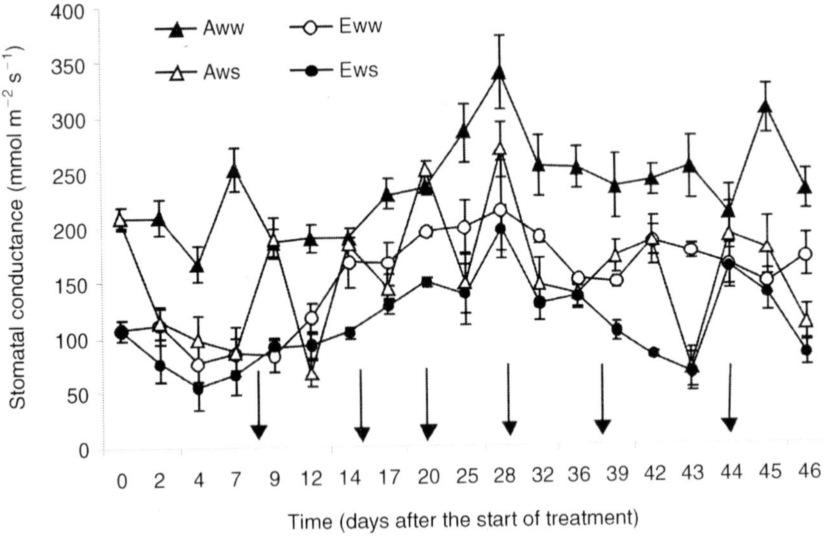

When stomatal conductance of *B. linariifolia* plants decreased it was positively correlated with the decrease in leaf water potential (Figure 6). Non-linear regression curves fitted to the data at the two CO_2 concentrations were found to follow two distinct lines (R^2 were 0.91 and 0.7 for ambient and elevated CO_2 plants, respectively), with the effect of elevated CO_2 concentration is indicated by the difference between the two lines (Figure 6).

Growth Measurements

Plant Dry Weight

The positive effect of elevated CO_2 concentration on the dry weights (shoot + root) of *B. linariifolia* plants appeared earlier in the experiment under water-stressed conditions than under well-watered conditions (Figure 7). At the second (4 WAT) and third (7 WAT) harvests, elevated CO_2 concentration caused an increase of 31% and 40% compared to well-watered plants under ambient CO_2 conditions, while under water stressed conditions elevated CO_2 concentration caused an increase of 28% compared to those at ambient CO_2 concentration. For the whole course of the experiment, the interaction be-

FIGURE 6. The relationship between stomatal conductance (g_s mmol m^{-2} s^{-1}) and leaf water potential of *B. linariifolia* plants that were grown under well-watered (ww) and water-stressed (ws) conditions under either ambient (A) or elevated CO_2 (E). The non-linear regression curves show the best fit for the relationship of plants grown at either ambient or elevated CO_2 concentration.

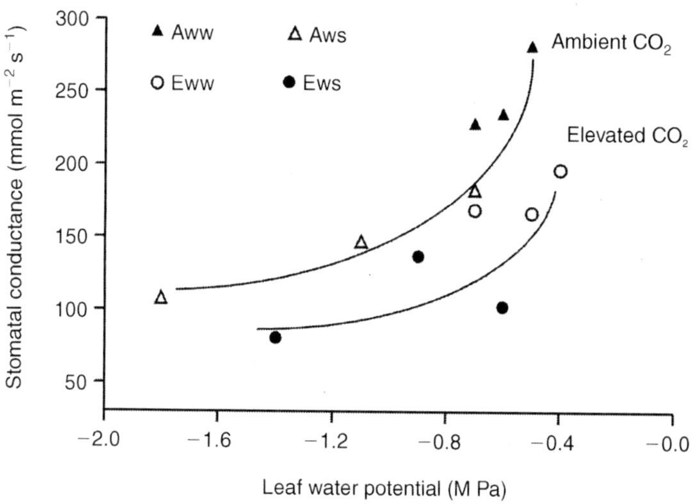

FIGURE 7. Total dry weights of *Blepharis linariifolia* plants grown at either ambient (A) or elevated (E) CO_2 under well-watered (ww) or water-stressed (ws) conditions. Values represent the mean ± (SE), n = 5. Time was weeks after the start of treatment.

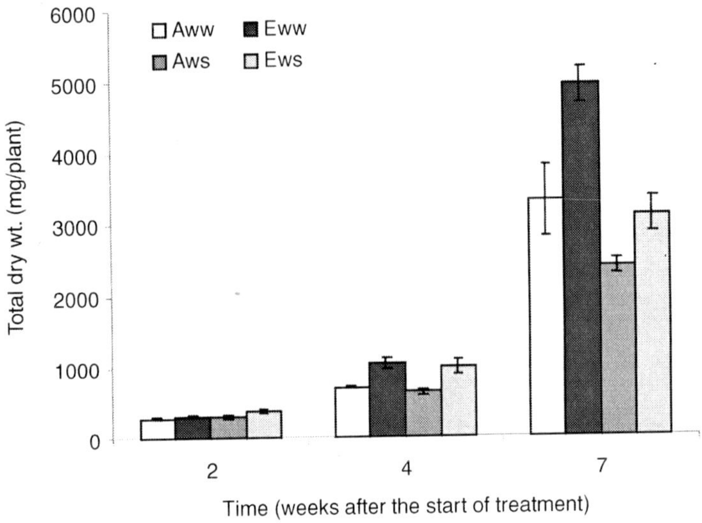

tween CO_2 and water stress was significant only at the second harvest. The comparisons of means (Scheffe´ Post Hoc test) showed a highly significant ($P < 0.001$) effect on well-watered plants under elevated CO_2 concentration.

Leaf Area

Elevated CO_2 concentration caused an increase in leaf area of 32% ($P < 0.001$) in well-watered plants compared to their counterparts at ambient CO_2 concentration and under water-stress conditions it led to an increase of 44% ($P < 0.001$) compared to drought-treated plants at ambient CO_2 over the 7 weeks of the experiment. Water stress caused a decrease of 33% ($P < 0.001$) in leaf area but the negative effect was not significant until 36 DAT ($P < 0.01$) (Figure 8). At this stage the decrease in leaf area was partly due to leaf rolling that was found to reduce the leaf width and consequently reduced the projected area. The water stress caused a decrease of 12 and 31% in leaf area of water-stressed plants at elevated and ambient CO_2 concentration, respectively, compared to control plants. Under water stress, elevated CO_2 concentration delayed leaf rolling in each water-stress cycle. Initially this was a two day difference but as

FIGURE 8. Leaf area (cm^2 $plant^{-1}$ of *Blepharis linariifolia* grown at either ambient (A) or elevated CO_2 (E) concentration under well-watered (ww) or water-stressed (ws) conditions. Values represent mean ± (SE), n = 5. Time interval was days after the start of treatment.

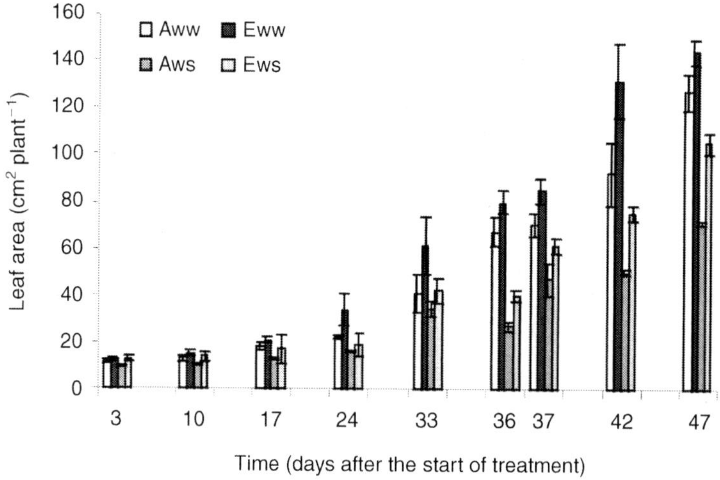

the plants grew the time difference decreased to one day and then half a day by the end of the experiment (7 WAT).

Photosynthesis

Single leaf photosynthesis of *B. linariifolia* grown at elevated CO_2 was higher than at ambient CO_2 concentration, under both the well-watered and water-stress conditions (Figure 9). Comparisons between assimilation rates (A) at CO_2 concentrations at which the plants were grown (either ambient (360 ppmv) or elevated (700 ppmv) CO_2) at intervals during the whole treatments periods were made to compare the combined effect of water-stress and elevated CO_2 concentration during wet and dry periods (Figure 9). When *B. linariifolia* plants were at the end of drought cycles, water stress was found to cause a general decrease of photosynthesis regardless of the CO_2 concentration ($P < 0.001$). Under water-stress condition elevated CO_2 concentration caused an increase ($P < 0.01$) of A from 14 to 19 μmol m^{-2} s^{-1} compared to ambient CO_2 grown plants. Also, even under well-watered conditions, elevated CO_2 caused an increase in A compared to the plants grown under ambient CO_2 concentration (25 vs. 20.3 μmol m^{-2} s^{-1}, respectively). When the water-stressed plants were re-watered, the overall effect of water-stress was

FIGURE 9. Assimilation rate (A, μmol m^{-2} s^{-1}) of *Blepharis linariifolia* plants that were grown under (a) well-watered (ww) and (b) water-stressed (ws) conditions under either ambient (A) or elevated CO_2 (E). Each point represents mean of 3 measurements ± (SE). Time was days after treatments. Arrows show measurements that were made at peak of water stress (end of stress cycle).

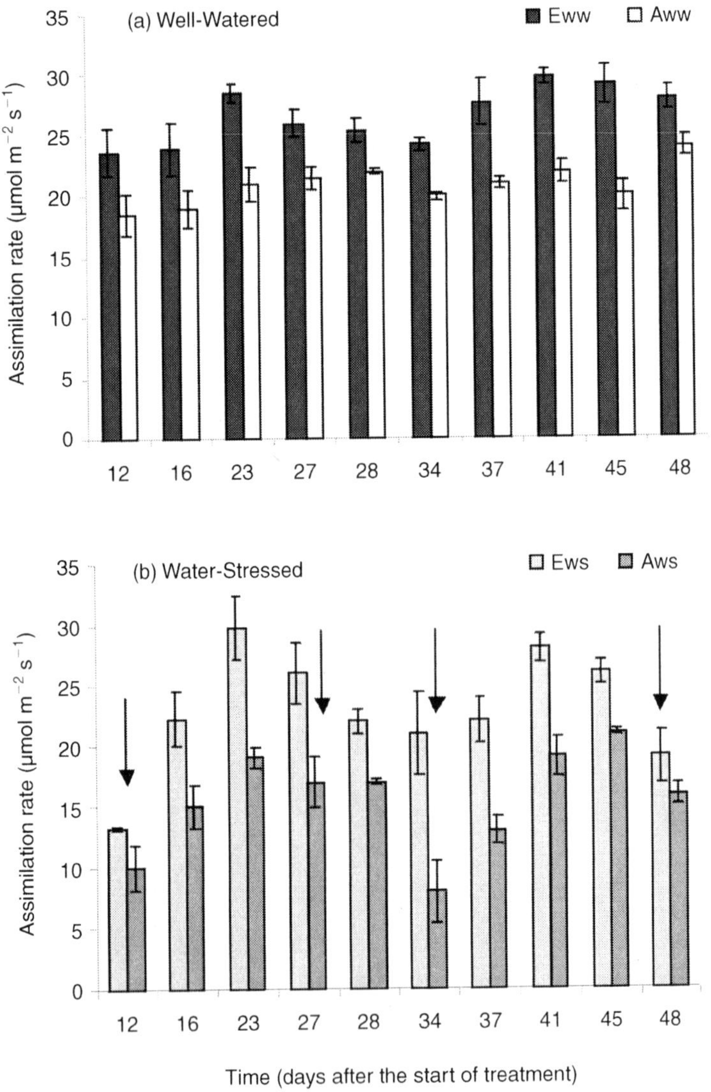

less significant $P < 0.05$, and the assimilation rate of the water-stressed plants under both ambient and elevated CO_2 concentration was found to show large recovery, so that no difference was found between them and the well-watered plants at the same CO_2 concentration. During this period, elevated CO_2 concentration also increased A of *B. linariifolia* plants.

Analysis of the A/C_i curves show that photosynthesis was not saturated at a cuvette CO_2 concentration of 360 ppmv (Figure 10). Under both well-watered and water-stressed conditions the saturation point was higher in elevated CO_2 plants. However, under water-stress treatment, the saturation point was even

FIGURE 10. A/C_i curves [assimilation, A μmol m^{-2} s^{-1} vs. intercellular CO_2 concentration (C_i)] of *Blepharis linariifolia* plants that were grown at either ambient or elevated CO_2 under (a) well-watered or (b) water-stressed conditions. Each point represents mean of 4 measurements ± (SE). Arrows point to measurements made at cuvette CO_2 concentrations of 360 and 700 ppm.

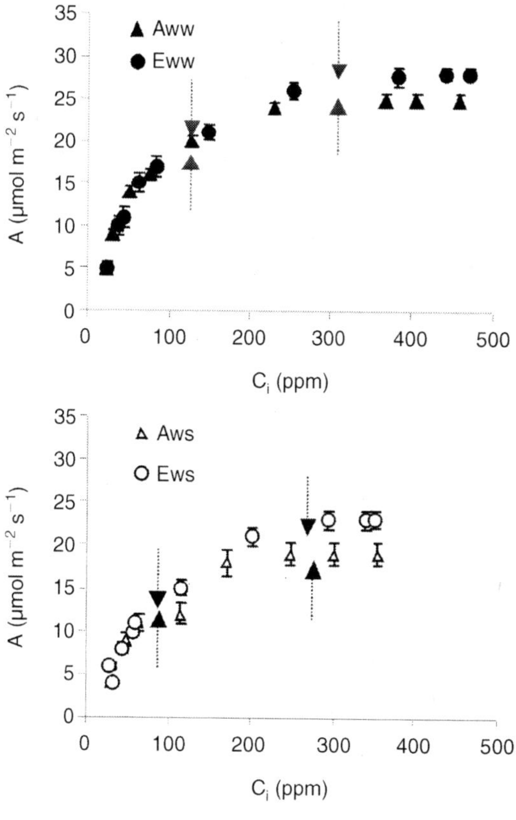

higher, as drought was found to reduce the plateau and consequently decrease the CO_2 saturation point under ambient CO_2 (Figure 10) compared to their counterparts at elevated CO_2 concentration. The ratio of intercellular CO_2 concentration (C_i) to ambient (leaf chamber) CO_2 concentration (C_a) (C_i/C_a) was affected by both elevated CO_2 concentration and water-stress. While elevated CO_2 had the effect of increasing the C_i/C_a ratio compared to that of the ambient CO_2 grown plants ($P < 0.01$) (from 0.42 to 0.49), water-stress on the other hand negatively affected the C_i/C_a ratio particularly by the end of the treatment period ($P < 0.01$).

Leaf Temperature

At elevated CO_2 concentration daytime leaf temperature of *B. linariifolia* was approximately 1°C higher than at ambient CO_2 under well-watered conditions, while under water-stressed conditions the increase was 0.7°C. At ambient CO_2 leaf temperature was also found to increase in the water-stress treatment ($P < 0.01$) and by the time of the second harvest (4 WAT, end of the 4th drought cycle) the leaf temperature of water-stressed plants under ambient CO_2, was even higher than that of water-stressed plants at elevated CO_2 concentration (31°C vs. 30°C).

DISCUSSION

At elevated CO_2 concentration water use by *B. linariifolia* was reduced because of the lower stomatal conductance (g_s) to water vapour from the leaves. Many previous studies have shown that elevated CO_2 concentration reduces g_s in leaves of both C_3 and C_4 species (Chaves & Pereira, 1992; Eamus, 1991; Ferris & Taylor, 1995; Murray, 1997a). An increase in CO_2 concentration to twice ambient (700 ppm) resulted in a 34% decline in g_s of *Schizachyrium scoparium* (C_4) similar to *B. linarifolia* while g_s of *Panicum laxum* (C_3) and *P. antidotale* (C_4) grasses was reduced by more than 50% (Ghannoum et al., 1997).

Under water stress, whole plant transpiration of *B. linariifolia* was reduced by elevated CO_2 concentration from the start of the treatment but by the end of the experiment it had no effect, whereas under well-watered conditions the negative effect on transpiration continued throughout. The decline in effect of elevated CO_2 concentration in water-stressed plants is explained by both the increase in leaf area and the significant increase in leaf temperature in the plants at elevated CO_2. As water stress developed, only small differences in stomatal conductance were found between the water-stressed plants at the two CO_2 levels. Similar results have been reported in the literature. For example,

Knapp et al. (1996) found that when water availability was high, elevated CO_2 concentration resulted in decreased stomatal conductance in 10 of the 12 species measured, but during a dry period reductions in stomatal conductance at elevated CO_2 concentration were found in only two species. Furthermore Kirkham et al. (1991) reported that elevated (2 × ambient) CO_2 concentration reduced transpiration rate of big bluestem (*Andropogon gerardii* Vitman) by only 25% under low water treatment, while the reduction was 35% at high-water availability, while Nie et al. (1992) found that elevated CO_2 concentration (700 ppmv) reduced canopy evapotranspiration of prairie species by 18% and 8% under the high and low waters levels, respectively.

In spite of the significant effect of elevated CO_2 concentration on transpiration of *B. linariifolia* under the two water levels early in the experiment there was no difference in the soil water content over the same period of time. This was presumably because the plants were too small to produce noticeable difference in soil moisture content. Thereafter, (2-4 WAT) the effect of elevated CO_2 concentration was mirrored by a similar effect on soil water content so that wetter soils ($P < 0.05$) were found under elevated CO_2 concentration grown plants under both water regimes. However, by the end of the experiment the effect of elevated CO_2 concentration on transpiration of water-stressed plants was no longer apparent and consequently no differences were found in soil water content. In the field, Kirkham et al. (1991) and Owensby et al. (1997) reported wetter soils under elevated CO_2 concentration in tallgrass prairie. However, the reduction in transpiration and soil drying is less than the reduction in leaf conductance due to leaf warming and greater leaf area (IPCC, 1996; Kirkham et al., 1992; Morrison & Gifford, 1984a). In fact Hunt et al. (1996) have demonstrated that the latter may offset or override the effect of smaller conductance on transpiration, resulting in no change in soil water or even in drier soil under elevated CO_2 concentration. This is what presumably had happened by the end of the treatment period in the present study.

Doubling of atmospheric CO_2 concentration increased the water use efficiency of the *B. linariifolia* under well-watered as well as under water-stressed conditions. This increase in WUE of the well-watered plants resulted from both an increase in dry weight and a reduction in transpiration in the earlier stages of the experiment (between 2-4 WAT), but by the end of the treatment period (4-7 WAT), the increase in WUE of the water-stressed plants was mainly caused by the increase in dry weight. Similar results were reported by Clark et al. (1999) for *Paspalum dilatatum* (C_4).

The positive responses in dry matter production under elevated CO_2 concentration and well-watered conditions in the current experiment does not support the original hypothesis. Elevated CO_2 concentration had an early positive ($P < 0.05$) effect on the total dry weights of water-stressed plants of *B. linariifolia* increasing dry weight by 28%, but under well-watered conditions the increase ($P < 0.001$) was even larger (40%). It has also been found that under the well-watered conditions doubling of atmospheric CO_2 concentration caused significant increase in the number of leaf pairs, number of branches, and number of inflorescences of *B. linariifolia* (Zaki-Eldeen, 2001), as well as changing the nature of growth of lateral branches from prostrate to erect. Similar results to the current experiment have been reported for desert communities, where elevated CO_2 concentration was found to have a significant influence on biomass production during high rainfall years, whereas during drought years no significant response to CO_2 enrichment was found (Smith et al., 2000). In contrast to these results Rozema et al. (1991) found no positive effect of elevated CO_2 concentration (580 ppmv) on the two C_4 grass species *Spartina angilica* C.E. Hubbard and *Spartina patens* (Ait). However, it should be noted that these plants were grown at a temperature of 21°C/day and 15°C/night, which was below their optimum (Eheringer et al. (1997).

Stimulation of dry weight production in *B. linariifolia* under well-watered conditions indicates that CO_2-induced growth enhancement occur as a direct result of stimulation of photosynthesis, as well as improved water relations resulting from partial stomatal closure. The carbon concentrating mechanism in C_4 species provides elevated CO_2 concentrations in the bundle sheath cells, which greatly reduces the rate of oxygenation by Rubisco and resultant photorespiration (Wand et al., 1999). Thus the direct effect of rising atmospheric CO_2 concentrations above present day concentrations should be negligible for C_4 plants. However, recent work has suggested that the photosynthetic rate of certain C_4 species can be stimulated by increasing CO_2 concentration, even under optimal water and nutrients (Ziska et al., 1999a). The current study confirms this observation for *B. linariifolia*. The A/C_i response curves under the two water regimes, show an increase in A with increasing C_i. Under drought conditions g_s and C_i/C_a ratio were both more stable under elevated than ambient CO_2 concentration. Under ambient CO_2 concentration, g_s was found to decrease at the end of each drought cycle, and this was associated with a similar decrease in the assimilation rate. During these periods A was 32% higher at elevated than at ambient CO_2 concentrations. This suggests that elevated CO_2 concentration largely protected the plants against the negative effects of drought on gas exchange.

It has been shown that *B. linariifolia* belongs to the NAD-ME C_4 subgroup (Zaki-Eldeen, 2001). The results reported here of a significant increase in dry weight at elevated CO_2 support the proposal of Hattersley (1984) that plants of the NAD-ME subgroup are the most likely to respond to CO_2 enrichment. However, in the literature, contrasting findings amongst C_4 subgroups have been reported. While, LeCain and Morgan (1998) found that the growth response of well-watered NADP-ME species to elevated CO_2 was bigger than that of NAD-ME subtype, Ziska et al. (1999a) found that the stimulation of leaf photosynthesis at elevated CO_2 was not associated with values of CO_2 leak rates from the bundle sheath. Also, elevated CO_2 was found to increase the plant dry mass of all C_4 subtypes by a similar amount of about 20% when soil water content and other environmental factors were optimised (Seneweera et al., 1998).

The effect of elevated CO_2 concentration on the assimilation rate of *B. linariifolia* were found to be largest when the plants were at the end of a drought cycle. This might be explained by the positive effect of elevated CO_2 concentration on water relations as it reduced transpiration significantly. This suggests that elevated CO_2 concentration, through its positive effect on water relations and increased C_i/C_a ratio, increased photosynthesis compared to the plants grown at ambient. Ham et al. (1995) and Owensby et al. (1999) made similar observations when they reported that the net carbon exchange and water vapour fluxes of plants grown in open-top chambers in the native tallgrass prairie (dominated by C_4 plants, *Adndropogon gerardii*) were reflected in greater net CO_2 flux at the ecosystem level under elevated CO_2 concentration. These authors speculated that this was caused primarily by continued photosynthesis in the elevated CO_2 concentration plots when water stress reduced leaf gas exchange in the ambient CO_2 treatment. They concluded that the response to elevated CO_2 was found to be primarily related to water stress.

It is also possible that some of the response to elevated CO_2 concentration in *B. linariifolia* is associated with a change in the rate of leaf ageing. Leaf age affects the sensitivity of both photosynthesis and growth of C_4 plants to elevated atmospheric CO_2 (Dai et al., 1995; Ziska et al., 1999a). It has been shown that the bundle sheath cell walls of both young and senescent maize leaves have a relatively high conductance, which could result in lower CO_2 concentration at the active sites of Rubisco (Dai et al., 1995). As a consequence, young and senescent C_4 leaves may be more responsive to CO_2 enrichment, and contribute to growth stimulation (Ghannoum et al., 1997). In addition, one of the unique features of *B. linariifolia* was the large number of meristems and high percentage of young developing leaves under both well-watered and water-stressed condition (Zaki-Eldeen, 2001). It is possible therefore, that this may have contributed to the overall growth stimulation at elevated CO_2 concentration observed in this species. Ziska et al. (1999a) reported that elevated CO_2 concentration increased significantly the leaf production of a C_4 dicot, *Flaveria trinervia*, which shown to have similar responses to those *B. linariifolia* when it has been grown under elevated CO_2 concentration.

Chaves (1992) found that photosynthetic capacity is normally affected only in severely stressed leaves when the RWC is below 60%, and when RWC falls to 30%, then an irreversible damage may occur. In *B. linariifolia*, the water status of water-stressed plants grown under elevated CO_2 concentration did not decline to the same degree as for those grown in ambient CO_2, as the RWC values were 52% and 35%, respectively. However, the potential of photosynthesis in *B. linariifolia* plants to recover was always very high as water stress was found to have had a relatively small effect on photosynthesis when the plants were re-watered. This might also suggests that the plants can adapt to severe environmental conditions as well as efficiently use the available resources.

B. linariifolia is native to Kordofan region in western Sudan, a region that is already influenced by great environmental changes mainly desertification and drought. Ecologists described the region as desertification vulnerable, with its typical landscape of the margin of the arid and semi-arid zone in Africa (IES, 1997a). The rainy season is very short and the uncertainty bounds both the time and amount of the rainfall. However, the demand on the available resources is still very high, and this places significant burdens on such a vulnerable system. The present study suggested some important responses of *B. linariifolia* to growth under elevated CO_2 concentration.

The reported positive effect of elevated CO_2 concentration on photosynthesis and biomass production under both well-watered and water-stressed conditions may suggest an increase in the potential productivity of this species as atmospheric CO_2 concentrations increase. Furthermore, the improvement of shoot water relations under elevated CO_2 concentration, through reduced g_s and leaf transpiration rates might be of a very significant value for a plant that grows in such dry region. The increase in productivity and water-use efficiency of *B. linariifolia* under elevated CO_2 concentration may increase the success of any efforts to re-introduce the species in degraded lands, since transplanted or sown annuals may grow faster and be more tolerant of seasonal water stress under elevated CO_2.

REFERENCES

Beadle, C. L., Ludlow, M. M. & Honeysett, J. L. (1993). Water Relations, pp. 93-128. In *Photosynthesis and production in a changing environment: a field and laboratory manual*. (eds. D. O. Hall, J. M. O. Scurlock, H. R. Bolhar Nordenkampf & S. P. Long). London: Chapman and Hall.

Beale, C. V., Morrison, I. L. & Long, S. P. (1999). Water use efficiency of C_4 perennial grasses in a temperate climate. *Agricultural and Forest Meteorology* 96, 103-115.

Boyer, J. S. (1995). Pressure Chamber. In *Measuring the water status of plants and soils* (ed. J. S. Boyer), pp. 178. San Diego: Academic Press.

Chaves, M. M. & Pereira, J. S. (1992). Water stress, CO_2 and climate change. *Journal of Experimental Botany* 43, 1131-1139.

Clark, H., Newton, P. C. D. & Barker, D. J. (1999). Physiological and morphological responses to elevated CO_2 and a soil moisture deficit of temperate pasture species growing in an established plant community. *Journal of Experimental Botany* 50, 233-242.

Clifford, S. C., Stronach, I. M., Mohamed, A. D., Azam-Ali, S. N. & Crout, N. M. J. (1993). The effects of elevated atmospheric carbon dioxide and water stress on light interception, dry matter production and yield in stands of ground nut (*Arachis hypogaea* L.). *Journal of Experimental Botany* 44, 1763-1770.

Conroy, J. P., Ghannoum, O., Jitla, D., Rogers, G. & Seneweera, S. (1998). Plant responses to elevated CO_2 and climate stress, pp. 181-191. In *Responses of plant metabolism to air pollution and global changes* (eds. DeKok et al.). Leiden The Netherland: Backhuys Publishers.

Dai, Z. Y., Ku, M. B. S. & Edwards, G. E. (1995). C_4 Photosynthesis-the effects of leaf development on the CO_2-concentrating mechanism and photorespiration in maize. *Plant Physiology* 107, 815-825.

Eamus, D. (1991). The interaction of rising CO_2 and temperatures with water use efficiency. *Plant, Cell and Environment* 14, 846-852.

Ehleringer, J. R., Cerling, T. E. & Helliker, B. R. (1997). C_4 Photosynthesis, atmospheric CO_2, and climate. *Oecologia* 112, 285-299.

Elamin, W. A. M. & Mustafa, M. I. (1992). Germination requirement of *Blepharis persica* (Burm) Kuntze, a hot desert forage. In *Environmental science and natural Resources*. Wad Medani: Gezira, Sudan.

Ferris, R. & Taylor, G. (1995). Contrasting effects of elevated CO_2 and water deficit on two native herbs. *New Phytologist* 131, 491-501.

Ghannoum, O., Caremmerer, S. V., Barlow, E. W. R. & Conroy, J. P. (1997). The effect of CO_2 enrichment and irradiance on the growth, morphology and gas exchange of C_3 (*Panicum laxum*) and C_4 (*Panicum antidotale*) grass. *Australian Journal of Plant Physiology* 24, 227-237.

Ghannoum, O. & Conroy, J. P. (1998). Nitrogen deficiency precludes a growth response to CO_2 enrichment in C_3 and C_4 *Panicum* grasses. *Australian Journal of Plant Physiology* 25, 627-636.

Ham, J. M., Owensby, C. E., Coyne, P. I. & Bremer, D. J. (1995). Fluxes of CO_2 and water-vapor from a prairie ecosystem exposed to ambient and elevated atmospheric CO_2. *Agricultural and Forest Meteorology* 77, 73-93.

Hattersley, P. W. (1984). Characterization of C_4 type leaf anatomy in grasses (Poaceae). Mesophyll: bundle sheath area ratios. *Annals of Botany* 53, 163-179.

He, H., Kirkham, M. B., Lawlor, D. J. & Kanemasu, E. T. (1992). Photosynthesis and water relations of Big Bluestem (C_4) and Kentucky Bluegrass (C_3) under high concentration carbon dioxide. *Transactions of The Kansas Academy of Science* 95, 139-152.

Hunt, H. W., Elliott, E. T., Detling, J. K., Morgan, J. A. & Chen. (1996). Responses of a C_3 and C_4 perennial grass to elevated CO_2 and temperature under different water regimes. *Global Change Biology* 2, 35-47.

IES. (1997). Trends of recovery in biomass and soil organic matter of Sudan's Sahel region: A case study of Northern Kordofan, pp. 46. Khartoum, Sudan: University of Khartoum, Institute of Environmental Studies (IES).

IPCC. (1996). Ecological, and soil processes in terrestrial ecosystems: A primer on general concepts and relationships, pp. 59-74. In *Climate change 95* (eds. R. T. Watson, M. C. Zinyowera & R. Moss). Cambridge: Cambridge University Press.

Kirkham, M. B., Ballou, L. K., Lawlor, D. J. & Kanemasu, E. T. (1992). Changes in prairie vegetation under elevated carbon dioxide levels and two soil moisture regimes. *Journal of Vegetation Science* 3, 673-678.

Kirkham, M. B., He, H., Bolger, T. P., Lawlor, D. J. & Kanemasu, E. T. (1991). Leaf photosynthesis and water-use of Big Bluestem under elevated carbon dioxide. *Crop Science* 31, 1589-1594.

Knapp, A. K., Hamerlynck, E. P., Ham, J. M. & Owensby, C. E. (1996). Responses in stomatal conductance to elevated CO_2 in 12 grassland species that differ in growth form. *Vegitatio* 125, 31-41.

LeCain, D. R. & Morgan, J. A. (1998). Growth, gas exchange, leaf nitrogen and carbohydrate concentrations in NAD-ME and NADP-ME C_4 grasses grown in elevated CO_2. *Physiologia Plantarum* 102, 297-306.

Lemon, E. R. (1984). The response of plants to rising levels of atmospheric carbon dioxide, pp. 65-106. In *CO_2 and plants*. (ed. E. R. Lemon). Westview Press, Inc. Boulder, Colorado.

Morgan, J. A., LeCain, D. R., Read, J. J., Hunt, H. W. & Knight, W. G. (1998). Photosynthetic pathway and ontogeny affect water relations and the impact of CO_2 on *Bouteloua gracilis* (C_4) and *Pascopyrum smithii* (C_3). *Oecologia* 114, 483-493.

Morrison, J. I. (1998). Stomatal response to increased CO_2 concentration. *Journal of Experimental Botany* 49, 443-452.

Morrison, J. I. & Gifford, R. M. (1984a). Plant growth and water use with limited water supply in high CO_2 concentrations. (1) Leaf area, water use and transpiration. *Australian Journal of Plant Physiology* 11, 361-374.

Mott, K. A. (1988). Do stomata respond to CO_2 concentrations other than intercellular? *Plant Physiology*. 86, 200-203.

Murray, R. D. (1997). Effects of elevated atmosphere CO_2 on the leaf surface, stomatal function and water-use efficiency. In *Carbon dioxide and plant responses*. (ed. R. D. Murray), pp. 129-155. New York: Research Studies Press Ltd.

Nie, D., Kirkham, M. B., Ballou, L. K., Lawlor, D. j. & Khanemasu, E. T. (1992). Changes in prairie vegetation under elevated carbon dioxide levels and two soil moisture regimes. *Journal of Vegetation Science* 3, 673-678.

Owensby, C. E., Ham, J. M., Knapp, A. K. & Auen, L. M. (1997). water vapour fluxes and their impact under elevated CO_2 in a C_4 Tallgrass Prairie. *Global Change Biology*. 3, 189-195.

Owensby, C. E., Ham, J. M., Knapp, A. K. & Auen, L. M. (1999). Biomass production and species composition change in a tallgrass prairie ecosystem after long-term exposure to elevated atmospheric CO_2. *Global Change Biology* 5, 497-506.

Poorter, H. (1993). Interspecific variation in the growth response of plants to an elevated ambient CO_2 concentration. *Vegetatio* 104/105, 77-79.

Read, J. J. & Morgan, J. A. (1996). Growth and partitioning in *Pascopyrum smithii* (C_3) and *Bouteloua gracilis* (C_4) as influenced by carbon dioxide and temperature. *Annals of Botany* 77, 487-497.

Rozema, J., Dorel, F., Janissen, R., Lenssen, G., Broekman, R., Arp, W. & Drake, B. G. (1991). Effect of elevated atmospheric CO_2 on growth, photosynthesis and water relations of salt marsh grass species. *Aquatic Botany* 39, 45-55.

Seneweera, S. P., Ghannoum, O. & Conroy, J. (1998). High vapour pressure deficit and low soil water availability enhance shoot growth responses of a C_4 grass (*Panicum coloratum* cv. Bambatsi) to CO_2 enrichment. *Austalian Journal of Plant Physiology* 25, 287-292.

Sionit, N. & Patterson, D. T. (1984). Responses of C_4 grasses to atmospheric CO_2 enrichment. I. Effect of irradiance. *Oecologia* 65, 30-34.

Smith, S. D., Humxan, T. E., Zitzer, S. F., Charlet, T. N., Housman, D. C., Coleman, J. S., Fenstermaker, L. K., Seemann, J. R. & Nowak, R. S. (2000). Elevated CO_2 increases productivity and invasive species success in an arid ecosystem. *Nature* 408, 79-81.

Wand, S. E., Midgley, G. F., Jones, M. H. & Curtis, P. (1999). Responses of wild C_4 and C_3 grass (*Poaceae*) species to elevated atmospheric CO_2 concentration: a meta-analytic test of current theories and perceptions. *Global Change Biology* 5, 723-741.

Woodward, F. I. (1987). Stomatal numbers are sensitive to increases in CO_2 from pre-industrial levels. *Nature* 327, 617-618.

Zaki-Eldeen, S. A. (2001). The effects of elevated CO_2 and water stress on two C_4 pasture species (*Blepharis linariifolia* and *Cenchrus biflorus*) native to Sudan. *Ph.D. thesis.* Dublin: University of Dublin, Trinity College.

Ziska, L. H., Sicher, R. C. & Bunce, J. A. (1999). The impact of elevated carbon dioxide on the growth and gas exchange of three C_4 species differing in CO_2 leak rates. *Physiologia Plantarum* 105, 74-80.

Increase in Atmospheric CO_2 and Ambient Temperatures in the North: Promises and Drawbacks for Crop Production

K. Hakala
H. Laurila
T. Mela

SUMMARY. Crop production in the North is constrained by a short growing season, spring and autumn frosts, long winters with deep snow cover and low temperatures during the growing season. Increase in concentration of atmospheric CO_2 and other greenhouse gases, and the subsequent increase in ambient temperatures could benefit crop production in countries located in the North, including Finland. However, longer growing seasons, milder winters and higher growing season temperatures could increase the occurrence of pests and pathogens, and ulti-

---

K. Hakala is Senior Researcher, MTT Agrifood Research Finland, Plant Production Research, FIN-31600 Jokioinen, Finland (E-mail: kaija.hakala@mtt.fi).

H. Laurila is Researcher, University of Helsinki, Department of Applied Biology, Crop Husbandry Section, FIN-00014 Helsingin Yliopisto, Finland.

T. Mela is Professor (Emeritus), MTT Agrifood Research Finland, Plant Production Research, FIN-31600 Jokioinen, Finland.

Address correspondence to: Dr. K. Hakala at the above address.

[Haworth co-indexing entry note]: "Increase in Atmospheric CO_2 and Ambient Temperatures in the North: Promises and Drawbacks for Crop Production." Hakala, K., H. Laurila, and T. Mela. Co-published simultaneously in *Journal of Crop Improvement* (Food Products Press, an imprint of The Haworth Press, Inc.) Vol. 13, No. 1/2 (#25/26), 2005, pp. 239-255; and: *Ecological Responses and Adaptations of Crops to Rising Atmospheric Carbon Dioxide* (ed: Zoltán Tuba) Food Products Press, an imprint of The Haworth Press, Inc., 2005, pp. 239-255. Single or multiple copies of this article are available for a fee from The Haworth Document Delivery Service [1-800-HAWORTH, 9:00 a.m. - 5:00 p.m. (EST). E-mail address: docdelivery@haworthpress.com].

http://www.haworthpress.com/web/JCRIP

Digital Object Identifier: 10.1300/J411v13n01_11

mately decrease crop yields and the net income of farmers. Benefits to be gained from increased tillering and higher numbers of grains per unit area resulting from elevated CO_2 levels might be smaller than expected in countries like Finland, where the uniculm growth habit of cereals is maintained by the long days that characterise the growing season.

KEYWORDS. Cereals, climate change, CO_2, crops, grasses, modelling, simulation, North

INTRODUCTION

In the northernmost agricultural areas, including Finland (lat. 60°N to 70°N), crop growth is constrained by the short growing season, low growing season temperatures and frosts that occur early and late in the growing season. The mean annual temperature in Finland is 4-5°C, and the mean temperature in the summer is 15°C in southern Finland, the main crop production area (Finnish Meteorological Institute, www.fmi.fi). The low mean growing season temperature is near the optimum for the production of C_3 grasses (17-19°C), but much below the optimum for cereals (28-30°C). In addition, rain early in the autumn often makes conditions for harvest and tillage very unfavourable.

Increase in the concentrations of greenhouse gases such as CO_2, CH_4, O_3, CO, N_2O, nitrogen oxides (NO_x), hydrofluorocarbons (HFCs), perfluorocarbons (PFCs), and sulphur hexafluoride (SF_6) may lead to enhanced greenhouse effect and changes in ambient temperatures and climate in the future (IPCC 2001a). Climate warming could improve crop production in the North, but it is not possible to predict accurately the effects of climate change on agricultural production in areas such as northern Europe. The enhanced greenhouse effect could affect climate in two ways. Firstly, increase in global temperatures could bring about increases in the temperatures in the North. Secondly, climate cooling is also possible as a result of submergence of the warm Gulf Stream in the northern Atlantic Ocean. Currently, the most probable scenario is that of climate warming (IPCC 2001b). For example, in Finland, the mean annual temperature is estimated to rise by 0.4°C per decade during this century (Carter et al. 1996a), which is greater than the predicted rise in global temperatures. As scenarios for climate change are much less accurate than those for increase in CO_2 concentration, there is considerable room for speculation as to what will occur.

Increase in annual temperatures could lengthen the growing season and increase the growing season temperatures, thus making conditions more favourable for production of current crop varieties, and even for introduction of new, more productive crop types (Carter et al. 1996a, 1996b, Mela 1996). Together with the increase in CO_2 concentration, crop production potential in the North could increase considerably. However, increased precipitation, as is predicted for Finland (Carter et al. 1996a), would make the conditions for harvest even more difficult than they are currently. Warmer temperatures might also increase the occurrence of pests and pathogens (Carter et al. 1996b).

This article summarises the results of some climate change studies for Finland, including those from a specific Finnish research project (Mela et al. 1996). Special emphasis is placed on the impacts of climate change in the North and on possible reasons why expected yield and biomass increases may or may not be attained with increase in CO_2 concentrations. The general plant physiological basis for the effects of climate change is also addressed.

DIRECT EFFECTS OF CO_2 INCREASE

Effect on Photosynthesis

Under the current conditions of 360 ppm CO_2 in the atmosphere, 20-25% of assimilated carbon is lost through the oxygenation of Rubisco and photorespiration (Bowes 1991). The predicted two-fold increase (by the end of this century) in the atmospheric concentration of CO_2 gives hope for a marked (from 25 to 100%, Cure and Acock 1986, Lawlor and Mitchell 1991, Stitt 1991, Drake et al. 1997) increase in CO_2 assimilation. The effect of elevated CO_2 concentration might even be greater at higher temperatures, as an increase in temperature decreases the aqueous solubility of CO_2 and the affinity of Rubisco for CO_2 relatively more than the aqueous solubility and affinity for Rubisco of O_2 (see reviews by Bowes 1991, Long 1991, and Drake et al. 1997).

Effect on Water, Nitrogen, and Radiation Use Efficiency

Stomatal conductance decreases with increase in the concentration of atmospheric CO_2 (Farquhar and Sharkey 1982, Tuba et al. 1994, Pearson et al. 1995, Drake et al 1997, Bunce 1998, Hakala et al. 1999). However, the relationship between ambient and intercellular CO_2 concentration remains the same, around 0.7 (Drake et al. 1997). Thus, even though the stomata close and evapotranspiration decreases at elevated CO_2 concentration, the CO_2 supply and assimilation by the plant remain on a higher level than at a lower CO_2 concentration. Thus, the water use efficiency of the leaf improves (Wong 1979, Tuba et al. 1994, Drake et al. 1997). Improved nitrogen use efficiency has also been recorded in an atmosphere of elevated CO_2 levels as a result of a reduction in the quantity of Rubisco in the leaves (Wong 1979, Schmitt and Edwards 1981, Conroy and Hocking 1993, Nie et al. 1995, Drake et al. 1997). Due to reduced photorespiration, the net number of CO_2 molecules assimilated per unit number of photons increases and thus radiation use efficiency of leaves increases.

Effects on Biomass Production

While there is strong evidence for increased CO_2 levels increasing photosynthesis, information on changes in biomass accumulation is more variable. In the first reviews, the yield increase under doubled CO_2 of several C_3 crops ranged from 33% (Kimball 1983) to 41% (Cure and Acock 1986). Later, more experiments were done covering larger areas of cultivation, and with increasing applicability to natural conditions. In the latest reviews, the increase in biomass varied from zero to more than 100% (Morison and Lawlor 1999), depending on the plant species and experimental conditions. However, most studies show at least some increase in crop production resulting from increased atmospheric CO_2. We will discuss later the conditions under which increase in CO_2 fails to increase biomass accumulation, especially in the North.

SECONDARY EFFECTS OF AN INCREASE IN CO_2

The secondary effects of increased CO_2 concentration include, in part, direct changes in plant physiology and composition. For example, a decrease in stomatal conductance increases internal and canopy temperatures, with effects on plant phenology and microclimate, even in the absence of increases in global temperatures and climate change (Morison and Lawlor 1999). Increased nitrogen use efficiency might decrease the nitrogen content of plants and thus decrease their nutritional value (Hocking and Meyer 1991, Thompson and Woodward 1994). Pests feeding on plants with lower nitrogen content may need more plant material to satisfy their nitrogen demand, thus causing destruction of a larger photosynthesising area than would be the case at current atmospheric CO_2 concentrations (Coley and Markham 1998). However, the number of pests could decrease with decreasing quality of leaf material caused by reduction of nitrogen content or increase in secondary substances such as tannins (Brooks and Whittaker 1998, Mansfield et al. 1999). Increase in water use efficiency and radiation use efficiency compensate for drought and low light conditions and could make crop production possible in harsh environments. The main secondary effect of increased concentrations of CO_2 and other greenhouse gases is an increase in global temperatures and the possible climatic changes that may follow from the increase.

CONDITIONS LIMITING THE EFFECT OF CO_2 ON BIOMASS

According to recent review articles, the stimulation of biomass accumulation and yield by doubling of CO_2 ranges between zero and 100% (e.g., Morison and Lawlor 1999). It is now clear that growing conditions characterised by favourable temperatures, good water and nutrient availability and high light intensity, all increase the production of plant biomass at high levels of atmospheric CO_2. Because of the increased water use efficiency, nitrogen use efficiency, and radiation use efficiency reported for plants growing at increased CO_2 concentrations, higher CO_2 may improve plant growth at low nutrient levels or when water is limiting. However, the expected increases in biomass and yield might not be realised under limiting growth conditions (e.g., Goudriaan and de Ruiter 1983, Idso et al. 1987, Rawson 1995, Wolf 1996).

Low Ambient Temperatures

Results from several studies have shown that yield and biomass of plants increases at high CO_2 only when growing temperatures are moderately high (Idso et al. 1987, McKee and Woodward 1994). For example, for every degree above a basic 10°C wheat yield increases by 1.8% (Rawson 1995). The temperatures would have to be high enough, particularly during critical phases of development, such as the tillering, for the CO_2 increase to have an effect (Grashoff et al. 1995). In experiments with meadow fescue in Finland there was no increase in yield at elevated CO_2 unless temperature also increased (Hakala and Mela 1996). Moreover, in experiments done on the south-west coast of Norway, the yield of some grasses did not change when CO_2 levels were elevated. This was explained by the possible effect of cool temperatures at the site nullifying the effect (Saebo and Mortensen 1995, 1996). Thus, increase in global temperatures in the North and other areas with normally low growing season temperatures, is crucial to benefit from the increased CO_2. Under elevated CO_2 levels the temperature of plant leaves may increase through the effect of CO_2 on stomatal aperture, thereby increasing the effect of CO_2 on photosynthesis. However, maintaining increased photosynthetic activity at elevated CO_2 is only possible if photosynthetic products are continuously transported out of the chloroplasts to the growing plant parts (Farrar and Williams 1991). This is possible only under conditions favourable for growth, including moderately high temperatures and adequate nutrient levels.

Low Nutrient Levels

Even though nitrogen use efficiency of plants improves when CO_2 levels are increased, the demand for soil nitrogen can also increase as a result of increased biomass production and greater nitrogen uptake (Hocking and Meyer

1991, Bowes 1993). Soil nutrient deficiency, especially of nitrogen and phosphorus, has been shown to limit the biomass gain under CO_2 enrichment (Goudriaan and de Ruiter 1983, Wolf 1996, Poorter 1998). In the worst case, elevated CO_2 can decrease the yield of cereals at low nitrogen levels. This may occur when tillering is initially stimulated by high CO_2, but because of nitrogen deficiency, the tillers fail to develop productive ears (Mitchell et al. 1993). Temperature conditions also affect the availability of nutrients to plants, both through changes in mineralisation of nutrients and through effects on root growth. If during winter soils remain frozen for long periods and cool and wet conditions follow in the spring, as occurs in the North, conditions for soil microbes are less favourable than in areas with a warmer climate. Higher mean annual temperatures and more favourable conditions for microbe and root growth would enhance the effect of CO_2 on biomass and yields in the North, not only through providing better conditions for biomass accumulation, but also through improving nutrient availability.

Determinate Crops with Strict Growth Habit

Cereals grow in distinct phases, which are reached according to a temperature sum. When the temperatures increase, the developmental phases are reached earlier, and as a consequence of the shorter growing time the grain number and the grain filling time both decrease with the inevitable consequence that grain yield also decreases (Wheeler et al. 1996a, 1996b). An average increase in growing season temperatures of 1.8°C by 2050 would result in a 4-6 days shorter grain filling period in barley, and a 5-11 days shorter grain filling period in wheat, in the most productive areas of Finland (Carter and Saarikko 1996, Carter et al. 1996b). Because of this, the yield of the Finnish barley variety 'Pomo', for example, would decrease by up to 1000 kg per hectare (Carter et al. 1996b). New crop genotypes adapted to prevailing conditions are being bred continuously and therefore the yield decrease described is to some extent hypothetical. Moreover, increased CO_2 can stimulate the production of carbohydrates through increased CO_2 assimilation and thus compensate for the shorter grain filling time (Carter et al. 1996b, Wheeler et al. 1996b). Only when temperatures are high enough to cause abortion or infertility of flowers, and thus severely reduce grain number, has an increase in CO_2 not been able to compensate for yield loss (Wheeler et al. 1996a, 1996b, Hakala 1998b).

Climatic conditions at the growth phases when number of tillers, ears per plant and number of grain in ears are determined affect the potential sink size of cereals. The survival of tillers and grains also depends on the conditions during their growth. Finally, grain filling depends on the conditions following anthesis (e.g., Kontturi 1979, review by Morison and Lawlor 1999). CO_2 enrichment can have an effect on the final biomass and grain yield at any of the growth phases. However, the main effect of CO_2 is to increase the number of organs (tillers, ears and grains per square meter) rather than their size (Krenzer and Moss 1975, Fischer and Aguilar 1976, Sionit et al. 1981, Goudriaan and de Ruiter 1983, Havelka et al. 1984, McKee and Woodward 1994, Hakala 1998a). If elevated CO_2 fails to have an effect at tillering or when the grain number is determined in cereals, the number of structures (future sinks) is not increased, and the total effect remains small. Increased CO_2 also increases the survival of tillers and grain, which makes favourable climatic conditions under high concentrations of CO_2 crucial during all developmental phases for an effect of CO_2 to manifest itself (see Morison and Lawlor 1999).

In Finland, where the growing season is short, the most profitable way to cultivate cereals is by sowing a dense stand, with little space for lateral shoots (Mela and Paatela 1974). A long tillering period would decrease the crop maturation time, since only a limited number of days are available for growth in the short growing season in the North. Because of this, formation of increased numbers of sinks through increased tillering under high CO_2 would probably not result in higher yields under the present climatic conditions of Finland. In a four year experiment with spring wheat at Jokioinen, southern Finland, it was shown that even when tillering increased under CO_2 enrichment, the yield gain was very limited due to the small size and poor yield of the lateral shoot ears (Hakala 1998a, 1998b). An increase in mean annual temperatures and growing season length would allow cultivation of crop varieties with longer growing time and thus allow CO_2 to affect crop yields also in Finnish conditions. The long summer days, however, may represent a problem, as long days are known to promote apical dominance and asynchronous growth of main shoots and lateral shoots (Peltonen-Sainio and Peltonen 1997).

Canopy Closure and Self-Shading Lead to Lowered Growth Rate

The effect of CO_2 is usually greatest during initial growth stages as radiation, nutrition and moisture conditions in a canopy are at their most favourable. Rapid growth early in the season is beneficial, because then the soil surface is quickly covered by leaves, which in turn enhances the interception of light and canopy photosynthesis per unit area (Poorter et al. 1996, Badger 1992). As the canopy develops, the relative growth enhancement decreases as a smaller proportion of leaves intercept incident radiation and the lower leaves in the canopy become increasingly shaded. At the same time the leaf area relative to the total biomass becomes smaller and the relative growth rate decreases. As the biomass accumulation is higher at elevated CO_2, these changes occur faster and the relative growth rate decreases earlier (Poorter et al. 1996, Badger 1992). As a result of this, the increase in biomass under elevated CO_2

may not be realised as anticipated according to increase in photosynthesis at saturating light intensity (Nijs et al. 1989, Hakala and Mela 1996, Hakala 1998a, Hakala et al. 1999).

A grass crop can be considered indeterminate as each cut removes a filled sink and creates a new one in which carbon can accumulate (Badger 1992). While the potential increase in yield of cereals in high CO_2 may be restricted by limited sink size, the increase in biomass of grasses can be limited when microclimatic conditions become unfavourable for growth and photosynthesis. Such conditions develop when the grass canopy reaches a leaf area index too high for light to penetrate completely (canopy closure). Then, the photosynthesising leaf area becomes small relative to the total biomass, and the carbon assimilated by the top layer is lost through respiration by the rest of the canopy (Nijs et al. 1989, Hakala 1998b). An increased number of cuts is crucial in gaining the full benefit from the grass crop in conditions of increased CO_2, especially when temperatures are elevated as well (Badger 1992, Hakala and Mela 1996, Hakala 1998b, Hakala et al. 1999).

IDEAL PLANT TYPE FOR THE FULL BENEFIT FROM ELEVATED CO_2

To realise the full benefit of increased CO_2, the ideal plant would be one with a fast growth rate (Poorter et al. 1996, Poorter 1998). Fast growth and a large sink capacity would help in accumulating more of the available carbon. A fast growth rate also helps the plant to compete in the canopy. When a crop plant is a fast grower, it will outgrow the weeds, especially in a high CO_2 environment. This would be economically attractive to the farmer, both because of increased yields and because of the reduced requirement for plant protection inputs. However, in some cases weeds may benefit more from the elevated CO_2 and out-compete the crops in a CO_2-enriched environment (Thompson and Woodward 1994, Ziska 2000).

One reason for different responses of crops to elevated CO_2 (Morison and Lawlor 1999) may be their different growth rates (Poorter 1998). However, different timing of critical growth stages during the year may have a significant effect. If a growth stage such as tiller initiation occurs at too low a temperature and light level, the CO_2 effect on the formation of the sinks may be lower than in plants with better timing (Grashoff et al. 1995). This may be crucial also in a field crop canopy, with weeds and crop plants competing for light and nutrients in a high CO_2 environment.

Lengthening of the growing season in the spring might be beneficial for crop production in the North (Mela 1996), as new crop species and varieties with a long growing time would be able to take advantage of the plentiful solar radiation in April-May. However, this benefit may be reduced through occurrence of spring or early summer frosts. Lengthening of the warm season in the autumn would be less beneficial because the rapidly shortening days in the autumn could limit growth despite warm temperatures.

For increased CO_2 to have a maximum effect, an ideal crop plant would have the following features:

- Flexibility to increase sink size according to the increase in the source when more CO_2 is available.
- Fast canopy closure at the tillering stage (Badger 1992), but a long period of grain filling (high temperature sum requirement from anthesis to maturity (Wolf 1993).
- Down-regulation of Rubisco for the benefit of growth (nitrogen from Rubisco directed to increased biomass-increased nitrogen use efficiency).
- Efficiency in water use through the ability to take the full benefit from the decrease of stomatal conductance in elevated CO_2 (Badger 1992).

A plant lacking these features would not be ideally adapted to the anticipated conditions of climate change. Because of the new and partly unpredictable growing conditions prevailing in the future, conservation of genetic variation and resources is increasingly important. Fortunately work is in progress to conserve crop genetic resources, including landraces on a global basis. It is crucial to retain as much genetic variability as possible in order to be prepared for any eventuality represented by future climate.

IMPLICATIONS OF THE SCENARIOS AND PHYSIOLOGICAL KNOWLEDGE FOR NORTHERN CONDITIONS

Positive Effects: Better Conditions for Crop Production

Countries in the far North, including Finland, currently suffer from very short growing seasons, harsh winter conditions and low growing season temperatures. The short growing season requires cultivation of fast-maturing crops. Moreover, even though solar radiation occurs at very high levels and daylength is sufficient already in early April, farmers have to wait until the beginning of May before the danger of late frosts is past and the fields are dry enough for agricultural machinery to be used on them. In the autumn, rain and cool weather begin early, which means that the crops have to be harvested early. This requires that crops mature in 3-4 months. If the temperatures increase according to the scenarios presented for Finland, the crops could be sown earlier, and they would benefit more from the high radiation intensities early in the growing season (Carter and Saarikko 1996, Carter et al. 1996a). The growing season might become longer and warmer and the winters milder (Carter et al. 1996a). Under these conditions the cultivation of the present crops would be more successful, and it would be possible to take new crops (e.g., maize) into cultivation, as well as more productive varieties of the present crops (Carter et al. 1996b, Mela 1996).

Uncertainties and Negative Effects: Pests and Pathogens, Climate Variability

The benefits for Finnish agriculture of an increase in temperature seem indisputable. However, it is possible that the average growing season temperatures will not increase as much as the average minimum temperatures. In the scenario for Finland of Carter et al. (1996a), growing season temperatures show the smallest increase (3°C per century), and winter temperatures the highest (6°C per century). Increase in winter temperatures may be positive for agriculture if they result in longer growing seasons. However, if the increase results in higher precipitation and extended periods of wet fields, the benefit is lost. Mild winters may also cause new problems for overwintering crops. If snow falls on unfrozen soil, the pathogens thriving under it have ideal conditions for infecting and destroying autumn-sown crops and grass stands (Nissinen 1996, Matsumoto and Nissinen 2001). Another problem follows if the temperatures during winter vary, with phases of freezing and thawing. Then the crops may be covered by ice, with consequent problems with anaerobic respiration resulting in poisonous concentrations of CO_2 and ethanol (Andrews 1996). Long spells of moderately high temperatures, either under snow cover or at the low light intensities during late autumn and midwinter, could decrease the carbohydrate reserves of overwintering crops by respiration (Niemeläinen 1990). Low carbohydrate levels may affect the resistance of overwintering crops to pathogens and slow their regrowth in the spring (Gaudet et al. 1999).

Another threat introduced by higher annual temperatures is represented by the effect of milder climate on pests and pathogens. Potato yields, for example, may increase under warmer growth conditions (Kaukoranta1996), but this benefit may not be realised if pest and disease incidence increases as well (Kaukoranta 1996, Carter et al. 1996b). One such pest is the potato cyst nematode, which currently thrives only south of the Arctic Circle, but could spread North and affect production in most of Lapland by 2050, if temperatures increase according to the highest temperature increase scenario of Carter et al. (1996b). Another pest, Colombian root-knot nematode, which is absent from Finland at present, would be able to produce two full generations in the best agricultural area of Finland under the most likely scenario of climatic change for 2050 (Carter et al. 1996b). Similar developments would apply to diseases as well (Carter et al. 1996b, Kaukoranta 1996). However, crop growth under higher temperatures could result in smaller and more fibrous leaves, containing more phenolic compounds and tannins (Morison and Lawlor 1999). Together with the effects of CO_2 (decrease in nitrogen concentration) and the rising UV-B radiation (increase in phenolics and tannins, see, e.g., Caldwell et al. 1995), an additional effect of climatic change could be an increase in the resistance of plants to pests and pathogens.

In addition to the effects of milder temperatures in general, extreme temperatures at flowering could cause flower abortion or infertility, which cannot always be compensated for by higher CO_2 levels (Wheeler et al. 1996a and b, Hakala 1998b). With perturbations of climate it is also unclear whether other biota would behave as expected of crops. If crops such as canola are sown early in the season they may flower before the pollinators have become active and may consequently yield less.

Higher Yields in Higher CO_2 Also in the Future?

If the climate becomes warmer in Finland, as predicted, it might be possible to cultivate more productive varieties of crops with longer growing times and higher tillering capacity. Higher tillering capacity would also allow elevated CO_2 to have a greater effect. However, as the light regime would remain unchanged, with very long days and high radiation intensities during the growing season, the prospects for higher production may not be realised. For example, the uniculm growth habit of cereals might not be easy to break (Peltonen-Sainio and Peltonen 1997). Currently, a dense stand with only a single head per plant represents the optimal cultivation method (Mela and Paatela 1974), and this might well remain so in the future unless new methods for breaking the uniculm growth habit are found.

It is clear that in the best case, excluding uncertainties, increase in annual temperatures alone would bring great benefits for agriculture in Finland. With a four degree higher annual temperature, as predicted in the scenarios for the climate of Finland a hundred years from now (Carter et al. 1996a), the climate might be similar to that of central Europe today (Mela 1996). The yield increases brought about by the cultivation of crop varieties with a higher production potential would be considerable. In addition to this there would be effects of elevated CO_2 and earlier sowing, with better use made of the high spring radiation intensities. In Table 1 selected effects of different conditions, simulated with the CERES-wheat crop model (Godwin et al. 1989), are shown for the spring wheat variety Polkka, currently widely cultivated in Finland. The simulation results suggest that without any increase in the CO_2 levels, earlier sowing alone would increase the grain yield of spring wheat considerably. Sowing 15 days earlier under current conditions of CO_2 (350 ppm) and tem-

TABLE 1. The effects of earlier sowing of spring wheat (*Triticum aestivum* L. cv. Polkka) under conditions of elevated atmospheric CO_2 and temperature on phenology, biomass, and yield components. Model CERES-Wheat potential model (without any limiting nutrient or water stresses).

Scenario	Sowing date (Julian day)	CO_2 (ppm)	Temp. change (°C)	Anthesis (Julian day)	Full maturity (Julian day)	Yield (t/ha)	Above ground biomass (t/ha)	1000 seed weight (g)	Grains/ ear	Tiller/ m^2
current reference*1)	135	350	0	192	244	4.7	9.7	35.4	18.9	574.5
reference sowing, doubled CO_2	135	700	0	190	238	8.77	17.22	37.8	23.5	718.4
earlier sowing, reference CO_2	120	350	0	182	230	7.7	14	37.9	28.1	612
earlier sowing, doubled CO_2	120	700	0	182	230	10.8	19.6	38.6	30.8	768.3
earlier sowing,+3° elev. temp., doubled CO_2	120	700	3	171	213	8.4	15.8	37.3	29	652.6

*1) current reference = sowing 15th May with ambient CO_2 at 350 ppm

perature, would produce an almost equal increase in grain yield as doubling CO_2 (700 ppm) at the reference sowing date (15th May) and current temperature. The results presented in Table 1 are aggregated from simulation studies of Laurila (2001) using the spring wheat cultivar Polkka.

The yield increase resulting from earlier sowing (Table 1) is brought about by increased tiller and kernel number and increased kernel size. The effect of doubling ambient CO_2 concentration is similar. It is not completely clear whether and to what extent an increase in CO_2 would increase crop yields under future climatic conditions. Experiments in countries at lower latitudes than Finland, including England and Germany, have shown that CO_2 has the most significant effect on yield when full yield potential is not otherwise reached. For instance, at very high tillering rates, when growth at normal CO_2 levels is strong, an increase in CO_2 does not always stimulate production of much extra biomass and yield (Weigel et al. 1994, Wheeler et al. 1996a). Thus, if the future climate of Finland resembles that of the present day central Europe, and grain yields increase as a result of a longer growing season and advances in plant breeding, it is still not clear that the conditions will necessarily promote the effects of increased CO_2. If the future climate favours more productive crop varieties that are sown earlier and at lower seed rates, the crop potential might be realised already at a lower ambient CO_2 concentration. However, under conditions where the full potential would not be fully manifested otherwise, increased CO_2 might "fill the gap." In the case of grasses, both the higher temperatures and elevated CO_2 levels would increase yields, at least if the grass cuts were adjusted to growth rates to prevent self-shading of the canopy (Nijs et al. 1989, Hakala and Mela 1996).

CONCLUSIONS

Although there are several scenarios for the effects of global change on crop production, it can probably be expected to increase in the North both for reasons of higher temperature and increased atmospheric CO_2 concentration. However, increased risks to production, represented by intensified pest and pathogen pressure, as well as unfavourable climate changes, including increase in autumn rain, could nullify any resultant benefits. Effects of climatic change on the environment as a whole are even more difficult to predict than effects on crop monocultures. Negative effects of climate change, such as drought and inundation, could override the positive ones many-fold. Conservation of the genetic resources of both cultivated and wild plants and animals will be crucial for developing crops adapted to any future changed environment.

REFERENCES

Andrews, C.J. (1996). How do plants survive ice? *Annals of Botany* 78:529-536.

Badger, M. (1992). Manipulating agricultural plants for a future high CO_2 environment. *Australian Journal of Botany* 40:421-429.

Bowes, G. (1991). Growth at elevated CO_2: photosynthetic responses mediated through Rubisco. *Plant, Cell and Environment* 14:795-806.

Bowes, G. (1993). Facing the inevitable: plants and increasing atmospheric CO_2. *Annual Review of Plant Physiology and Plant Molecular Biology* 44:309-332.

Brooks, G.L. and J.B. Whittaker. (1998). Responses of multiple generations of *Gastrophysa viridula* feeding on *Rumex obtusifolius*, to elevated CO_2. *Global Change Biology* 4:63-75.

Bunce, J.A. (1998). Effects of humidity on short-term responses of stomatal conductance to an increase in carbon dioxide concentration. *Plant, Cell and Environment* 21:115-120.S

Caldwell, M.M., A.H. Teramura, M. Tevini, J.F. Bornman, L.O. Björn and G. Kulandaivelu. (1995). Effects of increased solar ultraviolet radiation on terrestrial plants. *Ambio* 24:166-173.

Carter, T.R., M. Posch and H. Tuomenvirta. (1996a). The SILMU scenarios: specifying Finland's future climate for use in impact assessment. *Geophysica* 32:235-260.

Carter, T.R. and R.A. Saarikko. (1996). Estimating regional crop potential in Finland under a changing climate. *Agricultural and Forest Meteorology* 79:301-313.

Carter, T.R., R.A. Saarikko and K.J. Niemi. (1996b). Assessing the risks and uncertainties of regional crop potential under a changing climate in Finland. *Agricultural and Food Science in Finland* 5:329-350.

Coley, P.D. and A. Markham. (1998). Possible effects of climate change on plant/herbivore interactions in moist tropical forests. *Climatic Change* 39:455-472.

Conroy, J. and P. Hocking. (1993). Nitrogen nutrition of C_3 plants at elevated atmospheric CO_2 concentrations. *Physiologia Plantarum* 89:570-576.

Cure, J.D. and B. Acock. (1986). Crop responses to carbon dioxide doubling: a literature survey. *Agricultural and Forest Meteorology* 38:127-145.

Drake, B.G., M.A. Gonzàlez-Meler and S.P. Long. (1997). More efficient plants: a consequence of rising atmospheric CO_2? *Annual Review of Plant Physiology and Plant Molecular Biology* 48:609-639.

Farquhar, G.D. and T.D. Sharkey. (1982). Stomatal conductance and photosynthesis. *Annual Review of Plant Physiology* 33:317-345.

Farrar, J.F. and M.L. Williams. (1991). The effects of increased atmospheric carbon dioxide and temperature on carbon partitioning, source-sink relations and respiration. *Plant, Cell and Environment* 14:819-830.

Fischer, R.A. and M.I. Aguilar. (1976). Yield potential in a dwarf spring wheat and the effect of carbon dioxide fertilisation. *Agronomy Journal* 68:749-752.

Gaudet, D.A., A. Laroche and M. Yoshida. (1999). Low temperature-wheat-fungal interactions: A carbohydrate connection. *Physiologia Plantarum* 106:437-444.

Godwin, D., J.T. Ritchie, U. Singh and T. Hunt. (1989). A User's Guide to CERES-wheat (v. 2.10). Michigan State University.129 p.

Goudriaan, J. and H.E. de Ruiter. (1983). Plant growth in response to CO_2 enrichment, at two levels of nitrogen and phosphorus supply. 1. Dry matter, leaf area and development. *Netherlands Journal of Agricultural Science* 31:157-169.

Grashoff, C., P. Dijkstra, S. Nonhebel, A.H.C.M. Schapendonk and S.C. Van de Geijn. (1995). Effects of climate change on productivity of cereals and legumes; model evaluation of observed year-to-year variability of the CO_2 response. *Global Change Biology* 1:417-428.

Hakala, K. (1998a). Growth and yield potential of spring wheat in a simulated changed climate with increased CO_2 and higher temperature. *European Journal of Agronomy* 9:41-52.

Hakala, K. (1998b). Effect of climate change in the North on growth, yield formation and photosynthesis of spring wheat and meadow fescue. Doctoral thesis, 160 p. *Annales Universitatis Turkuensis SER. AII, TOM. 110.*

Hakala, K., R. Heliö, E. Tuhkanen and T. Kaukoranta. (1999). Photosynthesis and Rubisco kinetics in spring wheat and meadow fescue under conditions of simulated climate change with elevated CO_2 and increased temperatures. *Agricultural and Food Science in Finland* 8:441-457.

Hakala, K. and T. Mela. (1996). The effects of prolonged exposure to elevated temperatures and elevated CO_2 levels on the growth, yield and dry matter partitioning of field-sown meadow fescue (*Festuca pratensis*, cv. Kalevi). *Agricultural and Food Science in Finland* 5:285-298.

Havelka, U.D., V.A. Wittenbach and M.G. Boyle. (1984). CO_2-enrichment effects on wheat yield and physiology. *Crop Science* 24: 1163-1168.

Hocking, P.J. and C.P. Meyer. (1991). Effects of CO_2 enrichment and nitrogen stress on growth, and partitioning of dry matter and nitrogen in wheat and maize. *Australian Journal of Plant Physiology* 18:339-356.

Idso, S.B., B.A. Kimball, M.G. Anderson and J.R. Mauney. (1987). Effects of atmospheric CO_2 enrichment on plant growth: the interactive role of air temperature. *Agriculture, Ecosystems and Environment* 20:1-10.

IPCC. (2001a). *Climate Change 2001. The Scientific Basis.* The Third Assessment Report of Working Group I of the Intergovernmental Panel on Climate Change (IPCC), http://www.grida.no/climate/ipcc_tar/wg1/. (8 November 2002)

IPCC. (2001b). *Climate Change 2001. Impacts, Adaptation and Vulnerability.* The Third Assessment Report of Working Group II of the Intergovernmental Panel on Climate Change (IPCC), http://www.grida.no/climate/ipcc_tar/wg2/. (8 November 2002)

Kaukoranta, T. (1996). Impact of global warming on potato late blight: risk, yield loss and control. *Agricultural and Food Science in Finland* 5:311-327.

Kimball, B.A. (1983). Carbon dioxide and agricultural yield: an assemblage and analysis of 430 prior observations. *Agronomy Journal* 75:779-788.

Kontturi, M. (1979). The effect of weather on yield and development of spring wheat in Finland. *Annales Agriculturae Fenniae* 18:263-274.

Krenzer, E.G., Jr. and D.N. Moss. (1975). Carbon dioxide enrichment effects upon yield and yield components in wheat. *Crop Science* 15:71-74.

Laurila, H. (2001). Simulation of spring wheat responses to elevated CO_2 and temperature by using CERES-wheat crop model. *Agricultural and Food Science in Finland* 10:175-196.

Lawlor, D.W. and A.C. Mitchell. (1991). The effects of increasing CO_2 on crop photosynthesis and productivity: a review of field studies. *Plant, Cell and Environment* 14:807-818.

Long, S.P. (1991). Modification of the response of photosynthetic productivity to rising temperature by atmospheric CO_2 concentrations: Has its importance been underestimated? *Plant, Cell and Environment* 14:729-739.

Mansfield, J.L., P.S. Curtis, D.R. Zak and K.S. Pregitzer. (1999). Genotypic variation for condensed tannin production in trembling aspen (*Populus tremuloides*, Salicaceae) under elevated CO_2 and in high- and low-fertility soil. *American Journal of Botany* 86:1154-1159.

Matsumoto, N. and O. Nissinen. (2001). Environmental conditions that affect snow mold development. In: *Low temperature plant microbe interactions under snow*, ed. Norio Iriki, D. A. Gaudet and A.-M. Tronsmo: Hokkaido National Agricultural Experiment Station, Japan, pp. 13-21.

McKee, I.F. and F.I. Woodward. (1994). CO_2 enrichment responses of wheat: interactions with temperature, nitrate and phosphate. *New Phytology* 127:447-453.

Mela, T., T. Carter, K. Hakala, A. Hannukkala, T. Kaukoranta, H. Laurila, K. Niemi, R. Saarikko and K. Tiilikkala. (1996). The effects of climate change on crop production: results of a five-year research project. In: *The Finnish Research Programme on Climate Change: Final report*, ed. Jaana Roos, Helsinki, Finland: Publications of the Academy of Finland 4/96, pp. 324-336.

Mela, T.J.N. (1996). Northern agriculture: constraints and responses to global climate change. *Agricultural and Food Science in Finland* 5:229-234.
Mela, T. and J. Paatela. (1974). Grain yield of spring wheat and oats as affected by population density. *Annales Agriculturae Fenniae* 13:161-167.
Mitchell, R.A.C., V.J. Mitchell, S.P. Driscoll, J. Franklin and D.W. Lawlor. (1993). Effects of increased CO_2 concentration and temperature on growth and yield of winter wheat at two levels of nitrogen application. *Plant, Cell and Environment* 16:521-529.
Morison, J.I.L. and D.W. Lawlor. (1999). Interactions between increasing CO_2 concentration and temperature on plant growth. *Plant, Cell and Environment* 22:659-682.
Nie, G.Y., S.P. Long, R.L. Garcia, B.A. Kimball, R.L. Lamorte, P.J. Pinter, Jr., G.W. Wall and A.N. Webber. (1995). Effects of free-air CO_2 enrichment on the development of the photosynthetic apparatus in wheat, as indicated by changes in leaf proteins. *Plant, Cell and Environment* 18:855-864.
Niemeläinen, O. (1990). Factors affecting panicle production of cocksfoot (*Dactylis glomerata* L.) in Finland. III. Response to exhaustion of reserve carbohydrates and to freezing stress. *Annales Agriculturae Fenniae* 29:241-250.
Nijs, I., I. Impens and T. Behaeghe. (1989). Leaf and canopy responses of *Lolium perenne* to long-term elevated atmospheric carbon-dioxide concentration. *Planta* 177:312-320.
Nissinen, O. (1996). Analyses of climatic factors affecting snow mould injury in first-year timothy (*Phleum pratence* L.) with special reference to *Sclerotinia borealis*. *Acta Universitatis Ouluensis A 289.* Dissertation, University of Oulu, 115 p.
Patterson, D.T. and E.P. Flint. (1980). Potential effects of global atmospheric CO_2 enrichment on the growth and competitiveness of C_3 and C_4 weed and crop plants. *Weed Science* 28:71-75.
Pearson, M., W.J. Davies and T.A. Mansfield. (1995). Asymmetric responses of adaxial and abaxial stomata to elevated CO_2: impacts on the control of gas exchange by leaves. *Plant, Cell and Environment* 18:837-843.
Peltonen-Sainio, P. and J. Peltonen. (1997). Breaking uniculm growth habit of spring cereals at high latitudes by crop management. I. Leaf area index and biomass accumulation. *Journal of Agronomy and Crop Science* 178:79-86.
Poorter, H. (1998). Do slow-growing species and nutrient-stressed plants respond relatively strongly to elevated CO_2? *Global Change Biology* 4:693-697.
Poorter, H., C. Roumet and B.D. Campbell. (1996). Interspecific variation in the growth response of plants to elevated CO_2: a search for functional types. In: *Carbon Dioxide, Populations, and Communities*, eds. C. Körner and F.A. Bazzaz: Academic Press, pp. 375-412.
Rawson, H.M. (1995). Yield responses of two wheat genotypes to carbon dioxide and temperature in field studies using temperature gradient tunnels. *Australian Journal of Plant Physiology* 22:23-32.
Saebo, A. and L. Mortensen. (1995). Growth and regrowth of *Phleum pratense, Lolium perenne, Trifolium repens* and *Trifolium pratense* at normal and elevated CO_2 concentration. *Agriculture, Ecosystems and Environment* 55:29-35.
Saebo, A. and L. Mortensen. (1996). The influence of elevated CO_2 concentration on growth of seven grasses and one clover species in a cool maritime climate. *Acta Agriculturae Scandinavica Section B, Soil and Plant Science* 46:49-54.

Schmitt, M.R. and G.E. Edwards. (1981). Photosynthetic capacity and nitrogen use efficiency of maize, wheat and rice: a comparison between C_3 and C_4 photosynthesis. *Journal of Experimental Botany* 32:459-466.

Sionit, N., B.R. Strain and H. Hellmers. (1981). Effects of different concentrations of atmospheric CO_2 on growth and yield components of wheat. *Journal of Agricultural Science, Cambridge* 97:335-339.

Stitt, M. (1991). Rising CO_2 levels and their potential significance for carbon flow in photosynthetic cells. *Plant, Cell and Environment* 14:741-762.

Thompson, G.B. and F.I. Woodward. (1994). Some influences of CO_2 enrichment, nitrogen nutrition and competition on grain yield and quality in spring wheat and barley. *Journal of Experimental Botany* 45:937-942.

Tuba, Z., K. Szente and J. Koch. (1994). Response of photosynthesis, stomatal conductance, water use efficiency and production to long-term elevated CO_2 in winter wheat. *Journal of Plant Physiology* 144:661-668.

Weigel, H.J., R. Manderscheid, H.-J. Jäger and G.J. Mejer. (1994). Effects of season-long CO_2 enrichment on cereals. I. Growth performance and yield. *Agriculture, Ecosystems and Environment* 48:231-240.

Wheeler, T.R., G.R. Batts, R.H. Ellis, P. Hadley and J.I.L. Morison. (1996a). Growth and yield of winter wheat *(Triticum aestivum)* crops in response to CO_2 and temperature. *Journal of Agricultural Science, Cambridge* 127:37-48.

Wheeler, T.R., T.D. Hong, R.H. Ellis, G.R. Batts, J.I.L. Morison and P. Hadley. (1996b). The duration and rate of grain growth, and harvest index, of wheat (*Triticum aestivum* L.) in response to temperature and CO_2. *Journal of Experimental Botany* 47:623-630.

Wolf, J. (1993). Effects of climate change on wheat production potential in the European Community. *European Journal of Agronomy* 2:281-292.

Wolf, J. (1996). Effects of nutrient supply (NPK) on spring wheat response to elevated atmospheric CO_2. *Plant and Soil* 185:113-123.

Wong, S.C. (1979). Elevated atmospheric partial pressure of CO_2 and plant growth. I. Interactions of nitrogen nutrition and photosynthetic capacity in C_3 and C_4 plants. *Oecologia* 44:68-74.

Ziska, L.H. (2000). The impact of elevated CO_2 on yield loss from a C_3 and C_4 weed in field-grown soybean. *Global Change Biology* 6:899-905.

Climate Change and CO_2 Effects on Productivity of Danish Agricultural Systems

Jørgen E. Olesen

SUMMARY. Scenarios of climate change for Denmark suggest increases in annual mean temperature of 1 to 4°C by the end of the 21st century with an associated increase in rainfall of approximately 10%. The climatic warming and the associated increase in atmospheric CO_2 concentration will increase the productivity of agricultural crops. The increase may be slightly higher for indeterminate species, such as grass and other fodder crops, compared with cereals and other determinate species, where the duration of growth depends on temperature and daylength. The full benefits of the climatic warming requires adaptation in crop management, which at the individual crop level means changes in sowing dates, i.e., later sowings for winter cereals and earlier sowings for spring cereals. Some crop substitution will probably occur. On dairy farms more cereals will be grown due to higher productivity of the grasslands, which frees up some land for grain production. On sandy loam and loam soils spring cereals may become slightly more favorable and winter cereals slightly less favorable. *[Article copies available for a fee from The Haworth Document Delivery Service: 1-800-HAWORTH. E-mail address: <docdelivery@haworthpress.com> Website: <http://www.HaworthPress.com> © 2005 by The Haworth Press, Inc. All rights reserved.]*

Jørgen E. Olesen is affiliated with the Department of Crop Physiology and Soil Science, Danish Institute of Agricultural Sciences, P.O. Box 50, 8830 Tjele, Denmark (E-mail: JorgenE.Olesen@agrsci.dk).

[Haworth co-indexing entry note]: "Climate Change and CO_2 Effects on Productivity of Danish Agricultural Systems." Olesen, Jørgen E. Co-published simultaneously in *Journal of Crop Improvement* (Food Products Press, an imprint of The Haworth Press, Inc.) Vol. 13, No. 1/2 (#25/26), 2005, pp. 257-274; and: *Ecological Responses and Adaptations of Crops to Rising Atmospheric Carbon Dioxide* (ed: Zoltán Tuba) Food Products Press, an imprint of The Haworth Press, Inc., 2005, pp. 257-274. Single or multiple copies of this article are available for a fee from The Haworth Document Delivery Service [1-800-HAWORTH, 9:00 a.m. - 5:00 p.m. (EST). E-mail address: docdelivery@haworthpress.com].

http://www.haworthpress.com/web/JCRIP

Digital Object Identifier: 10.1300/J411v13n01_12

KEYWORDS. Atmospheric CO_2, temperature, rainfall, cereals, winter wheat, spring barley, potato, grass, dairy farm

INTRODUCTION

Climate change is expected to affect agriculture very differently in different parts of the world (Parry et al., 1999). The resulting effects depend on current climatic and soil conditions, the direction of change and the available resources and infrastructure to cope with change. Even within Europe large regional differences in response are expected with predominant positive effects on productivity in Northern areas and negative effects in the Mediterranean area (Olesen and Bindi, 2002). Denmark is located in a part of Northern Europe, where effects of anticipated climatic warming are estimated to be largely beneficial and the agricultural sector in general has a large adaptive capacity (Fenger et al., 1993).

Denmark is a member of the European Union (EU), and the trends in Danish agriculture are dominated by the EU Common Agricultural Policy (CAP). The agricultural area of the current 15 EU countries is about 135 million ha, and that of the 10 countries seeking EU membership is about 60 million ha. Agricultural land constitutes 63% of the total land area in Denmark, which is a typical proportion for Western Europe. However, 58% of the agricultural area is used for growing cereals, which is a considerably higher percentage than in other European countries. Denmark also has the highest proportion of land area suited for wheat-growing in the EU (van Lanen et al., 1992).

Danish agriculture is renowned for its large and competitive exports of food products, especially dairy products and pork. About 80% of the production of both dairy and meat products is exported, but only about 20% of the grain production is exported. The crop production is thus largely used domestically for animal feed. This dependency of Danish agriculture on dairy cows and pigs is expected to continue, possibly with an even higher focus on pig production. Under climate change, the crops in Danish agriculture will therefore also have to sustain a large animal sector, which means a large area for production of cereals for feed for pigs and of fodder crops for cattle. This paper examines the possible effects of the anticipated global warming and increasing atmospheric CO_2 concentrations on productivity of these crops.

CLIMATE AND CROP PRODUCTION

The length of the growing season in Denmark is limited by both temperature and solar radiation (Figure 1a). Low temperatures limit growth in spring,

FIGURE 1. Annual variation in mean temperature, global radiation (a), and rainfall in Denmark (b). The monthly means for temperature and rainfall are shown for two 30-year periods, 1880-1909 and 1970-1999.

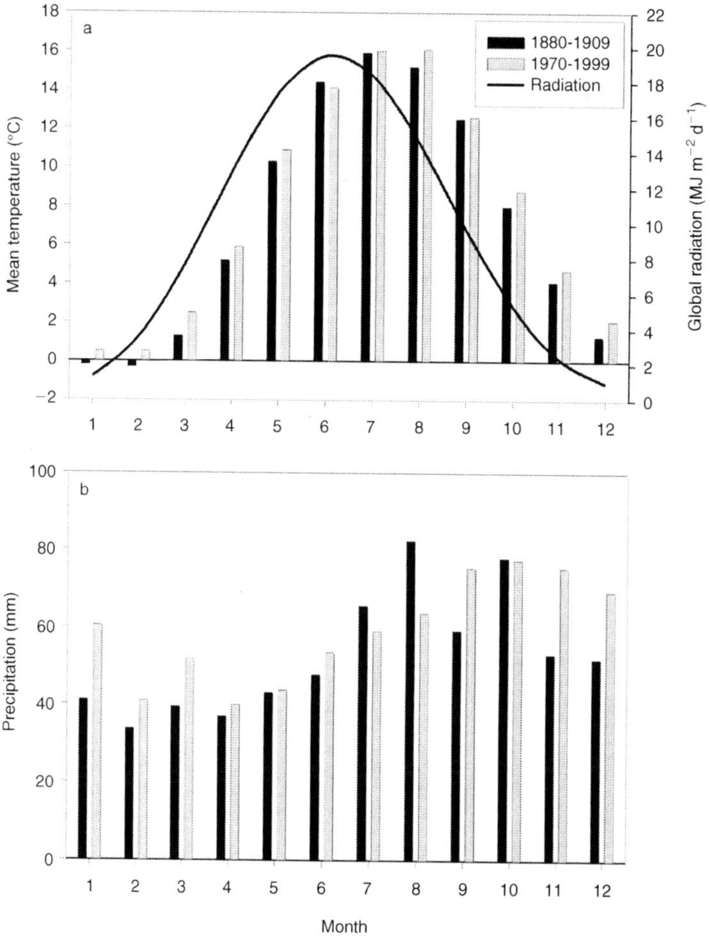

whereas solar radiation is the main limitation in autumn. A temperature increase will therefore mainly improve crop resource capture through an earlier growth start in spring.

Total annual rainfall normally exceeds potential evapotranspiration in Denmark. However, soil water deficits often develop during the summer season, where the monthly evapotranspiration is about 100 mm (Olesen, 1991). This is compensated by a water surplus during winter (Figure 1b). In general, increasing temperatures will increase the risk of drought on lighter soils during the summer season because of higher evapotranspiration. The main effects of changes in precipitation on most soil types will be on runoff during the winter season, because the capacity of soils to store water provides a buffer on many soil types against changes in seasonal water balance.

The climate in Denmark during the 20th century has become slightly warmer and wetter (Figure 1). The annual temperature has increased by 0.6°C and the precipitation by 78 mm. The increases in temperature and precipitation have mainly occurred during the autumn and winter seasons. The total duration of the growing season taken as the period with average daily temperatures above 5°C has increased by 11 days over the same period.

The global temperature increase at the end of the 21st century is projected at 1.4 to 5.8°C from Global Circulation Models (GCM) (Houghton et al., 2001). The range represents an uncertainty that mainly originates from different tracks of economic, technological, social and political development during the coming century. Scenarios for climate change in Denmark based on GCM runs indicate a similar warming, which will be most pronounced during the winter (Parry, 2000). The annual precipitation is projected to increase by about 10%. The increase in precipitation will mainly occur during the winter. The projected seasonal changes in climate thus resemble the changes that have occurred during the 20th century, albeit at a different rate.

CEREAL RESPONSES TO TEMPERATURE

Climate and CO_2 effects on agricultural crops depend on the type and use of the crop. Cereals, oilseed and protein crops including pulses are generally determinate species, and the duration to maturity depends on temperature and in many cases daylength. A temperature increase will therefore shorten the duration of the growing period and reduce yields, if management is not altered (Porter and Gawith, 1999; Tubiello et al., 2000). In contrast many forage crops are indeterminate species, where the duration of the growing season depends on the duration of periods of unfavorably hot, cold, dry or wet conditions. In Northern Europe the cold conditions during winter provide the main restriction on the growing season of forage crops.

The combined effect of changes in climatic parameters and CO_2 concentration on crop production can be evaluated using crop growth models (Rosenzweig and Hillel, 1998). Olesen et al. (2000) developed the generic CLIMCROP model to explore the effects of climate on crop production and yield of winter wheat in Denmark. The model includes a crop model and a soil model that is updated in daily time steps, and the model is also calibrated for other crop species.

Winter Wheat

The effect of methodology for simulating crop development on yield response of winter wheat to climate change was estimated by including three different crop developmental routines in the CLIMCROP model. The first method is the one used originally in CLIMCROP, which is based on the approach used in the SIRIUS model (Jamieson et al., 1998). The second method is a simplified method used in the FASSET model (Olesen et al., 2002), and the third method is based on the one used in the ARC-Wheat model (Weir et al., 1984). Methods 1 and 2 were originally calibrated for Danish conditions, and the parameters of ARC-Wheat were adjusted to give approximately similar dates of anthesis and yellow ripeness for normal sowing dates as the other methods. Both CLIMCROP and ARC-Wheat include a vernalisation response, which was set to similar duration at optimal vernalisation temperature. The FASSET model does not include a vernalisation response (Table 1).

The effect of changes in mean temperature was analysed using sets of 200 years of climate data generated by the LARS-WG stochastic weather generator (Semenov and Barrow, 1997). The LARS-WG was calibrated using observed daily climate data from Ødum (56°18′N, 10°08′E) for 1961 to 1997. A change in mean temperature was simulated by altering the present-day daily temperatures by between −2°C and +5°C in steps of 0.5°C. The response of simulated yield was calculated under rain fed conditions for two different soil

TABLE 1. Main characteristics of the three developmental models for winter wheat tested.

Phase	1–CLIMCROP	2–FASSET	3–ARC-Wheat
Sowing to emergence	T-sum (T_b = 0°C)	T-sum (T_b = 0°C)	T-sum (T_b = 0°C)
Vernalisation	Optimal at 11°C	None	Optimal at 3 to 10°C
Emergence to flag leaf	T-sum (T_b = 0°C) Final leaf number depends on daylength	Sum of temperature times daylength (T_b = 4°C)	Sum of temperature times daylength (T_b = 1°C)
Flag leaf to anthesis	T-sum (T_b = 0°C)	Sum of temperature times daylength (T_b = 4°C)	Sum of temperature times daylength (T_b = 1°C)
Anthesis to maturity	T-sum (T_b = 6°C)	T-sum (T_b = 6°C)	T-sum (T_b = 9°C)

T-sum is temperature sum or thermal time. T_b is the base temperature used in thermal time calculations.

types (coarse sand and sandy loam). The mean yield and mean date of selected growth stages were calculated for each temperature change on the basis of 200 individual years of model runs. The results of the change in temperature can be affected by the sowing date used. All scenarios were therefore run for sowing dates varying from 16 August to 31 October.

Figure 2 shows that the response to sowing date of simulated date of flag leaf emergence and anthesis were unaffected by estimation method for sowings after 15 September under the baseline climate. The ARC-Wheat method showed almost no response in either flag leaf emergence or anthesis throughout the range of sowing dates. The largest response was obtained with the FASSET model giving very early dates of flag leaf emergence with early sowings. The FASSET routine lacks a vernalization function, which for early

FIGURE 2. Response of simulated mean day of flag leaf emergence (a, b) and anthesis (c, d) in winter wheat to changes in sowing date at baseline climate (a, c) and for 2°C temperature increase (b, d). The simulations were performed with three different methods of simulating crop development.

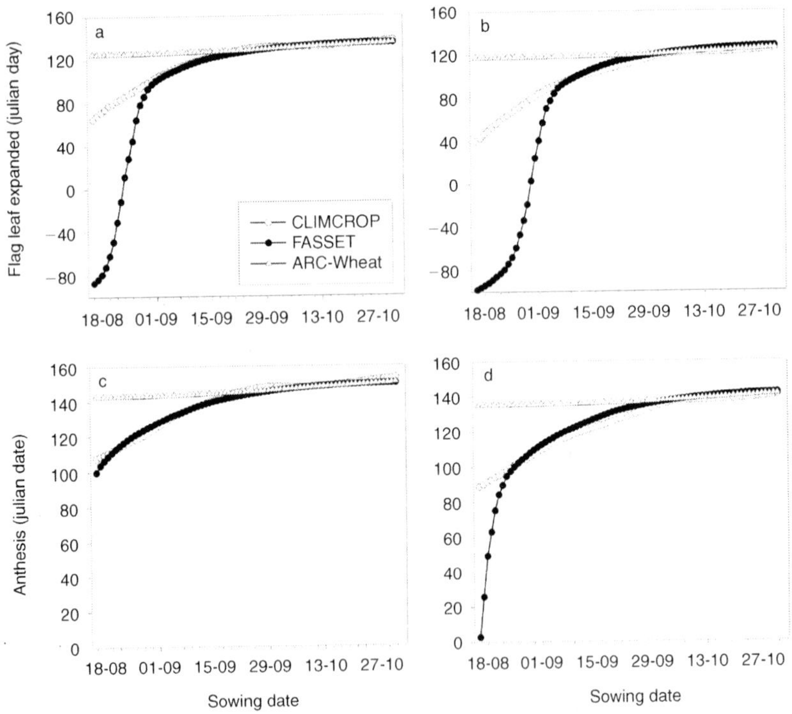

sowings results in considerable contributions to effective thermal time already in autumn. The response of the CLIMCROP development routine was intermediate to those of ARC-Wheat and FASSET. The duration of the vernalization phase in CLIMCROP and ARC-Wheat were set to be identical at optimal vernalization temperature. However, the optimal temperatures for vernalization were lower in the ARC-Wheat model. It is unlikely that the differences in the models in their description of temperature and daylength responses after fulfillment of the vernalization requirement would have caused the differences in response in Figure 2, in particular because similar dates of anthesis were obtained with the CLIMCROP and FASSET models over a wide range of sowing dates, and these models differ considerably in methodology of simulating the daylength response.

There was an optimal sowing date for high grain yield for all methods of simulating crop development (Figure 3). However, the response of grain yield to changes in sowing date was most marked for the FASSET method and least so for the ARC-Wheat method, reflecting mainly the differences in simulated dates of flag leaf emergence and anthesis. Early anthesis thus reduced yield, because the duration of resource capture by the crop was reduced. On the other hand late sowings reduced the period of crop establishment and thus the size and duration of the crop canopy, again affecting yield.

The grain yield of winter wheat at the optimal sowing date decreased with increasing temperature for all methods of estimating crop development (Figure 4). However, the average change varied between methods with a decrease on sandy soils of 0.24, 0.28, and 0.42 t DM $ha^{-1}°C^{-1}$ for FASSET, CLIMCROP and ARC-Wheat, respectively. The corresponding values for a loamy sand was 0.30, 0.34, and 0.46 t DM $ha^{-1}°C^{-1}$. The larger yield reduction for ARC-Wheat was mainly caused by the higher base temperature for grain filling used in ARC-Wheat, 9°C as opposed to 6°C for the two other methods (Table 1). The effect of a temperature increase on the duration of the grain filling period was thus a reduction of 1.8 $d°C^{-1}$ for ARC-Wheat and 1.1 $d°C^{-1}$ for the other two methods.

The optimal sowing date increased with increasing temperature for all three methods (Figure 4). However, the sowing date increased by about 4.7 $d°C^{-1}$ for CLIMCROP and ARC-Wheat, but only by 2.4 $d°C^{-1}$ for FASSET.

The results demonstrate that the response of crop development is of key importance when evaluating the effect of climate change on crop production of winter cereals. One of the most important processes to consider is the vernalization response, one of the weak points in the current understanding of crop development in winter wheat (Kirby et al., 1999; Gonzalez et al., 2002). In addition the cold tolerance of winter wheat appears to be linked to vernalization (Mahfoozi et al., 2001).

FIGURE 3. Response of simulate mean dry matter grain yield in winter wheat to changes in sowing date at baseline climate (a, c) and for 2°C temperature increase (b, d), and for a coarse sandy soil (a, b) and a sandy loam (c, d). The simulations were performed with three different methods of simulating crop development.

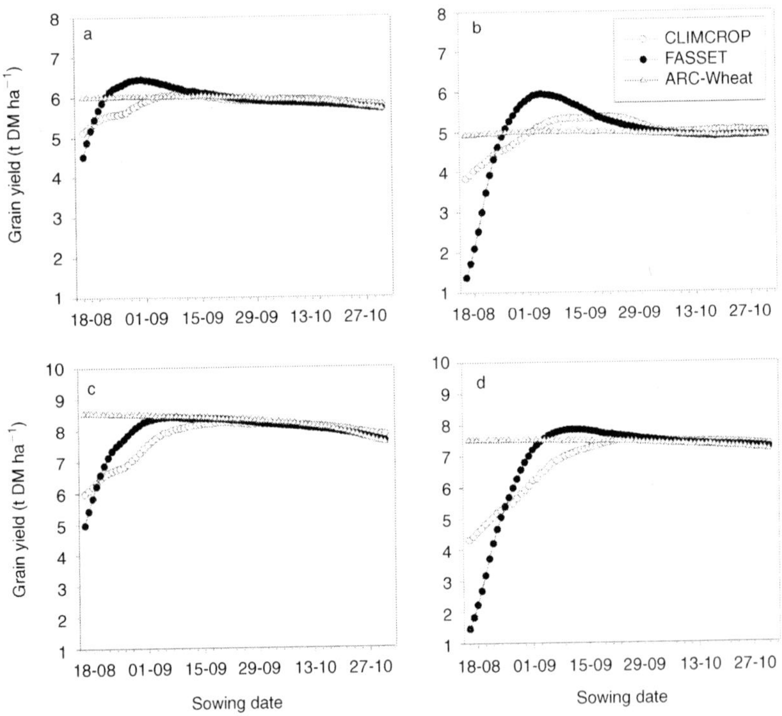

The results show that a higher vernalisation response reduces the magnitude of the yield decline with rising temperature. This indicates that varieties of winter wheat under a warmer climate should have a moderate to strong vernalization response. It would also be beneficial to increase the duration of other developmental phases, in particular the grain filling period. There is some genetic variability in these traits, and this should be exploited in the breeding of new cultivars.

Spring Barley

The effect of sowing date and temperature change on growth and development of spring barley was analysed using the CLIMCROP model and the same

FIGURE 4. Response of simulated mean dry matter grain yield, sowing date and anthesis date in winter wheat to changes in temperature for a coarse sandy soil (a, c, e) and a sandy loam (b, d, f). The results are shown for the sowing date giving the highest mean grain yield. The simulations were performed with three different methods of simulating crop development.

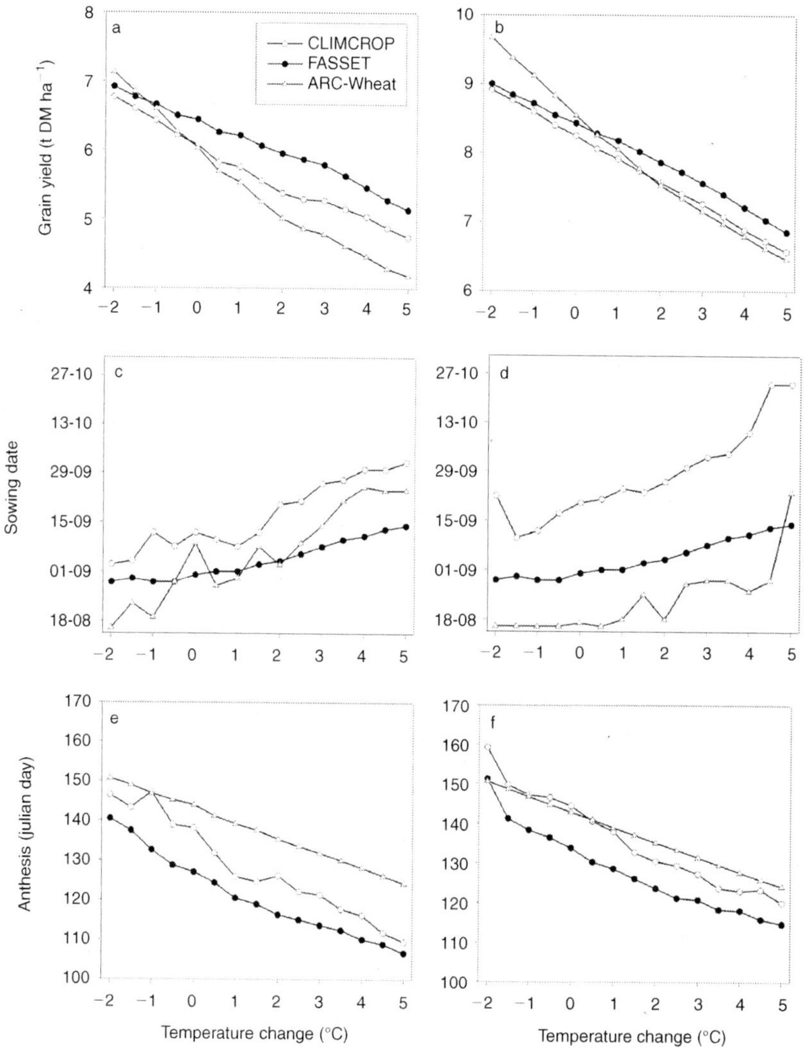

methodology as described above for winter wheat. Only the generic developmental model for spring barley of CLIMCROP was tested, but with changes in mean temperature of −2°C to +5°C in steps of 0.5°C. All temperature scenarios were run for sowing dates varying from 1 February to 30 April. Crop development of spring barley during vegetative growth in CLIMCROP is affected by both thermal time and daylength (Jensen and Rudemo, 1991).

Grain yield in spring barley declined when sowing was delayed after mid March (Figure 5). The effect was most marked on the sandy loam soil. The yield decline occurred at earlier sowing dates with increasing temperature. The decline was caused primarily by a reduction in the duration of the active growth period.

The grain yield at the optimal sowing date decreased with increasing tem-

FIGURE 5. Response of simulated mean dry matter grain yield (a, b), anthesis date (c) and maturity date (d) in spring barley to changes in sowing date at three different climate scenarios. The results for grain yield are shown for a sandy soil (a) and a sandy loam (b).

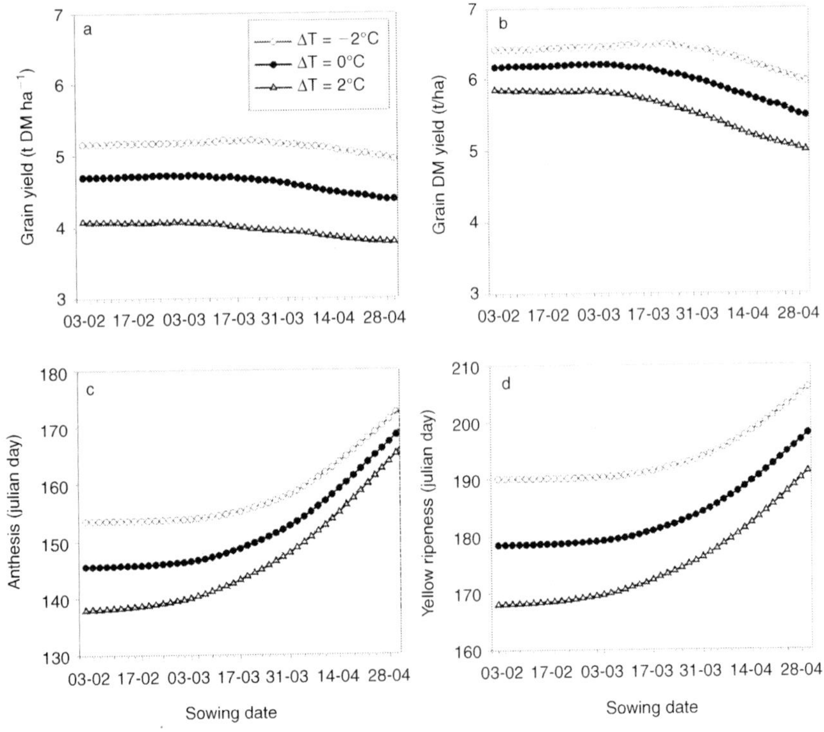

perature (Figure 6). The decrease was 0.27 and 0.19 t DM ha^{-1}°C^{-1} for coarse sand and sandy loam, respectively. The larger decrease on the coarse sandy soil is an effect of increased drought risk on this soil type caused by higher potential evapotranspiration at higher temperatures. The optimal sowing date was reduced by about 6.7 d°C^{-1} for both soil types.

The results show that a temperature increase in spring barley would cause a larger yield reduction on the coarse sandy soil than on the sandy loam. The effect is opposite for winter wheat as shown above. The yield reduction for an increase in temperature of 1°C is thus the same in winter wheat and spring barley on the coarse sandy soil. A temperature increase will therefore be more favourable for spring barley than winter wheat on the sandy loam soil.

FIGURE 6. Response of simulated mean dry matter grain yield (a) and optimal sowing date (b) in spring barley to changes in temperature for two soil types. The results are shown for the sowing date giving the highest mean grain yield.

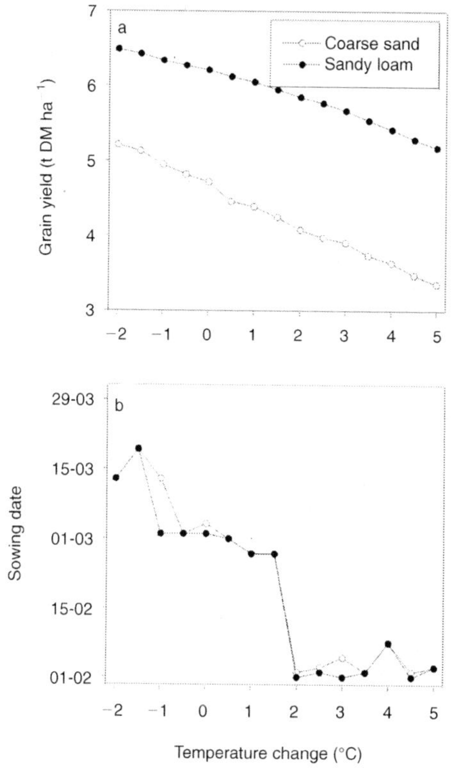

CO_2 AND TEMPERATURE RESPONSES

The yield reduction from temperature increase in cereal crops will be at least partially compensated by the beneficial effects of CO_2 enhancement. The combined effect of temperature increase and CO_2 enhancement on yield of four different crop types is shown in Figure 7. The graph is based on runs of

FIGURE 7. Response of simulated mean yield of four agricultural crops (winter wheat, spring barley, potato and grass) to changes in mean air temperature and atmospheric CO_2 concentration on a loamy sand soil in Denmark. The lines indicate contours of the percentage yield changes. The yields under baseline conditions in 1990 were 8.9 t ha^{-1} for winter wheat, 5.7 t ha^{-1} for spring barley, 63.2 t ha^{-1} for potato and 11.9 t DM ha^{-1} for grass.

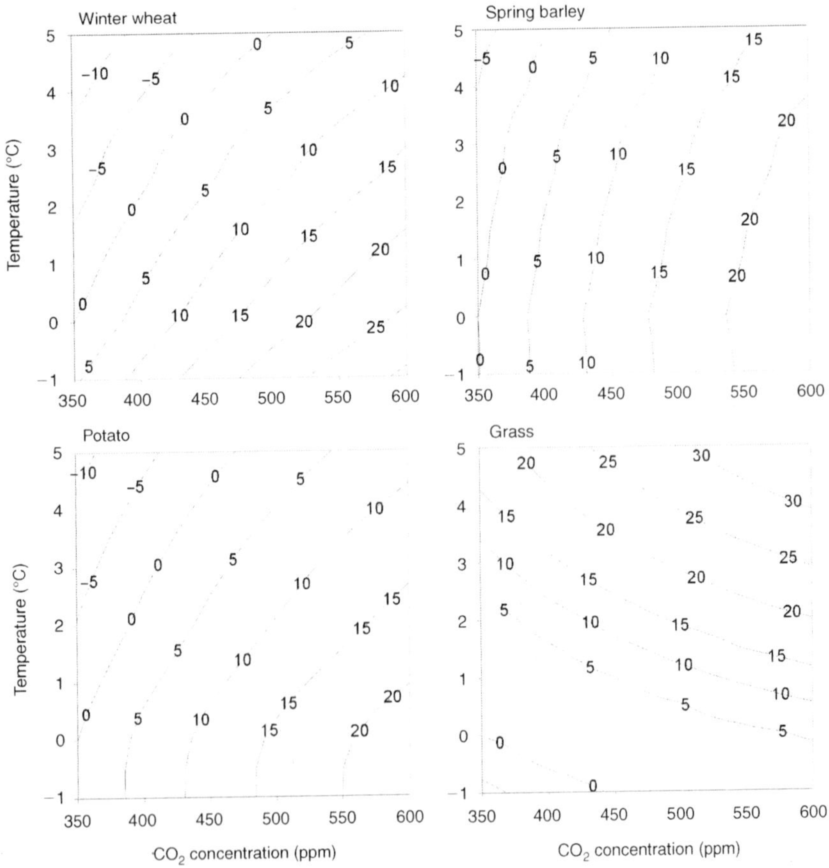

the CLIMCROP model using 200 years of generated climate data for Ødum as described above. The crops represent three determinate species (winter wheat, spring barley and potato) and one indeterminate species (grass) that is taken as an example of a forage crop. The sowing time for the determinate species was adjusted to match the optimal sowing time, i.e., later sowing for winter wheat and earlier sowing for spring barley and potato under warmer conditions.

In most cases the effect of CO_2 enhancement more than compensated for the yield reduction due to increasing temperature in all crops in Figure 7. Only in the unlikely case of a small CO_2 increase in combination with a high temperature increase was the result a net yield decline. The yield reduction caused by temperature increase was much smaller in spring barley and potatoes compared with winter wheat. The reason is that earlier sowing reduces the effective temperature increase as experienced by the crop. The earlier sowing also results in a better utilization of incoming radiation in spring (Figure 1). Grass production is favored by higher temperatures, mainly due to a longer growing season. A possible negative effect for grass is the higher associated risk of summer drought, which will affect grazing animals and may increase demands for irrigation.

Table 2 shows the sensitivity of yield of these four crops to change in either temperature or rainfall. The effect of two GCM scenarios with and without associated change in atmospheric CO_2 concentration is also shown. The results were simulated with the CLIMCROP model as described above. A temperature increase of 2°C gave a larger response in all crops than a reduction in rainfall of 20%. The low response to reduced rainfall is a consequence of the humid climate in Denmark (Figure 1), where summer drought on many loamy sand and sandy loam soils is buffered by the soil water reserves.

The two GCM scenarios in Table 2 represent transient runs with the HadCM2 GCM model for year 2050 with (HCGS) and without (HCGG) sulphate aerosols for grid cells that included the Ødum site in Denmark (Barrow et al., 2000). Both GCM scenarios resulted in a small yield decrease for all determinate crops with baseline CO_2 concentration. The projected increase in atmospheric CO_2 concentration to 515 ppm in 2050 resulted in a net yield increase of 9 to 17% depending on crop type.

CROP PRODUCTION ON A DAIRY FARM

Climate change also affects management at the farm scale, especially on farms involving livestock systems. Such farms may be influenced through different pathways, including effects on availability and price of grains for feeding, changes in productivity of pastures and forage crops, effects on animal health, growth and reproduction, and changes in requirements for housing and

TABLE 2. Simulated absolute and relative changes in mean yield of four crops for climate change compared with baseline climatic conditions in 1961-97 for a loamy sand soil in Denmark. The climatic changes were either arbitrary changes in temperature and precipitation alone, or changes in both, based on outputs from transient runs up to 2050 of the HadCM2 general circulation model including (HCGS) or excluding (HCGG) the effects on climate of sulphate aerosols. The effects on grain yield were simulated using the CLIMCROP model using both baseline CO_2 concentration (353 ppm) for the year 1990 and projected CO_2 concentration for the year 2050 (515 ppm). Yields of cereals are shown with 15% water, grass as dry matter and potato as fresh weight.

Scenario	CO_2 ppm	Winter wheat		Spring barley		Potato		Grass	
		t ha^{-1}	%	t ha^{-1}	%	t ha^{-1}	%	t ha^{-1}	%
Baseline	353	8.9		5.7		63		11.9	
Temperature +2°C	353	8.3	−7	5.6	−2	60	−5	12.3	3
Precipitation −20%	353	8.6	−4	5.7	−1	63	0	11.9	0
HCGG, 2050	353	8.5	−4	5.7	−1	59	−7	12.2	2
	515	10.1	13	6.7	17	69	9	13.4	12
HCGS, 2050	353	8.4	−6	5.5	−4	61	−4	12.2	2
	515	10.0	12	6.5	14	71	11	13.4	12

feed storage. The effects at farm level of changes in grain production or of feed-grain prices will be greatly moderated by market forces (Reilly, 1994). Effects may be larger on dairy farms, where climate effects on grasslands and forage crops will affect farm output through altered land use.

The effect of climate change on crop choice and overall productivity on a dairy farm was studied using the CLIMCROP model. The farm was assumed to have a total rotational area of 100 ha with a stocking rate of 1 LU ha^{-1} (73 dairy cows plus 80 young stock). This is slightly lower than the current average stocking rate on dairy farms in Denmark. The soil type was assumed to be coarse sand, representative of the majority of dairy farms in Denmark. The crop rotation was based on grass for grazing and silage, and on spring barley and winter wheat for whole crop silage and grain production. The animal production has an annual requirement of 240 t DM in grass, 126 t DM in cereal whole crop, 87 t cereal grain and 61 t cereal straw. In addition the farm imports molasses and rapeseed cake corresponding to 26% of the total energy requirement of the livestock. The area with grass was calculated from the simulated average grass production. The barley area was assumed to correspond to half the grass area, as a two-year rotational grass crop was established as an under-sown grass crop in spring barley. The rest of the farm area was used for winter wheat. Spring barley was preferred over the other crops for making whole crop silage. The grain production in excess of the feed requirements was exported from the farm. The climate data used was generated using the LARS weather generator as described above for winter wheat based on climatic data from Ødum in Denmark.

The response of the crop area and of the grain export to changes in temperature and precipitation and to two GCM scenarios was analyzed. The GCM scenarios were based on runs of the HADCM2 model for the year 2050 with (HCGS) and without (HCGG) sulphate aerosols. The crops were irrigated, but a maximum annual average irrigation capacity of 60,000 m^3 was applied. The irrigation water was preferentially used for the grass and forage crops.

The scenarios of temperature increase and precipitation decrease resulted in rather large decreases in grain export, because of lack of irrigation water for the cereal crops (Table 3). There were considerable differences in grain export between the two GCM scenarios. The HCGS scenario gave an increase in grain export of 24% and the HCGG scenario gave an increase of 61%. This was mainly an effect of differences in rainfall between the two scenarios.

TABLE 3. Simulated effect of climate change compared with baseline conditions in 1990 on crop area and grain export for a 100 ha dairy farm with 1 LU/ha. The climatic changes were either arbitrary changes in temperature and precipitation alone, or changes in both, based on outputs from transient runs up to 2050 of the HadCM2 general circulation model including (HCGS) or excluding (HCGG) the effects on climate of sulphate aerosols. The effects on grain yield were simulated using the CLIMCROP model. A CO_2 concentration of 515 ppm was assumed for the year 2050 (adapted from Olesen et al., 2000a).

Climate	CO_2 concentration (ppm)	Crop area (ha)			Grain export ($t\ yr^{-1}$)
		Grass	Cereal whole crop	Cereal grain	
Baseline	353	27	15	59	205
Temperature +2°C	353	26	15	59	166
Temperature +4°C	353	24	15	61	155
Precipitation +20%	353	27	15	59	316
Precipitation −20%	353	27	14	59	140
HCGG, 2050	515	24	12	64	331
HCGS, 2050	515	24	12	63	255

CONCLUSIONS

Scenarios of climate change for Denmark suggest increases in annual mean temperature of 1 to 4°C by the end of the 21st century depending on socio-economic development (Parry, 2000). Winter rainfall may increase up to 20% (Christensen and Christensen, 2000). This implies a wide range of possible consequences. The results of the HadCM2 scenarios shown in Tables 2 and 3 are within the expected range of climate change for the year 2050. Agricultural productivity in Denmark may therefore be expected to increase under increasing temperature and increasing CO_2 concentration.

The increase in productivity will increase the grain production on both arable and dairy farms. This may therefore be expected to reinforce the current trends in Danish agriculture of declining cattle population and increasing pig and cereal production. Apart from an anticipated continued decline in total agricultural area, land use will probably not be greatly affected, although spring cereals may become slightly more competitive crop and winter cereals slightly less competitive on sandy loams and loamy soils.

There are still many uncertainties associated with the estimates of climate change on crop production. The major uncertainties are concerned with the possible range of climate changes considered and the possible technological progress over the coming century. However, there are also uncertainties in the models used to simulate climate change impacts, e.g., the way crop growth and development is simulated, as illustrated here for crop development in winter wheat. An even larger uncertainty is, however, associated with the effect of secondary climate change factors, such as soil fertility, weeds, pests and diseases, that also affect crop production. All these factors need to be considered in future research on climate change impacts on agricultural crop production.

REFERENCES

Barrow, E.M., M. Hulme, M.A. Semenov, and R.J. Brooks. (2000). Climate change scenarios. In *Climate change, climate variability and agriculture in Europe: An integrated assessment*, eds. T.E. Downing, P.A. Harrison, R.E. Butterfield, and K.G. Lonsdale (eds.). Environmental Change Unit, University of Oxford, UK, pp. 11-27.

Christensen, J.H. and O.B. Christensen. (2000). Regional climate scenarios-A study in precipitation. In *Climate Change Research. Danish Contributions*, eds. A.M.K. Jørgensen, J. Fenger, and K. Halnæs, Danish Climate Centre, Danish Meteorological Institute, Denmark, pp. 151-166.

Fenger, J., A.M.K. Jørgensen, H.E. Mikkelsen, and M. Philipp. (1993). Greenhouse effect and climate change-implications for Denmark. *Ambio* 22: 378-385.

Gonzalez, F.G., G.A. Slafer, and D.J. Miralles. (2002). Vernalization and photoperiod responses in wheat pre-flowering reproductive phases. *Field Crops Research* 74: 183-195.

Houghton, J.T., Y. Ding, D.J. Griggs, M. Noguer, P.J. van den Linden, X. Dai, K. Maskell, and C.A. Johnson (eds.). (2001). *Climate change 2001: The scientific basis. Contribution of Working Group I to the Third Assessment Report of the Intergovernmental Panel on Climate Change.* Cambridge University Press, Cambridge, UK.

Jamieson, P.D., M.A. Semenov, I.R. Brooking, and G.S. Francis. (1998). Sirius: a mechanistic model of wheat response to environmental variation. *European Journal of Agronomy* 8: 161-179.

Jensen, J.E., and M. Rudemo. (1991). First passage time models for prediction of developmental stages. *Biometrics Journal* 1: 3-16.

Kirby, E.J.M., J.H. Spink, D.L. Frost, R. Sylvester-Bradley, R.K. Scott, M.J. Foulks, R.W. Clare, and E.J. Evans. (1999). A study of wheat development in the field: analysis by phases. *European Journal of Agronomy* 11: 63-82.

Mahfoozi, S., A.E. Limin, and D.B. Fowler. (2001). Influence of vernalization and photoperiod responses on cold hardiness in winter cereals. *Crop Science* 41: 1006-1011.

Olesen, J.E. (1991). *Jordbrugsmeteorologisk årsoversigt 1990.* Tidsskrift for Planteavls Specialserie S2130, Research Centre Foulum, Denmark.

Olesen, J.E., and M. Bindi. (2002). Consequences of climate change for European agricultural productivity, land use and policy. *European Journal of Agronomy* 16: 239-262.

Olesen, J.E., T. Jensen, and J. Petersen. (2000). Sensitivity of field-scale winter wheat production in Denmark to climate variability and climate change. *Climate Research* 15: 221-238.

Olesen, J.E., B.M. Petersen, J. Berntsen, S. Hansen, P.D. Jamieson, and A.G. Thomsen. (2002). Comparison of methods for simulating effects of nitrogen on green area index and dry matter growth in winter wheat. *Field Crops Research* 74: 131-149.

Parry, M.L. (ed.). (2000). *Assessment of potential effects and adaptations for climate change in Europe: The Europe ACACIA project.* Jackson Environment Institute, University of East Anglia, Norwich, UK.

Parry, M.L., C. Rosenzweig, A. Iglesias, G. Fischer, and M. Livermore. (1999). Climate change and world food security: a new assessment. *Global Environmental Change* 9: 51-67.

Porter, J.R., and M. Gawith. (1999). Temperatures and the growth and development of wheat: a review. *European Journal of Agronomy* 10: 23-36.

Reilly, J. (1994). Crops and climate change. *Nature* 367: 118-119.

Rosenzweig, C., and D. Hillel. (1998). *Climate change and the global harvest. Potential impacts of the greenhouse effect on agriculture.* Oxford University Press, New York.

Semenov M.A., and E.M. Barrow. (1997). Use of stochastic weather generator in the development of climate change scenarios. *Climate Change* 35: 397-414.

Tubiello, F.N., M. Donatelli, C. Rosenzweig, and C.O. Stockle. (2000). Effects of climate change and elevated CO_2 on cropping systems: model predictions at two Italian locations. *European Journal of Agronomy* 13: 179-189.

van Lanen, H.A.J., C.A. van Diepen, G.J. Reinds, G.H.J. de Koning, J.D. Bulens, and A.K. Bregt. (1992). Physical land evaluation methods and GIS to explore the crop growth potential and its effects within the European Communities. *Agricultural Systems* 39: 307-328.
Weir, A.H., P.L. Bragg, J.R. Porter, and J.H. Rayner. (1984). A winter wheat crop simulation model without water and nutrient limitations. *Journal of Agricultural Science, Cambridge* 102: 371-382.

Interactive Effects of CO_2 and O_3 on the Growth of *Trisetum flavescens* and *Trifolium pratense* Grown in Monoculture or a Bi-Species Mixture

C. Haldemann
J. Fuhrer

SUMMARY. Golden oat grass (*Trisetum flavescens* L.) and red clover (*Trifolium pratense* L.) were grown as monocultures or bi-species mixtures under controlled conditions and exposed to ambient (350 ppm) or elevated (580 ppm) CO_2, with or without addition of O_3 (diel profile with 150 ppb maximum). Shoot biomass measurements after the initial growth and two re-growth periods were used to determine the specific responses of both species, and the difference in the specific response between monocultures and mixtures. *T. pratense* was much more responsive to CO_2, O_3, and their combination, compared to *T. flavescens*. In the case of O_3 but not of CO_2, the difference in sensitivity between species

---

C. Haldemann is affiliated with the Swiss Federal Research Station for Agroecology and Agriculture (FAL), Air Pollution/Climate Group, CH-8046 Zurich, Switzerland.

J. Fuhrer is affiliated with the Swiss Federal Research Station for Animal Production (RAP), CH-1725 Posieux, Switzerland (E-mail: juerg.fuhrer@fal.admin.ch).

Address correspondence to: J. Fuhrer at the above address.

The authors thank R. Perler for technical help during this study.

[Haworth co-indexing entry note]: "Interactive Effects of CO_2 and O_3 on the Growth of *Trisetum flavescens* and *Trifolium pratense* Grown in Monoculture or a Bi-Species Mixture." Haldemann, C., and J. Fuhrer. Co-published simultaneously in *Journal of Crop Improvement* (Food Products Press, an imprint of The Haworth Press, Inc.) Vol. 13, No. 1/2 (#25/26), 2005, pp. 275-289; and: *Ecological Responses and Adaptations of Crops to Rising Atmospheric Carbon Dioxide* (ed: Zoltán Tuba) Food Products Press, an imprint of The Haworth Press, Inc., 2005, pp. 275-289. Single or multiple copies of this article are available for a fee from The Haworth Document Delivery Service [1-800-HAWORTH, 9:00 a.m. - 5:00 p.m. (EST). E-mail address: docdelivery@haworthpress.com].

http://www.haworthpress.com/web/JCRIP

Digital Object Identifier: 10.1300/J411v13n01_13

was larger in mixture than in monoculture. In contrast to elevated CO_2, O_3 significantly reduced the root:shoot ratio in the mixture, which could explain the increasing negative effect of O_3 on clover with progressing harvests.

The relative CO_2 stimulation of *T. pratense* and of the cumulative mixture shoot biomass was larger in the presence than in the absence of O_3, which was due to an almost complete protection from O_3 stress by elevated CO_2. In the mixture, the fraction of *T. flavescens* was small and increased during the experiment; this increase was most pronounced with O_3, but any change in mixture biomass was dominated by the response of *T. repens*.

The results confirm that in grass/legume mixtures legumes are most sensitive to elevated CO_2 and O_3, but the magnitude of specific responses depends on canopy structure and of plant development. Elevated CO_2 minimizes the negative impacts of O_3 stress on above- and below-ground plant growth.

KEYWORDS. Carbon dioxide, ozone, golden oat, red clover, root and shoot growth

INTRODUCTION

The concentration of CO_2 in the atmosphere has risen from pre-industrial levels of about 270 ppm to currently 370 ppm, and a further rise is predicted for the twenty-first century (IPCC, 2001). This increase in CO_2 is expected to affect the biosphere in general, and more specifically agricultural production systems. The majority of studies with agricultural crops tend to suggest that effects of elevated CO_2 on crop growth and yield are positive (Bazzaz, 1990, Idso et al. 1994). However, the CO_2 effect may be altered by limiting environmental conditions (e.g., Meier & Fuhrer, 1997); thus, yield stimulations observed in experiments under non-limiting nutrient and water supply may not be attained under realistic field conditions. Tropospheric ozone (O_3) is an additional factor which at elevated concentrations may modify the CO_2 stimulation. Air pollution by O_3 is of prime importance in many regions, particularly in the northern hemisphere (Chameides et al., 1994). Current levels of O_3 in Europe, the USA and elsewhere are sufficiently high to cause crop losses (Fuhrer et al., 1997).

According to current understanding, plant biomass production is largely protected from toxic O_3 effects in a CO_2-rich atmosphere. This is mainly due to ozone exclusion (Allen, 1990), but there is some evidence that treatment with elevated CO_2 also increases the capacity for detoxification (Heagle et al. 1999). When applied in combination, elevated CO_2 can compensate at least for part of the negative effects of elevated O_3, and the positive effect of elevated CO_2 on yield can be larger in an atmosphere containing elevated levels of O_3 relative to the effect without O_3 (Heagle et al., 1998). Results from a number of studies underline the importance of the interaction between the effects of the two gases, but for quantitative risk assessments the possible variability in response patterns between species and cultivars must be considered.

In contrast to arable crops, grasslands have received little attention with respect to interactive effects of elevated CO_2 and O_3. The situation in multi-species grassland communities is more complex than in monocultures of arable crops; the effects of the two gases and their interaction may differ substantially between co-occurring species (Porter, 1993; Volin et al. 1998), which, in turn, may lead to a shift in the floristic composition of the plant community in an changed atmosphere. Shifts in species mixtures have been shown for pastures exposed to elevated O_3 (Fuhrer et al., 1994), or to elevated CO_2 (Hebeisen et al., 1997). A reduction in the clover fraction in pastures may have important negative consequences for forage quality (Muntefering et al., 2000). The issue is further complicated when grasslands are cut or grazed. In the case of cutting, effects of pollutants may differ between the initial growth and the re-growth periods. For clover it has been shown that O_3 effects on growth are larger during re-growth because of reduced storage of resources necessary for re-growth (cf. Fuhrer, 1997). Effects of O_3 may also depend on the cutting regime, for instance because of a change in O_3 penetration (Ashmore & Ainsworth, 1995). Overall, available information suggests that effects of a change in the atmospheric composition on grassland species depend on the specific physiological and morphological characteristics of the competing species, and also on management.

In order to gain a better insight into some of the possible interactions, the aim of this study was to investigate under controlled conditions the effects of CO_2 and O_3, singly and in combination, on two species grown in monoculture or in bi-species mixture during three growth periods. Red clover (*Trifolium pratense* L.) and golden oat grass (*Trisetum flavescens* L.) were chosen as typical species of managed permanent grasslands in pre-alpine and alpine regions; they also represent different functional groups, and their shoot architecture differs considerably.

MATERIALS AND METHODS

Plant Material and Growth Conditions

Seeds of red clover (*Trifolium pratense* L. cv. Temara) and golden oat grass (*Trisetum flavescens* L. cv. Trisett) were germinated in moist quartz sand. After emergence seedlings were transplanted as monocultures or mixtures into 16 cylindrical pots (monoliths with 0.2 m diameter, 0.8 m height) (Meier & Fuhrer, 1997). The lower 50 cm of the pots were filled with washed quartz sand, and the top 30 cm with top soil (60% humus, 23% sand, 10% peat; pH_{H_2O} 6.5-7.0). Monoliths were covered with plastic lids (2 mm) containing 46 holes (12mm diameter) in which the seedlings were planted. Plant density was equivalent to 1500 plants m^{-2}, and in the mixtures the clover:grass planting ratio was about 1:2. One and 10 days after transplanting the seedling to the pots, rhizobium bacteria were applied to clover roots.

All experiments were carried out in a controlled environment growth chamber with a 14-h photoperiod. The arrangement of the units followed a completely randomized design. Illumination was provided by halogen-metal lamps giving a PPFD of 360 $\mu mol\ m^{-2}\ sec^{-1}$ at the top of the cuvettes. Air temperature followed a diurnal cycle with a maximum of 25°C at noon and a minimum of 15°C at midnight. Each pot was irrigated by means of a peristaltic pump (Ismatec, Glattbrugg, Switzerland) with 100 ml deionized water per day. N-fertilization with ammonium nitrate (NH_4NO_3) was applied three times per growth period, with increasing amounts of N according to the to the estimated biomass. The total N application per growth period was equivalent to 100 kg N ha^{-1}. Solutions for irrigation and fertilization were distributed through a pierced tube placed on the surface of the soil column.

Treatments

Cylindrical plexiglass cuvettes (0.75 m height) were fitted to the top of the pots (Meier & Fuhrer, 1997). Horizontal tubes at the top and at the bottom were used as air inlet and outlet, respectively. Ozone-free air was passed through a CO_2 molecular-sieve, reduced to a pressure of 2 bar and flow-adjusted by a mass-flow controller before being distributed to four separate mixing boxes (DMP, Hegnau-Volketswil, Switzerland) for each of the following treatments:

Control:	CO_2 = 350 ppm	O_3 = 0 ppb
CO_2:	CO_2 = 580 ppm	O_3 = 0 ppb
O_3:	CO_2 = 350 ppm	O_3 = 150 ppb, with a diel pattern passing through the maximum at noon and with 0 ppb during the dark period
$O_3 \times CO_2$:	CO_2 = 580 ppm	O_3 = 150 ppb (same as in the O_3 treatment)

Pure CO_2 from a gas cylinder and O_3 generated from pure oxygen by electrical discharge (Fischer ozone generator 500M, Meckenheim-Bonn, Germany) were injected into the air inside the mixing box of the respective treatment using mass flow controllers. The air from each mixing box was sub-divided to obtain four replicates of each treatment. Air flow was measured with electronic flow meters. The pressure in the cuvettes was kept slightly above atmospheric pressure to reduce possible effects of leaks. To reduce boundary layer resistance inside the cuvettes the air was stirred with propellers (500 rpm) placed at the top of the canopy.

CO_2 concentration at the inlet of the cuvettes was measured with an IR CO_2 analyzer (Ultramat 5E, Siemens, Germany), and water content with a the dew-point hygrometer (Hygro-M2 monitor, General Eastern, Watertown, MA, USA). Concentrations of O_3 were measured in the inlet tubes with a Dasibi, 1008 RS ozone monitor (Glendale, CA, USA). Gas concentrations in each cuvette were measured using a time-share system with 5-min intervals. O_3 fumigation, monitoring, and data acquisition were computer-controlled (Labview 4, National Instruments, Austin, TX, USA).

Harvests

Plants were cut 4 cm above the soil surface 29, 45, and 61 days after transplanting the seedlings to the exposure units. The shoot material was separated by species. At the end of the experiment, roots were separated from the soil by a washing procedure (Meier & Fuhrer, 1997) to determine total root biomass. All biomass weights was determined after drying at 80°C for 24 h. Data are expressed in units of dry weight per experimental unit.

Statistical Analysis

Treatment mean differences in specific biomass data from individual harvests were tested for significance using a multiple comparison test (Fishers LSD). Normality of the data was tested prior to analysis. Analysis of variance (ANOVA) was used to test for the effects of treatments and their interactions on cumulative shoot and root biomass data. The statistics were calculated using NCSS 2000 statistical software (Kaysville, UT).

RESULTS

The two species produced similar amounts of shoot biomass when grown in monoculture and exposed to the control atmosphere, and the differences between the three harvests were small (Figure 1). The elevated O_3 treatment affected shoot biomass in both species to a different extent. Elevated O_3 caused a reduction in biomass in *T. pratense* as compared to the control which was significant in the case of harvest 2 and harvest 3. In *T. flavescens* the negative effect of O_3 was only significant for harvest 1. Elevated CO_2 had no effect on *T.*

FIGURE 1. Shoot biomass of *T. flavescens* and *T. pratense* grown in monoculture or mixture and harvested three times (mean ± SE, $n = 4$). Different letters indicate significant differences between treatments for each harvest separately (Fisher LSD contrasts, $P < 0.05$).

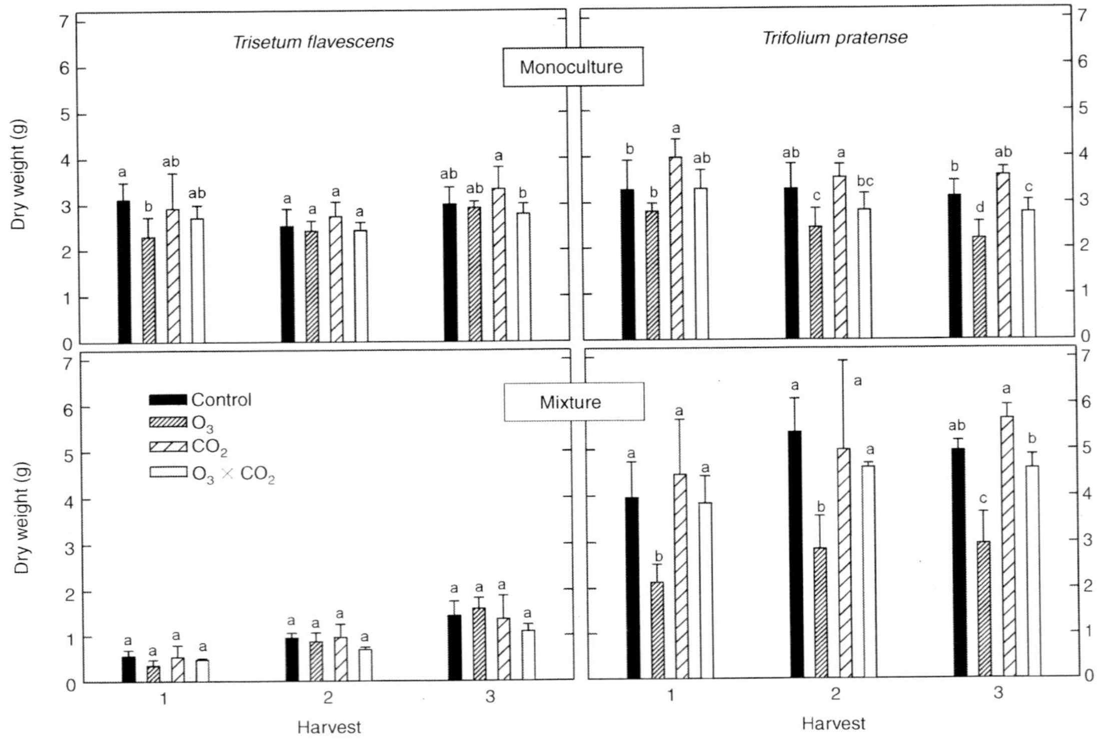

flavescens but caused a stimulation of biomass in *T. pratense* which was significant during the first growth period. In the $O_3 \times CO_2$ treatment, biomass production never differed significantly from the control, except for the last harvest in the $O_3 \times CO_2$ treatment.

In mixture with *T. pratense*, growth of *T. flavescens* was strongly reduced, compared to growth in monoculture, but increased with progressing harvests (Figure 1c). No significant treatment effects were detected. In contrast, *T. pratense* in the control treatment produced more biomass than in monoculture (Figure 1d), and the negative effect of O_3 was always significant and larger than in monoculture. Conversely, the effect of elevated CO_2 was never significant. Also, in the combined gas treatment, *T. pratense* biomass was not different from the control.

Total mixture shoot biomass increased with progressing harvests in all treatments (Figure 2a). Because of the dominance of *T. pratense* in the mixture, treatment effects on total were similar to those observed in *T. pratense*. The biomass ratio of *T. flavescens* to *T. pratense* in the mixture also increased from harvest 1 to harvest 3 (Figure 2b). This increase was largest in the O_3 treatment, thus indicating a stronger increase in *T. flavescence* relative to *T. pratense* with progressing duration of the experiment. This pronounced O_3 effect was absent in the presence of elevated CO_2, which by itself had no significant effect on the species ratio.

Over three growth periods, specific cumulative biomass in the monocultures for *T. flavescens* was similar to the biomass of *T. pratense*, whereas in the mixture, *T. pratense* was stimulated at the expense of *T. flavescens* (Figure 3). In the monocultures, O_3 reduced the biomass of both species, while CO_2 tended to have the opposite effect. In the $O_3 \times CO_2$ treatment, shoot biomass was equal to the control. In the mixture, no effects of CO_2 and O_3 on *T. flavescens* were found, in contrast to *T. pratense* in which O_3 reduced the cumulative biomass by 44%. The ratio of biomass in mixture to biomass in monoculture was unaffected by the treatments in *T. flavescens*, but significantly reduced in *T. pratense*.

Cumulative mixture shoot and root biomass, and their ratio, differed between treatments (Figure 4). The most pronounced effect was observed in the O_3 treatment which significantly reduced shoot and root biomass, and the root:shoot ratio. The effect of CO_2 was not significant for shoot and root biomass, and root:shoot ratio. ANOVA revealed a significant treatment effect for roots and shoots ($P < 0.001$), but no significant interaction of CO_2 and O_3 (Table 1).

DISCUSSION

The two species selected for this study differed in their sensitivity to elevated CO_2. In *T. pratense*, cumulative shoot biomass in monoculture in-

FIGURE 2. Total shoot biomass of mixtures of *T. flavescens* and *T. pratense* (a), and species biomass ratio (b) for each harvest (mean ± SE, $n = 4$). Different letters indicate significant differences between treatments for each harvest separately (Fisher LSD contrasts, $P < 0.05$).

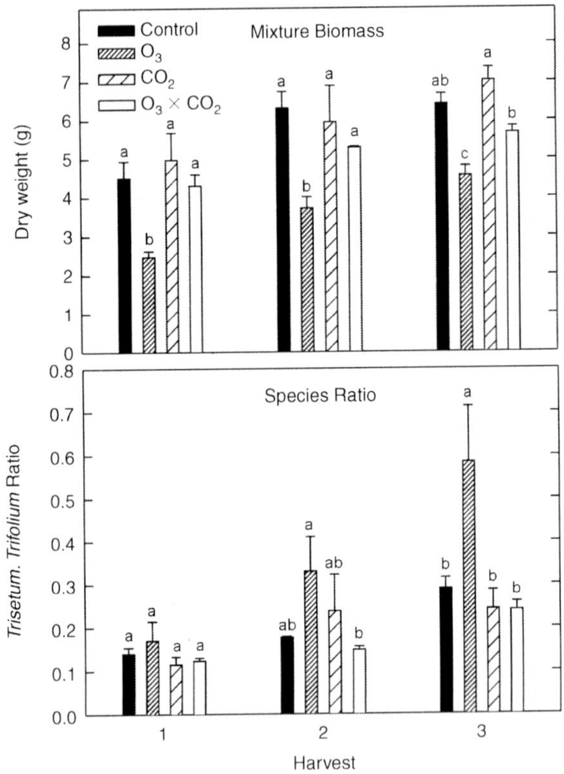

creased by about 14% at elevated CO_2, compared to the control treatment, while in *T. flavescens* the effect was only 4% and non-significant (Figure 3). It has been demonstrated that legumes generally respond more strongly to elevated CO_2 than grasses (Hebeisen et al., 1997; Lüscher et al., 1998). In the short run, this can be attributed to increased symbiotic fixation of N_2 in legumes which provides the extra N necessary to cope with additional C supply under elevated CO_2 (Zanetti et al., 1998). In contrast, grasses are restricted by N, and only in mixtures with legumes the necessary extra soil N becomes available through the N input by symbiotic N_2 fixation by the legumes. (Zanetti et al., 1997; Lüscher et al., 2000). In the mixture, biomass production in *T. pratense* was less stimulated by elevated CO_2 than in monoculture, al-

FIGURE 3. Cumulative shoot biomass of *T. flavescens* and *T. pratense* grown in monoculture or mixture (a), the ratio between the cumulative biomass produced in mixture and monoculture for each species (b) (mean ± SE, $n = 4$). Different letters indicate significant differences between treatments for each harvest separately (Fisher LSD contrasts, $P < 0.05$).

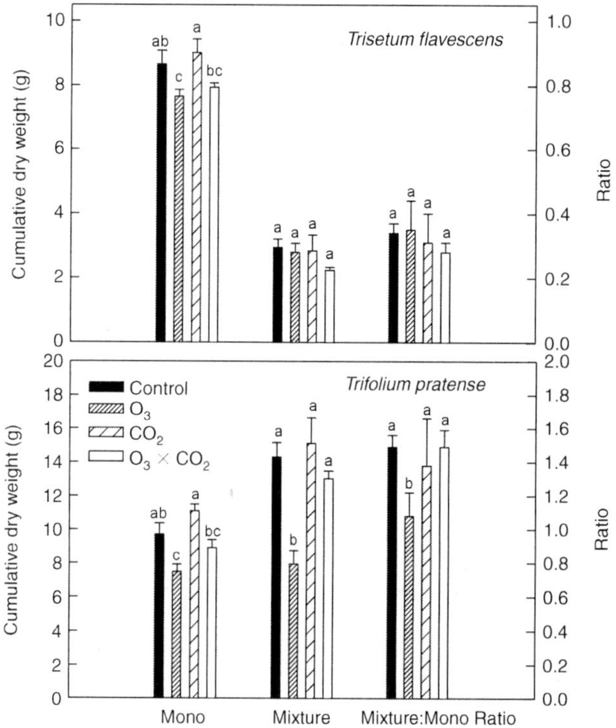

though the mixture response corresponded to the response of the dominating *T. pratense*, in agreement with data presented by Navas et al. (1999). It has been well been recognized that competition can alter the response of plant species to elevated CO_2 (Navas, 1998). Some species are more responsive to CO_2 when grown in monculture, whereas the reverse is true for other species (cf. Navas et al., 1999). Here, *T. pratense* in monoculture produces a denser canopy than in mixture with the tall-stature grass species. It could be expected that this would favor the CO_2 responsiveness of *T. pratense*. However, the opposite was found in the present experiment. Reduced competition for light in the less dense canopy of the mixture could have reduced the CO_2 responsiveness in *T. pratense*. This interpretation would agree with the notion that the relative

CO_2 stimulation is less under favorable conditions, than in situations with limiting resources (Idso & Idso, 1994).

In terms of the specific cumulative shoot biomass, *T. pratense* was more affected by elevated O_3 than *T. flavescens* (Figure 3). This difference in O_3 sensitivity between the two species confirms earlier findings. Bungener et al. (1999) reported that the growth response to O_3 was much stronger in *T. pratense* than *in T. flavescence*, and Nussbaum et al. (2001) related the inter-specific difference in sensitivity to differences in stomatal O_3 uptake. It is well known that the amount of O_3 absorbed by leaves determines the effects of O_3 on plants, and that species with lower O_3 uptake rates may be partially protected from O_3. The sensitivity to O_3 of *T. flavescens* observed here was even lower in the mixture than in monoculture, which could be attributed to the stronger shading or better protection from O_3 penetration in the presence of *T. pratense* plants. In agreement, *T. flavescens* responded more strongly in mixture with a low stature species (*Centaurea jacea*) than in mixture with *T. pratense* (Nussbaum et al., 2000). In contrast, the response of *T. pratense* to O_3 was larger in the mixture than in monoculture, which could be related to reduced self-shading and/or increased penetration of O_3 into the less dense canopy. However, in agreement with Nussbaum et al. (2000), *T. flavescens* could not benefit from the reduction in *T. pratense* biomass under O_3 stress. These observations underline that effects of CO_2 and O_3 on species mixtures not only depend on the characteristics of the individual species involved, but also on the canopy structure of the mixture. When extrapolating these results to real plant communities it could thus be concluded that the CO_2 response of *T. pratense* is stronger in a dense canopy, in contrast to the O_3 response. The latter would be favored by less self-shading and less restricted O_3 penetration into the canopy, and that effects of CO_2 and/or O_3 on grassland mixtures may be dominated by the response of the legumes.

The absence of a significant effect of CO_2 on the root:shoot ratio is in agreement with observations made earlier for grass/clover mixtures under non-limiting water and nutrient availability (Meier & Fuhrer, 1997), or native herbs (Ferris & Taylor, 1993), and it suggests that roots respond proportionally, and that biomass partitioning is not altered (Figure 4). However, the effect of elevated CO_2 on root:shoot ratio is highly variable among species and study conditions (Rogers et al., 1996). In one study with 27 herbaceous species, root:shoot ratio decreased in 14 species, increased in 6 species, and remained unchanged in 7 species under elevated CO_2 (Hunt et al., 1991). Thus, the results obtained here should not be generalized and extrapolated beyond the conditions of the experiment. In contrast to CO_2, O_3 caused a strong reduction in root:shoot ratio, suggesting preferential partitioning of biomass to shoots rather than roots. This is a common effect of long-term O_3 exposure (Cooley & Manning, 1987), and in functional terms it may reflect the shift in partitioning

FIGURE 4. Cumulative shoot biomass, final root biomass, and the root:shoot ratio in mixtures of *T. flavescens* and *T. pratense* after three growth periods (mean ± SE, n = 4). Different letters indicate significant differences between treatments for each harvest separately (Fisher LSD contrasts, $P < 0.05$).

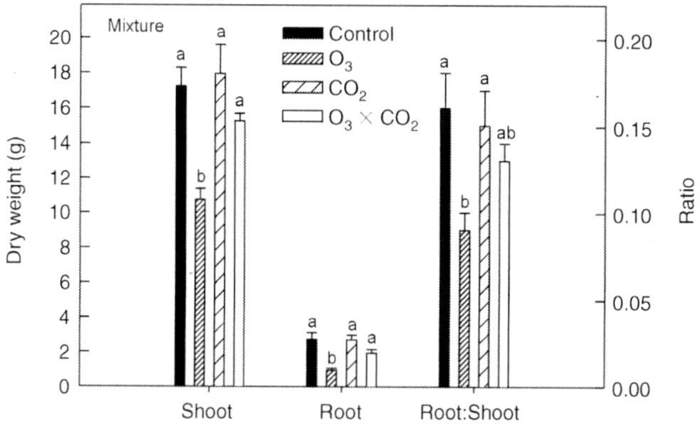

in response to reduced C supply due to O_3 impacts on photosynthesis. In the present study, the effect is dominated by the response of *T. pratense* because of the larger fractional biomass. A reduction in partitioning towards roots in this species is detrimental because of the reduction in stored resources (cf. Fuhrer, 1997). The negative impact of O_3 on *T. pratense* was small for harvest 1, and increased with progressing harvests (Figure 1). This confirms the notion that under O_3 stress reduced remobilization of reserves after a cut negatively affects re-growth and is the main mechanism leading to long-term changes in the grass:clover ratio in pastures (Fuhrer et al., 1994). In mixtures, this effect could favor less sensitive species, as indicated here by the stronger increase in the fractional biomass of *T. flavescens* during the re-growth periods as compared to the control. In the long run this could lead to shifts in species dominance (Fuhrer et al., 1994). In the $O_3 \times CO_2$ treatment, the negative effect of O_3 on root:shoot ratio was partly reversed, reflecting the protective effect of elevated CO_2 from O_3 effect.

The effect of elevated CO_2 on plant growth is modified in the presence of O_3. For instance, in the absence of O_3, cumulative shoot biomass in *T. pratense* was increased by 14% under elevated CO_2 relative to the control, but by some 20% in the $O_3 \times CO_2$ treatment relative to O_3 alone (Figure 3). In the mixture, stimulation of *T. pratense* and cumulative mixture biomass in the $O_3 \times CO_2$ treatment was 64% and 71%, respectively, relative to the treatment with O_3 alone. These results are in agreement with findings for wheat, cotton and soybean (Heagle et al., 1998, 1999, 2000). They clearly demonstrate that

TABLE 1. Statistical analysis[a] of the effects of CO_2 and O_3 treatments on cumulative shoot and root biomass of *T. flavescens*/*T. pratense* mixtures.

Factor	Shoot	Root	Root:Shoot
O_3	***	***	**
CO_2	*	n.s.	n.s.
$O_3 \times CO_2$	n.s.	n.s.	n.s.

[a] ANOVA: n.s. = not significant; * $P < 0.05$; ** $P < 0.01$; $P < 0.001$

in a sensitive species, such as *T. pratense*, effects of elevated CO_2 on growth and productivity depend on the level of O_3. Relative to the control with no O_3 and with ambient CO_2, cumulative mixture biomass was reduced by 38% by O_3 alone, but only by 12% in the $O_3 \times CO_2$ treatment, thus indicating that elevated CO_2 provides protection from O_3 stress (Figure 4). It is well known that elevated CO_2 may reduce O_3 sensitivity in a range of species by reducing stomatal conductance, and hence O_3 uptake (Fiscus et al., 1997; McKee et al., 2000). This was confirmed in the present experiment by canopy gas exchange measurements (data not shown). Although the possibility exists that elevated CO_2 increases the potential for cellular detoxification of absorbed O_3 (Polle & Pell, 1999), direct evidence is largely lacking. The results of the present study support the notion that reduced O_3 uptake in elevated CO_2 atmospheres is the more important mechanism.

In summary, the results of this study lead to the following conclusions:

- Effects of elevated CO_2 are modified by O_3 stress; the relative CO_2 stimulation of shoot and root biomass is larger in the presence of O_3, and the negative effects of O_3 are absent in the presence of elevated CO_2.
- Effects of CO_2 and O_3 on shoot biomass production are more pronounced in *T. pratense* than in *T. flavescens*; the low O_3 sensitivity of the latter species coincides with a lower specific responsiveness to elevated CO_2, particularly when grown in mixture with *T. pratense*. The mixture biomass response is determined by the response of the dominating species (*T. pratense*).
- O_3 stress strongly reduces the root:shoot ratio in the mixture, with possible consequences for re-growth of *T. pratense*, whereas elevated CO_2 has no effect.
- The results indicate that grassland species differ in their sensitivity to CO_2 and O_3, and the specific responses differ between monocultures and mixtures, and between harvests. CO_2 effects on both species tend to be larger in monocultural stands, whereas the opposite applies to effects of O_3.

REFERENCES

Allen, L.H. (1990). Plant responses to raising carbon dioxide and potential interactions with air pollutants. *Journal of Environmental Quality* 19:15-34.

Ashmore, M.R., Aisnworth, N. (1995). The effects of cutting on the species composition of artificial grassland communities. *Functional Ecology* 9:708-712.

Bazzaz, F.A. (1990). The response of natural ecosystems to the rising global CO_2 levels. *Annual Review of Ecology and Systematics* 21:167-196.

Bungener, P., Nussbaum, S., Grub, A., Fuhrer, J. (1999). Growth response of grassland species to ozone in relation to soil moisture condition and plant strategy. *New Phytologist* 142:283-294.

Chameides, W.L., Kasibhatla P.S., Yienger J., Levy H. (1994). Growth of continental-scale metro-agro-plexes, regional ozone pollution, and world food production. *Science* 264:74-77.

Cooley, D.R., Manning, W.J. (1987). The impact of ozone on assimilate partitioning in plants: a review. *Environmental Pollution* 47:95-113.

Ferris R., Taylor, G. (1993). Contrasting effects of elevated CO_2 on the root and shoot growth of four native herbs commonly found in chalk grassland. *New Phytologist* 125:855-866.

Fiscus, E.L., Reid, C.D., Miller, J.E., Heagle, A.S. (1997). Elevated CO_2 reduces O_3 flux and O_3-induced yield losses in soybeans: possible implications for elevated CO_2 studies. *Journal of Experimental Botany* 48:307-313.

Fuhrer, J. (1997). Ozone sensitivity of managed pasture. In *Ecological Advances and Environmental Impact Assessment* (P.N. Cheremisinoff, Ed.). Advances in Environmental Control Technology Series, Houston, TX: Gulf Publishing Company, pp. 681-706.

Fuhrer, J., Skärby, L., Ashmore, M. (1997). Critical levels for ozone effects on vegetation in Europe. *Environmental Pollution* 97:91-106.

Fuhrer, J., Shariat-Madari, H., Perler, R., Tschannen, W., Grub, A. (1994). Effects of ozone on managed pasture. II. Yield, species composition, canopy structure, and forage quality. *Environmental Pollution* 86:307-314.

Heagle, A.S., Miller, J.E., Pursley, W.A. (1998). Influence of ozone stress on soybean response to carbon dioxide enrichment: III. Yield and seed quality. *Crop Science* 38:128-134.

Heagle, A.S., Miller, J.E., Booker, F.L., Pursley, W.A. (1999). Ozone stress, carbon dioxide enrichment, and nitrogen fertility interactions in cotton. *Crop Science* 39: 731-741.

Heagle, A.S., Miller, J.E., Pursley, W.A. (2000). Effects of ozone and carbon dioxide interactions on growth and yield of winter wheat. *Crop Science* 40:1656-1664.

Hebeisen, T., Lüscher, A., Zanetti, S., Fischer, B.U., Hartwig, U.A., Frehner, M., Hendrey, G.R., Blum, H., Nösberger, J. (1997). Growth response of *Trifolium repens* L. and *Lolium perenne* L. as monocultures and bi-species mixture to free air CO_2 enrichment and management. *Global Change Biology* 3:149-160.

Hunt, R., Hand, D.W., Hannah, M.A., Neal, A.M. (1991). Response to CO_2 enrichment in 27 herbaceous species. *Functional Ecology* 5:410-421.
Idso, K.E., Idso, S.B. (1994). Plant responses to atmospheric CO_2 enrichment in the face of environmental constraints: a review of the past 10 years' research. *Agricultural and Forest Meteorology* 69:153-203.
IPPC (2001). *Climate Change 2001–The Scientific Basis.* Intergovernmental Panel on Climate Change. Cambridge UK: Cambridge University Press.
Lüscher, A., Hendrey, G.R., Nösberger, J. (1998). Long-term responsiveness to free air CO_2 enrichment of functional types, species and genotypes of plants from fertile permanent grassland. *Oecologia* 113:37-45.
Lüscher, A., Hartwig, U.A., Suter, D., Nösberger, J. (2000). Direct evidence that symbiotic N_2 fixation in fertile grassland is an important trait for a strong response of plants to elevated atmospheric CO_2. *Global Change Biology* 6:655-662.
McKee, I.F., Mulholland, B.J., Craigon, J., Black, C.R., Long, S.P. (2000). Elevated concentrations of atmospheric CO_2 protect against and compensate for O_3 damage to photosynthetic tissues of field-grown wheat. *New Phytologist* 146:427-435.
Meier M., Fuhrer J. (1997). Effect of elevated CO_2 on orchard grass and red clover grown in mixture at two levels of nitrogen or water supply. *Environmental and Experimental Botany* 38:251-262.
Muntifering, R.B., Crosby, D.D., Powell, M.C., Chappelka, A.H. (2000). Yield and quality characteristics of bahiagrass (*Paspalum notatum*) exposed to ground-level ozone. *Animal and Feed Science Technology* 84:243-256.
Navas, M.-L. (1998). Individual species performance and response of multi-specific communities to elevated CO_2: a review. *Functional Ecology* 12:721-727.
Navas, M.-L., Garnier, E., Austin, M.P., Gifford, R.M. (1999). Effect of competition on the responses of grasses and legumes to elevated atmospheric CO_2 along a nitrogen gradient; differences between isolated plants, monocultures and multi-species mixtures. *New Phytologist* 143:323-331.
Nussbaum S., Bungener, P., Geissmann M., Fuhrer J. (2000). Plant-plant interactions and soil moisture may be important in determining ozone impacts on grasslands. *New Phytologist* 147:327-335.
Nussbaum, S., Geissmann, M., Eggenberg, P., Strasser, R.J. Fuhrer, J. (2001). Ozone sensitivity in herbaceous species as assessed by direct and modulated chlorophyll fluorescence techniques. *Journal of Plant Physiology* 158:757-766.
Polle, A., Pell, E. (1999). Role of carbon dioxide in modifying the plant response to ozone. In *Carbon Dioxide and Environmental Stress*, eds. Y. Luo, H.A. Mooney, New York NY: Academic Press, pp. 193-213.
Porter, H. (1993). Interspecific variation in the growth response of plants to an elevated ambient CO_2 concentration. *Vegetatio* 104/105:77-97.
Rogers, H.H., Prior, S.A., Runion, G.B., Mitchell, R.J. (1996). Root to shoot ratio of crops as influenced by CO_2. *Plant and Soil* 187:229-248.
Volin, J.C., Reich, P.B. and Givnish, T.J. (1998). Elevated carbon dioxide ameliorates the effects of ozone on photosynthesis and growth: species respond similarly regardless of photosynthetic pathway or plant functional group. *New Phytologist*, 138:315-325.

Zanetti, S., Hartwig, U.A., vanKessel, C., Lüscher, A., Hebeisen, T., Frehner, M., Fischer, B.U., Hendrey, G., Blum, H., Nösberger, J. (1997). Does nitrogen nutrition restrict the CO_2 response of fertile grassland lacking legumes? *Oecologia*, 112:17-22.

Zanetti, S., Hartwig, U.A., Nösberger, J. (1998). Elevated atmospheric CO_2 does not affect per se the preference for symbiotic nitrogen as opposed to mineral nitrogen of *Trifolium repens* L. *Plant, Cell and Environment* 21:623-630.

The Potential Effect of Climate Change and Elevated Air Carbon Dioxide on Agricultural Crop Production in Central and Southeastern Europe

Vesselin A. Alexandrov
Josef Eitzinger

SUMMARY. The vulnerability and adaptation of major agricultural crops to different soils in Austria and Bulgaria under a changing climate and elevated air CO_2 were investigated. Several incremental and transient GCM climate change scenarios were created and applied. Warming will decrease the crop-growing duration of the selected crops in the regions of interest. All GCM scenarios, including the climate change effect only, projected reductions in grain yield of winter wheat and spring barley, caused by a shorter crop-growing period. However, when the direct effect of an increased CO_2 level was assumed, most GCM climate

---

Vesselin A. Alexandrov is Associate Professor in Meteorology, National Institute of Meteorology and Hydrology, BG-1874 Sofia, Bulgaria (E-mail: Vesselin. Alexandrov@ meteo.bg).

Josef Eitzinger is Assistant Professor in Agronomy, University of Agricultural Sciences (BOKU), Institute of Meteorology and Physics, A-1180 Vienna, Austria.

Bulgaria Research supported in part by the M486-GEO and M599-GEO projects subsidized by the Austrian Science Foundation (FWF) and Inco-Copernicus project ERBIC 15CT 98 0105.

[Haworth co-indexing entry note]: "The Potential Effect of Climate Change and Elevated Air Carbon Dioxide on Agricultural Crop Production in Central and Southeastern Europe." Alexandrov, Vesselin A., and Josef Eitzinger. Co-published simultaneously in *Journal of Crop Improvement* (Food Products Press, an imprint of The Haworth Press, Inc.) Vol. 13, No. 1/2 (#25/26), 2005, pp. 291-331; and: *Ecological Responses and Adaptations of Crops to Rising Atmospheric Carbon Dioxide* (ed: Zoltán Tuba) Food Products Press, an imprint of The Haworth Press, Inc., 2005, pp. 291-331. Single or multiple copies of this article are available for a fee from The Haworth Document Delivery Service [1-800-HAWORTH, 9:00 a.m. - 5:00 p.m. (EST). E-mail address: docdelivery@haworthpress.com].

http://www.haworthpress.com/web/JCRIP

Digital Object Identifier: 10.1300/J411v13n01_14

change scenarios projected an increase in wheat and barley yield and especially in soybean yield. An increased level of CO_2 alone had no significant impact on the simulated maize yield reductions under climate change.

KEYWORDS. Austria, Bulgaria, climate change scenarios, crops, DSSAT, WOFOST, vulnerability, climate variability, model comparison, adaptation

INTRODUCTION

Every year of the last decade was amongst the warmest years on a global scale since 1860 (e.g., WMO, 2000). Increases in greenhouse gas concentrations since pre-industrial times have led to a positive radiative forcing of the climate, tending to warm the surface and to produce other changes. The concentration of greenhouse gases in the atmosphere continues to increase (e.g., IPCC WG1 TAR, 2001). This is largely due to human activities, mostly fossil fuel use, land use change, and agriculture. In some regions there is also an increase in the concentration of aerosols, which has an opposite effect on the radiative balances and tend to cool the atmosphere. About 64% of the warming effect due to greenhouse gas increases during the last 200 years are caused by carbon dioxide. These changes in concentration of greenhouse gases and aerosols are projected to lead to regional and global changes in temperature, precipitation and other climate variables. This can ultimately result in global changes in soil moisture, an increase in global mean sea level, and prospects for more severe extreme high-temperature events, floods and droughts in many locations (e.g., Houghton et al., 1996; IPCC, 1997).

Climate change could have a significant impact on the world economy in the 21st century (e.g., Houghton et al., 1996). Global climate change will affect all economic sectors to some degree, but agricultural production is perhaps the most sensitive and vulnerable, as climate is the primary determinant of agricultural productivity (e.g., IPCC WG2 TAR, 2001; Watson et al., 1996). Agriculture in Europe accounts for only a small part of the GDP (gross domestic product), and although the vulnerability of the overall economy to changes that affect agriculture is therefore low (e.g., Reilly, 1996), the impact at a local level could be considerable. Europe as a whole is notable for its great output of both arable crops and animal production (e.g., FAOSTAT, 1998).

Many recent studies have assessed the effects of climate change on agricultural productivity in Europe on different scales (e.g., Downing et al., 2000; Dubrovsky et al., 2000, Harrison et al., 1995; Harrison and Butterfield, 1995; IPCC, 1997; Rosenzweig and Iglesias, 1994; Sathaye et al., 1997; Sirotenko et al., 1997; Watson et al., 1996; Wolf and van Diepen, 1995). We have also initiated studies on the climate change impacts on agricultural production in Central and Southeastern Europe (e.g., Alexandrov, 1999; Alexandrov and Hoogenboom, 2000; Alexandrov et al., 2002; Eitzinger et al., 2000, 2001).

The main goal of this paper is to summarize some of our up-to-date results on the potential effect of climate change and elevated air carbon dioxide on agricultural crop production in Central and Southeastern Europe using Austria and Bulgaria as case studies. Specific objectives included: to create new climate change scenarios for the 21st century in the selected regions in Austria (Lower Austria, Upper Austria and Styria) and Bulgaria; to simulate the impact of climate change variability for major crops in the regions of interest, including winter wheat, spring barley maize and soybean; and to assess possible adaptation measures for agriculture in Austria and Bulgaria under an expected climate change.

MATERIALS AND METHODS

Regions

Three Austrian agricultural regions were selected to represent the case study for Central Europe (Figure 1a). The region under consideration in Lower Austria is situated between 48.20°N and 48.77°N latitude and between 15.70°E and 16.90°E longitude; the region in Upper Austria–between 47.92°N and 48.60°N and between 13.30°E and 14.48°E; the region in Styria–between 46.77°N and 47.22°N and between 15.18°E and 16.08°E. The selected weather stations are located in relatively flat areas with minor variations in elevation, ranging between 153 and 242 m in Lower Austria; from 273 to 465 m in Styria; mainly between 260 and 545 m in Upper Austria.

Some of the Austrian weather stations are characterized by relatively different climatic conditions, especially by different monthly, seasonal and annual precipitation amounts. Annual average temperature in the stations in Lower Austria varies between 9.1° to 9.9°C and annual precipitation ranges between 426 and 546 mm. In Upper Austria the climate norms (1961-1990) of annual air temperature are lower–between 6.9° and 9.3°C, however, annual precipitation is higher–from 692 to 1056 mm. Annual temperature in the stations in Styria ranges from 8.5° to 9.5°C and annual precipitation varies between 765 and 1151 mm.

FIGURE 1. Locations of the weather stations in Austria (a) and experimental crop variety stations in Bulgaria (b) used in the study.

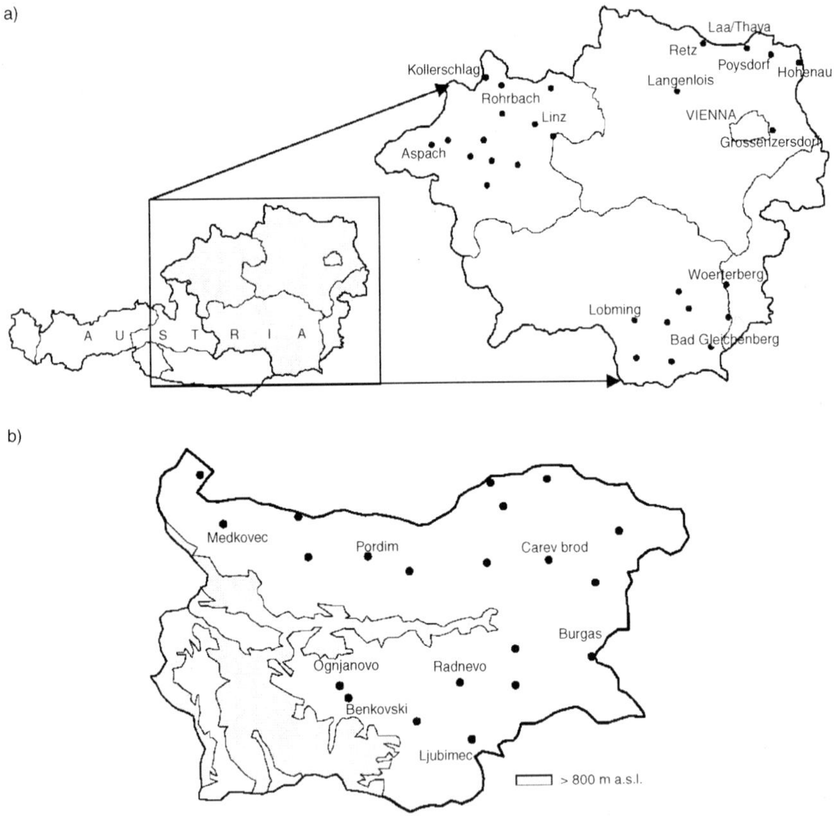

Bulgaria (Figure 1b) is located on the Balkan Peninsula in Southeastern Europe. The country includes 31% lowlands (with elevation 0-200 m), 41% hills (200-600 m), 25% highlands (600-1600 m), and 3% mountains (> 1600 m). The Balkan Mountains split the country into Northern and Southern Bulgaria, and they have a strong effect on the temperature regime. The annual mean air temperatures in Bulgaria vary from −3.0° to 14.0°C, depending on the location and elevation. Air temperature normally reaches a minimum in January, and a maximum in July. The monthly mean temperature varies from −10.9° to 3.2°C in January and from 5.0° to 25.0°C in July. Annual mean total precipitation is approximately 500-650 mm, with an annual variation ranging from 440 to 1020 mm. The highest monthly values are measured in June, and at some places in May, with the mean total varying between 55 and 85 mm. February, and sometimes March and September, are the driest months, with mean totals varying between 30 and 45 mm. Mean precipitation during the warm months, e.g., April through September, is 333 mm.

Data

Near 30 weather stations were selected to be representative for the Austrian simulation study (Figure 1a). The selection of the Austrian weather stations was based on the assumption that all main soil types in the regions of interest should be represented. For each of the weather station locations, at least three soil profiles were defined, representing the major soil types in the areas, surrounding the selected weather stations. The soils in the selected regions include light, medium to heavy soils with low, medium and high soil water storage capacity, respectively. For example, in the Austrian soil classification for the selected region in Lower Austria they are designated as "Braunerde," "Chernozem," "Chernozem on Loess," "kalkfreie Felsbraunerde," "kalkhaltiger Kulturrohboden," "Kolluvium," "Lockersediment Braunerde" and "Rohboden." Soil characteristics such as field capacity, wilting point, horizon designation, percentages of clay, silt, coarse fractions, organic carbon, saturated hydraulic conductivity, total nitrogen, aluminum saturation, bulk density, pH in water and buffer for every given soil layer were used in order to create the appropriate soil profiles.

Data from crop experiments, conducted during the last two decades in the selected Austrian regions were applied. The Austrian registered winter wheat cultivar "Perlo" was used for Lower Austria. "Perlo" is a well established cultivar, especially adapted for dry and warm regions such as the region of Marchfeld. Agrotechnological, "Perlo" phenological and yield data, as well as weather data from Grossenzersdorf (Marchfeld, Lower Austria) for the period 1985-1999 were gathered for the simulation study. Agrotechnological, phenological and yield data of two additional winter wheat cultivars ("Renan" and "Silvius") grown in Upper Austria and Styria for the period 1991-2000 were also used. For the same period and regions crop data of two spring barley cultivars namely "Meltan" and "Elisa" were collected, as well. Available agrometeorological data of soybean from Grossenzersdorf, for the period 1985 to 1999 and 1992 to 1999 were also collected.

Crop and agrotechnological data from 21 experimental variety stations across Bulgaria (Figure 1b) for the period 1980-1993 were also gathered. In the study we used the Bulgarian maize hybrid "Knezha 611" and winter wheat bread variety "Sadovo 1". The phenological and yield data were obtained from the Bulgarian National Variety Commission of the Ministry of Agriculture. Different soils in Bulgaria (e.g., typical chernozems; leached chernozemsmolnitza; grey forest, moderately loamy; leached cinnamonic forest; delluvial-meadow, sandy; alluvial, clayey-sandy; etc.) were used. Daily weather data, including precipitation, maximum and minimum temperature, and simulated solar radiation (Slavov and Georgiev, 1985), were obtained for the same period for the nearest weather stations. All observed Bulgarian meteorological data were provided by the weather network of the National Institute of Meteorology and Hydrology in Sofia.

Crop Growth Models

DSSAT

The IBSNAT (International Benchmark Sites Network for Agrotechnology Transfer) project has developed a computerized decision support system that integrates soil, weather and crop data bases with dynamic crop simulation models (e.g., IBSNAT, 1993; Tsuji et al., 1994). System analysis and validated crop simulation models provide an alternative method for representing agricultural production (e.g., Tsuji et al., 1998). Using models, crop growth and development can be readily evaluated for a wide range of management scenarios and environmental conditions. Strategy analyses to evaluate the change impacts of climate on agricultural production and agrotechnological adaptation are one of the effective applications of crop models (e.g., Thornton and Hoogenboom, 1994).

The CERES generic grain cereal and CROPGRO grain legume models of the DSSAT (e.g., Tsuji et al., 1998) were used to determine the vulnerability of current agricultural management scenarios in Austria and Bulgaria. These crop models are designed to use a minimum set of soil, weather, genetic and management information. The models are daily incrementing and require daily weather data, consisting of maximum and minimum temperature, solar radiation and precipitation as input. They calculate crop phasic and morphological development using temperature, daylength, genetic characteristics and vernalization, where appropriate. Leaf expansion, leaf growth and plant population provide information for determining the amount of light intercepted, which is assumed to be proportional to biomass production. The biomass is partitioned into various growing organs in the plant using a priority system. A water and nitrogen balance submodel provides feedback that influences various growth and development processes. A new soil water model was completed, which contains improved infiltration, redistribution and root uptake calculations. Restrictions to percolation are included in soil inputs so that perched water tables can be simulated along with oxygen stress effects on root and crop growth processes. An option has been added to compute potential evapotranspiration using the Penman equation, which uses humidity and wind speed as input if these data are available (e.g., Hoogenboom et al., 1999; Tsuji et al., 1994). The DSSAT seasonal analysis program (e.g., Thornton and Hoogenboom, 1994) was used to simulate possible adaptation measures, which can decrease the potential agricultural crop vulnerability under expected climate change conditions.

Most plants grown in atmospheric CO_2 conditions that are higher than ambient levels exhibit increased rates of net photosynthesis, i.e., total photosynthesis minus respiration. High CO_2 also reduces the stomatal opening and closure of some crop plants. By so doing, CO_2 reduces transpiration per unit leaf area while enhancing photosynthesis. This often improves water-use efficiency (the ratio of crop biomass accumulation or yield and the amount of water used in evapotranspiration) (e.g., Raschi et al., 1997; Schulze and Caldwell, 1994; Tsuji et al., 1998). All crops that are simulated with the DSSAT models are sensitive to carbon dioxide concentrations (e.g., Tsuji et al., 1998). The models simulate the changes in photosynthesis and evapotranspiration caused by higher levels of CO_2. Ratios have been calculated between measured daily photosynthesis and evapotranspiration rates for a canopy exposed to a range of high CO_2 values, based on published experimental results (Tsuji et al., 1998). Instantaneous midday values were then modified to determine daily integrated increases, allowing for lower light intensities in morning and evening. In the DSSAT crop models, the photosynthesis ratios were applied to the maximum amount of daily carbohydrate production, which is based on incoming solar radiation. To account for the effect of elevated carbon dioxide on stomatal closure and increased leaf area index, and hence on potential transpiration, the evapotranspiration formulation of the DSSAT models includes the ratio of transpiration under elevated CO_2 conditions to that under ambient conditions. Leaf resistance is calculated as a function of the different CO_2 concentrations. The ratio procedure results in a lower transpiration rate per unit leaf area for higher CO_2 levels on a daily basis. Seasonal evapotranspiration, however, may not change proportionately, and may even increase, because of the greater leaf area produced under elevated CO_2 conditions (e.g., Tsuji et al., 1998).

WOFOST

The WOFOST (WOrld FOod STudies) explanatory and dynamic crop model (e.g., Boogaard et al., 1998; Van Diepen et al., 1989) was also used for the Austrian regions. This model was developed by the DLO-Winand Staring Centre and Research Institute for Agrobiology and Soil Fertility in Wageningen (e.g., Boogaard et al. 1998). WOFOST is a member of the family of models developed in Wageningen by the school of C.T. de Wit. It is designed to simulate the growth and development of annual field crops and grass during the growing season, from sowing to maturity or harvest in daily increments. It simulates a cropping system defined by crop, the weather conditions and the soil parameters, including the plant and soil water balance. Outside the crop-growing period the soil water balance can be calculated for bare soil conditions. The major processes taken into account are phenological development, assimilation, respiration and evapotranspiration. WOFOST uses parameters and functions describing the effects of temperature, radiation and water stress on important physiological crop processes as a function of the development stage and crop status. For example, the photosynthesis response curve is limited by a maximum leaf CO_2 assimilation rate and initial light use efficiency of a single leaf. These parameters are further related to temperature at a specified carbon dioxide concentration. Biomass partitioning is a function of the development stage of the crop, while temperature determines the development rate of the crop.

The WOFOST model is designed for simulation of three production levels. The potential yield production level is limited only by temperature, solar radiation and the specific physiological plant characteristics. Such conditions are possible in greenhouses or in very intensive agricultural production systems (e.g., under field conditions with optimum irrigation and nutrition). At the water-limited production level, the soil and plant water balance is also included in the simulation of crop growth with the interactions between transpiration, stomata opening, CO_2 assimilation and water uptake being considered. The third production level is also limited by nutrients.

The above crop growth models were previously calibrated and verified at selected regions in Austria and Bulgaria (e.g., Alexandrov et al., 2001). There was found a good agreement between real phenological stages and yield of important crops in Austria and Bulgaria and phenology and yield, estimated by the CERES, CROPGRO and WOFOST models. Overall, the model test results were very good. The obtained results indicated a satisfactory performance of the applied simulation models for winter wheat, spring barley, maize, and soybean in the selected environments.

Climate Change Scenarios

Most climate change studies use estimates of regional climate change from global circulation models (GCMs) (e.g., IPCC, 1997; Tegart et al., 1990; Watson et al., 1996). The major advantage of using GCMs as the basis for creating climate change scenarios is that they offer a physically consistent way of estimating changes in climate due to increased greenhouse gases for a large number of climate variables. The 30-year averaged transient GCM monthly meteorological outputs were provided for this study by the IPCC DDC (Intergovernmental Panel on Climate Change Data Distribution Center) for the periods:

1961-1990, 2010-2039, 2040-2069, and 2070-2099. The last three periods are referred to as 2020s, 2050s and 2080s and the period 1961 to 1990 represents the baseline climate. The IPCC DDC data is regarded as one of the most recent GCM climate change scenarios currently recommended for climate change impact. The GCMs used in the study include the models from the Max-Planck Institute for Meteorology (ECHAM4), UK Hadley Centre for Climate Prediction and Research (HadCM2), Canadian Centre for Climate Modelling and Analysis (CGCM1), Australian Commonwealth Scientific and Industrial Research Organization (CSIRO-Mk2b) and Geophysical Fluid Dynamics Laboratory (GFDL-R15) (e.g., IPCC DDC, 1999). The outputs for air temperature, precipitation, and solar radiation of the GFDL-R15 model for the 2080s were not available and the ECHAM4 outputs for the 2050s and 2080s do not account the cooling aerosol effect.

In order to interpolate GCM model output data for a specific weather station, linear averaging based on the inverse distances between the specific point and the GCM grid points from the four nearest grid points was used (e.g., ANL, 1994). The climate change scenarios were created by combining average monthly output obtained by taking the difference between transient GCM runs with observed climate data from the 30-year baseline climate period, adding the change in temperature to the observed temperatures and multiplying ratio changes in precipitation and solar radiation by their observed daily values during the period 1961 to 1990 (e.g., ANL, 1994).

It is strongly recommended to provide an understanding of the response of the ecosystems to individually perturbed meteorological variables (e.g., Cramer et al., 2000). That is why several incremental climate change scenarios (e.g., from 0°C to +2°C with 0.5°C steps in air temperature are combined with 0%, ± 10%, ± 20% changes in precipitation) were created and used for the Austrian agricultural regions. A major reason for using incremental scenarios is that they capture a wide range of potential climate changes. In addition, because individual variables are changed independently of each other, incremental scenarios also help identify the relative sensitivities of agricultural ecosystems to changes in temperature and precipitation (e.g., IPCC-TGCIA, 1999).

The modified temperature, precipitation, and solar radiation databases, corresponding to each of the climate change scenarios, were used to run the CERES, CROPGRO and WOFOST simulation models for winter wheat, spring barley and soybean in Austria and for winter wheat and maize in Bulgaria. The WGEN weather generator of the DSSAT WeatherMan program (e.g., Pickering et al., 1994; Tsuji et al., 1994) was used to change the climate variability of expected climate in the Austrian selected regions. Crop management and land use were assumed to be constant. Planting dates were selected based on the latitude and current agrotechnological information applied in the selected regions. At planting the usual (for the current agrotechnology) amount of nitrogen fertilizer was applied for the selected crops.

The simulations were run with the climate change effect only as well as with the direct effect of increased atmospheric CO_2 levels (e.g., Watson et al., 1996). The climate change effect only included the impact of changes in only air temperature, precipitation and solar radiation. In this case the atmospheric CO_2 concentration for the DSSAT and WOFOST crop simulation models was assumed to be constant, i.e., 330 ppmv, for the baseline (1961-1990) and expected climate. The direct effect of increased CO_2 concentration incorporated both the influence of changes in climate and the enhancement of plant growth as a result of elevated atmospheric CO_2 concentrations. In this case the atmospheric CO_2 concentrations were assumed to be 330 ppmv for the baseline climate, 447 ppmv for the 2020s, 554 ppmv for the 2050s and 697 ppmv for the 2080s (e.g., IPCC DDC, 1999).

RESULTS AND DISCUSSION

Vulnerability Assessments Under Incremental Climate Change Scenarios for Austria

It is important to consider the changes in the duration of the crop-growing season under the climate change scenarios to analyze the possible impacts of global climate change on agricultural crops. Crop growth is very sensitive to change and variation in weather conditions during the normal growing season. Most physiological processes, starting as early as the germination process immediately after sowing, and lasting up to the ripening process during physiological maturity, are affected by air temperature. Warming will decrease crop growth duration as a result of the enhanced phenological development, expressed by a shorter period for air temperature accumulation to reach a defined temperature sum. An increase of 0.5°C in air temperature will shorten the growing season for winter wheat, spring barley and soybean by two to four days. In case of 2.0°C warming, soybean in Lower Austria is projected to mature eight to nine days earlier, while maturity dates of winter wheat and spring barley in Lower Austria, Upper Austria, and Styria are projected to occur by 11-15 days and 6-11 days earlier, respectively (Figure 2).

A sensitivity analysis was also conducted for each selected Austrian weather station, using different soil profiles-an evaluation of the effect of step changes in temperature and precipitation on winter wheat and spring barley yield was done. For winter wheat an increase in air temperature between 0.5° and 2.0°C resulted in a grain yield decrease of 1 to 7% and between 7 and 26%, respectively. Similarly, for spring barley a warming between 0.5° and 2.0°C resulted

FIGURE 2. Changes in growing season duration (in days) of winter wheat (a) and spring barley (b) under warming of 2°C in the selected Austrian regions, relative to 1961-1990; CERES model.

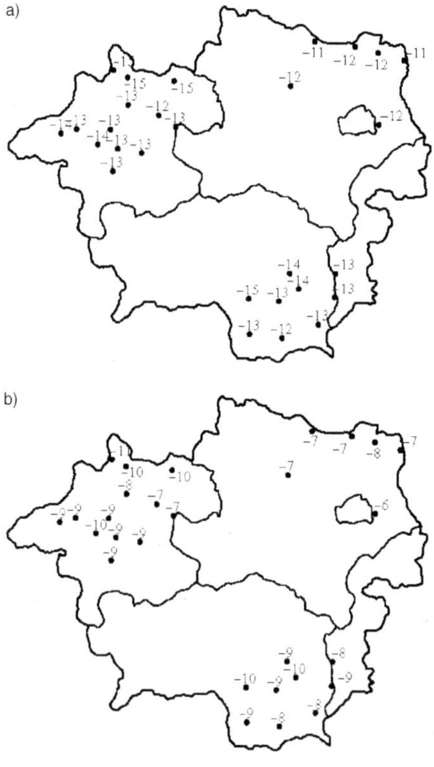

in a grain yield decrease of 2 to 6% and between 6 and 24%, respectively. Precipitation increases also projected grain yield reductions for medium and heavy soils under assuming the nitrogen effect. Actually, the only positive grain yield changes in Hohenau (Lower Austria, medium soil) were simulated through a warming by 1.0°C, combined with precipitation decreases (Figure 3a). However, for some light soils with low soil water storage capacity, an increase of precipitation has a positive effect on final grain yield (Figure 3b). Similar findings were located for spring barley (Figure 4a).

Warming, especially in combination with higher precipitation, produced a significant increase in soybean seed yield–more than 20% with an air temperature increase of 2°C and a 20% increase in precipitation (Figure 4b). At some locations even the negative impact of slight reductions in precipitation was

FIGURE 3. Changes in grain yield of winter wheat under incremental changes of air temperature and precipitation in Hohenau (a: Lower Austria, medium soil) and Rohrbach (b: Upper Austria, light soil); CERES model.

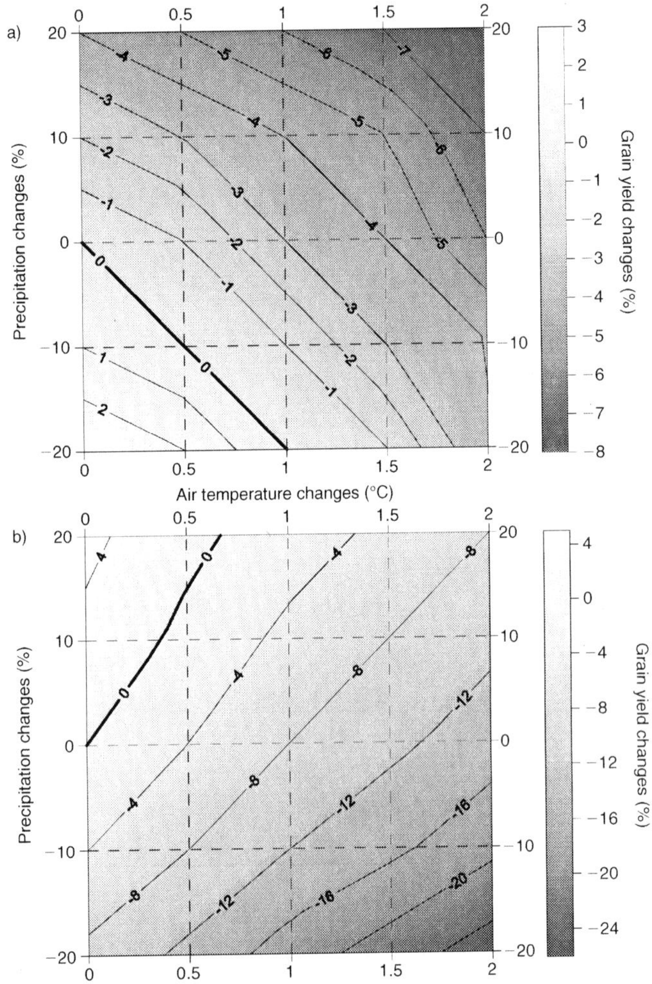

FIGURE 4. Changes in grain yield of spring barley in Lobming (a: Styria, medium soil) and soybean in and Retz (b: Lower Austria, medium to heavy soil) under incremental changes of air temperature and precipitation; CERES (a) and CROPGRO (b) models.

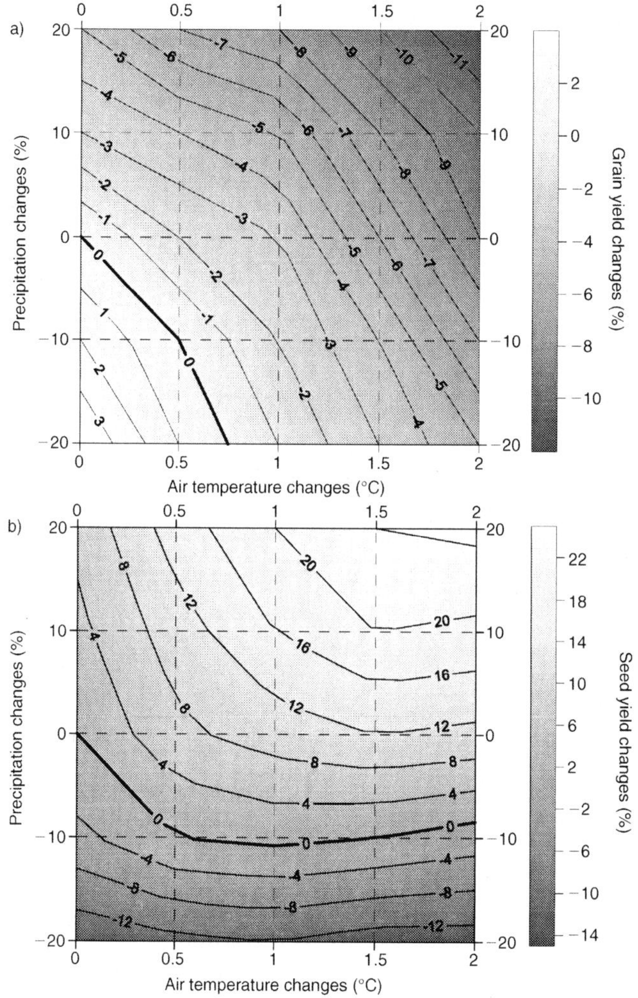

compensated by an air temperature increase. A drier climate than the current one will lead to soybean yield decreases–e.g., between 0 and 14% in Retz (Lower Austria, medium to heavy soils) (Figure 4b).

Soybean, especially on soils with low soil water storage capacity, showed the greatest sensitivity to changes in precipitation, in contrast to heavy soils (Table 1). The soybean yields on light soils in Grossenzersdorf, Langenlois, and Retz with low soil water storage capacity are especially vulnerable to precipitation deficit. The soil profiles at the last two locations are not very deep (50 cm) and they are situated on rocks. The light soil in Langenlois and Retz also showed a high sensitivity to increases in precipitation. There were thus significant fluctuations in soybean yield. The heavy to medium soil in Poysdorf and the medium to heavy soil in Hohenau were also affected by the decreased precipitation, depending on the soil water storage capacity.

The above results were obtained under the assumption of an unchanging atmospheric CO_2 concentration. However, winter wheat, spring barley and especially soybean, which are member to the C3 crop group, are sensitive to the increased/fertilising CO_2 effect. Figure 5 represents the simulated increase of winter wheat yield in Bad Gleichenberg (Styria), spring barley yield in Aspach (Upper Austria), and soybean yield in Poysdorf (Lower Austria), relative to the current climate, with different CO_2 concentration levels in the atmosphere. When the simulations were run with increased CO_2 levels, assumed both for the current climate conditions and incremental scenarios, there were no significant changes in the direction and magnitude of the yield changes simulated under different incremental air temperature and precipitation scenarios.

The results for the increased ambient CO_2 cases agree with other published studies (e.g., Goudriaan and Unsworth, 1990; Idso and Idso, 1994; Nonhebel, 1996). Most of the projected yield reductions will be compensated by the known fertilising effect (e.g., Cure and Acock, 1986; Parry, 2000) of ambient CO_2 levels and positive yield changes will be additionally enhanced. Beside the positive impact on wheat photosynthesis, elevated CO_2 was also found to stimulate the rate of individual wheat grain filling (e.g., Li et al., 2001).

GCM Climate Change Scenarios

According to the transient GCMs used in the study annual temperatures in the selected regions in Austria are expected to rise between 0.9° (HadCM2 in Lower Austria and Styria; CGCM1 in Lower Austria and Upper Austria) and 1.8°C (GFDL-R15 in Upper Austria and Styria) in the 2020s of the 21st century. Annual temperatures are projected to be higher in the 2050s and 2080s-between 1.6° (CGCM1 in all selected regions) and 3.2°C (ECHAM4 in Upper Austria) in the 2050s and between 2.9°C (CSIRO-Mk2b in Upper Austria) and 4.9°C (ECHAM4 in Upper Austria) in the 2080s, respectively. Most of the GCMs show higher increases of air temperature during winter and summer than in spring season. In general, precipitation in Austria for the 2020s is ex-

TABLE 1. Soybean yield changes (in %) under incremental scenarios and different soil types in Lower Austria.

Location in Lower Austria	Soil Type	SW (mm)	Temperature change (°C)									
			0.5					1.0				
			Precipitation change (%)									
			−20	−10	0	10	20	−20	−10	0	10	20
	l	240	−21	−7	3	12	19	−19	−7	3	13	20
	m	258	−30	−15	3	17	28	−32	−14	4	18	29
Grossenzersdorf	h	404	−2	3	5	6	7	−2	4	7	9	10
	l*	118	−32	−15	0	15	28	−32	−17	−1	14	27
	m*	416	−11	−4	4	8	11	−11	−3	5	10	13
	h*	200	−21	−9	2	12	20	−22	−9	2	12	22
	m	339	−5	4	8	11	13	−3	8	15	19	21
Langenlois	mh	351	−4	4	10	13	15	−2	8	15	20	23
	l	95	−48	−25	1	30	55	−50	−26	3	30	56
	m	308	−1	4	10	12	13	4	10	16	19	21
	m	339	−13	1	6	10	13	−14	0	10	14	18
Retz	mh	351	−13	0	7	10	14	−12	1	10	16	20
	l	95	−57	−33	−2	34	68	−60	−35	−5	31	67
	m	308	−6	2	7	11	13	−6	5	12	17	20
	l	310	−10	−1	6	11	14	−7	2	10	16	20
Poysdorf	m	318	4	7	9	10	10	9	12	15	16	17
	hm	254	−23	−9	4	15	24	−23	−8	6	18	28
	l	310	−6	1	6	10	13	−4	4	11	15	18
Hohenau	m	318	4	6	8	8	9	8	11	13	15	15
	hm	254	−23	−10	3	14	21	−23	−8	4	15	23
	mh	337	1	4	7	8	9	4	9	12	13	14
Laa/Thaya	mh	120	−16	−6	4	14	24	−15	−5	5	16	26
	m	332	0	3	6	7	8	3	8	11	13	13

TABLE 1 (continued)

Location in Lower Austria	Soil Type	SW (mm)	Temperature change (°C)									
			1.5					2.0				
			Precipitation change (%)									
			−20	−10	0	10	20	−20	−10	0	10	20
Grossenzersdorf	l	240	−20	−6	4	14	21	−21	−6	4	15	21
	m	258	−34	−17	2	18	30	−36	−20	0	17	30
	h	404	−2	4	8	10	12	−2	4	9	12	12
	l*	118	−33	−18	−2	12	26	−35	−19	−3	11	25
	m*	416	−11	−3	6	11	15	−11	−2	6	13	16
	h*	200	−24	−12	0	11	20	−25	−13	−3	9	19
Langenlois	m	339	−3	8	19	23	26	−3	9	20	26	29
	mh	351	−1	10	19	25	28	−1	10	20	27	32
	l	95	−51	−28	1	30	58	−52	−29	−2	27	55
	m	308	6	14	20	24	27	7	17	23	28	30
Retz	m	339	−15	−1	12	18	21	−16	−3	10	19	23
	mh	351	−14	0	12	20	24	−14	−2	11	19	25
	l	95	−60	−37	−8	26	64	−61	−39	−11	24	59
	m	308	−5	6	14	20	24	−6	6	15	21	26
Poysdorf	l	310	−5	4	12	18	22	−5	5	13	19	23
	m	318	11	15	18	21	22	11	16	20	23	25
	hm	254	−24	−8	6	18	30	−25	−9	6	20	30
Hohenau	l	310	−3	6	13	18	22	−3	7	14	20	24
	m	318	10	14	17	18	19	11	16	19	21	22
	hm	254	−24	−8	4	16	25	−25	−9	4	16	25
Laa/Thaya	mh	337	6	11	15	17	18	7	13	17	20	22
	mh	120	−15	−5	5	15	25	−16	−6	4	14	24
	m	332	5	10	13	16	17	5	11	15	18	20

Legend: SW–soil water storage capacity for the whole soil profile; l–light soils; m–medium soils; h–heavy soils; mh–medium to heavy soils; hm–heavy to medium soils;
*–different field capacity

FIGURE 5. Simulated yield increase of winter wheat in Bad Gleichenberg (Styria), spring barley in Aspach (Upper Austria) and soybean in Poysdorf (Lower Austria) under different CO_2, concentration levels in the atmosphere; CERES and CROPGRO models.

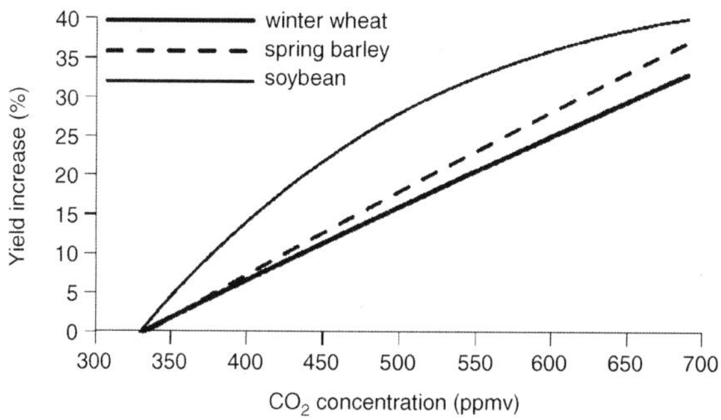

pected to be higher during the year, although some GCMs simulated some slight variations (both decreases/increases) during the summer season. Precipitation is expected to increase during the cold-half of the year in the 2050s and 2080s and slightly to decrease during the warm half of the year in the 2080s (CGCM1 model). The changes in monthly solar radiation are expected to be on the interval $-10 \div 10\%$ during the 21st century.

The transient GCMs predicted that annual temperatures in Bulgaria are to rise between 0.7° (HadCM2) and 1.8°C (GFDL-R15) in the 2020s. However, the HadCM2 model simulated a slight decrease in air temperature for November in the 2020s. A warmer climate is also predicted for the 2050s and 2080s, with an annual temperature increase ranging from 1.6° (HadCM2) to 3.1°C (GFDL-R15) in the 2050s, and 2.9° (HadCM2 and CGCM1 models) to 4.1°C (ECHAM4) in the 2080s. Warming is projected to be higher during the summer in the 2080s. The CGCM1 model predicted an increase in annual precipitation in Bulgaria for the 2020s and 2050s. The GFDL-R15 model projected a decrease in precipitation in May, June, and July in the 2020s and 2050s. The ECHAM4, HadCM2, and CSIRO-Mk2b models simulated a decrease in monthly, seasonal and annual precipitation in the 2080s. An increase of solar radiation is expected during the cold half of the year, based on the ECHAM4 model runs.

Vulnerability Assessments Under GCM Climate Change Scenarios

Crop Growth Duration

All transient GCM climate change scenarios projected a shorter growing season for the selected crops in Austria (Figure 6) and Bulgaria. These changes were driven by the predicted temperature increase of the scenarios. The duration of the regular crop-growing season for winter wheat and spring barley in Austria was simulated between four and fourteen days and between three to ten days shorter in the 2020s than the growing season during the current climatic conditions. Maturity dates for winter wheat, spring barley, and soybean were predicted to occur occur from seven to twenty one days, between five and fifteen and from ten to fourteen days earlier for the 2050s. The highest warming was simulated by the ECHAM4 model for the 2080s–that is why the crop-growing season for the above three crops was projected to occur, respectively, between twenty-five and thirty-two, from seven to nineteen and fourteen days earlier in Austria at the end of the 21st century (Figure 6).

The simulated duration of the regular crop-growing season for maize in Bulgaria was between five (HadCM2) and twenty (GFDL-R15) days shorter. Maturity dates for maize were expected to occur between eleven days and thirty days earlier in the 2050s. The predicted changes in the crop-growing duration for maize in the 2050s were less for the HadCM2, CGCM1, and CSIRO-Mk2b climate change scenarios than the changes predicted by the ECHAM4 and GFDL-R15 models. These last two models simulated a higher increase of air temperature in Bulgaria, especially the GFDL-R15 model, during the summer months July and August. The GCM climate change scenarios for the 2080s projected a decrease in maize growing season by seventeen (CSIRO-MK2b) to thirty-nine (ECHAM4 and CGCM1) days. This will cause a shift in harvest maturity dates for maize from September to August at the end of the next century. Winter wheat showed a decrease in growing season duration for the 2020s in Bulgaria, varying between three days (HadCM2) and fourteen days (GFDL-R15). The projected decreases in growing season for the 2020s, 2050s, and 2080s were less for the HadCM2 model, which predicted a smaller air temperature increase during November and December. Even a slight decrease in monthly air temperature in November was projected for the 2020s under the HadCM2 climate change scenario. The transient GCM climate change scenarios predicted that harvest maturity for winter wheat in the country would be approximately one to two weeks earlier in the 2050s, and between two to three weeks earlier in the 2080s.

Winter Wheat in Austria and Bulgaria and Spring Barley in Austria

All transient GCM climate change scenarios for the 21st century, including the climate change effect only, projected reductions of winter wheat and spring barley yield in the selected Austrian agricultural regions. The projected winter wheat yield reductions, for example in LaaThaya (Lower Austria) were between 3 and 11% in the 2020s and 2050s and between 10 and 15% in the 2080s (Figure 7a). The model reductions for spring barley were between 1 and

FIGURE 6. Changes in growing season duration of winter wheat in Kollerschlag (a: Upper Austria), spring barley in Woerterberg (b: Styria) and soybean in Grossenzersdorf (c: Lower Austria) CERES (a,b) and CROPGRO (c) models.

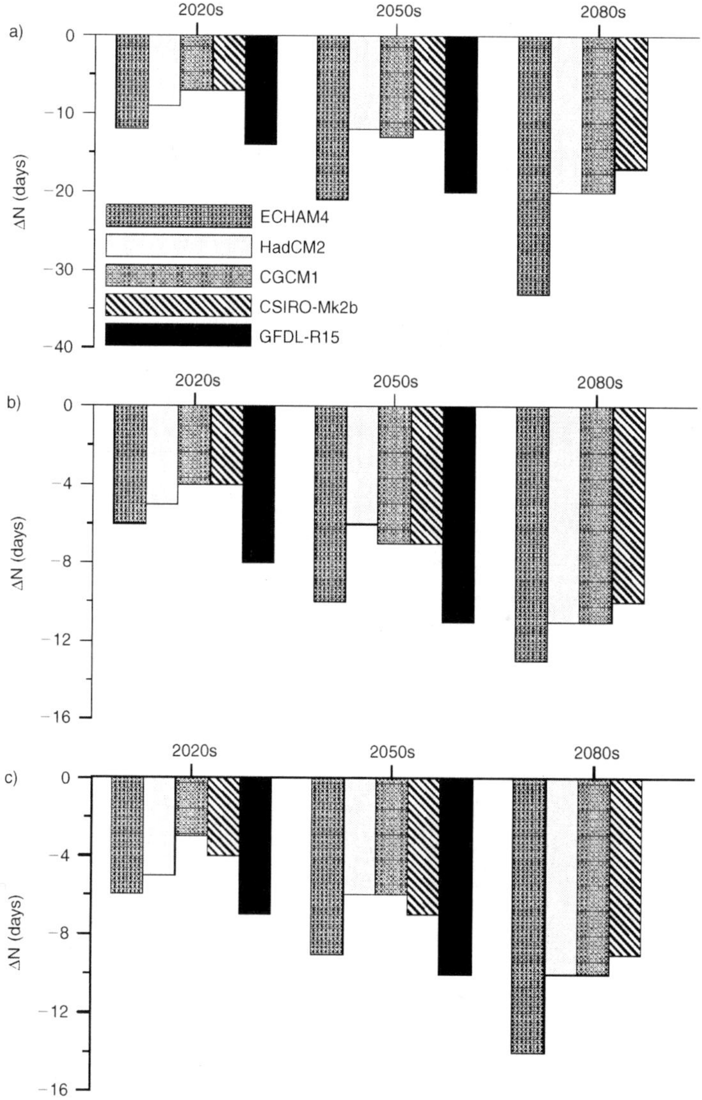

18% in the 2020s and 2050s and between 10 and 33% in the 2080s (Figure 7b). The decrease in simulated grain yield of wheat and barley for the 21st century was primarily caused by a shortened growing season due to warming and some simulated increases in precipitation by the GCMs during the crop-growing season. The previous sensitivity analysis showed that winter wheat and spring yield are affected under warmer and wetter (in most cases) climate. The HadCM2 model simulated less increase of air temperature and reduced autumn precipitation in the 2020s and 2050s–that is why, the projected winter wheat yield decreases were less than the others, simulated by the rest of the GCMs, used in the study (Figure 7a).

FIGURE 7. Winter wheat (a) and spring barley (b) yield changes, relative to the current yields, under GCM scenarios for the 2020s, 2050s and 2080s in Lower Austria (a: LaaThaya; b: Hohenau), assuming climate change effect only and direct CO_2 effect; CERES model.

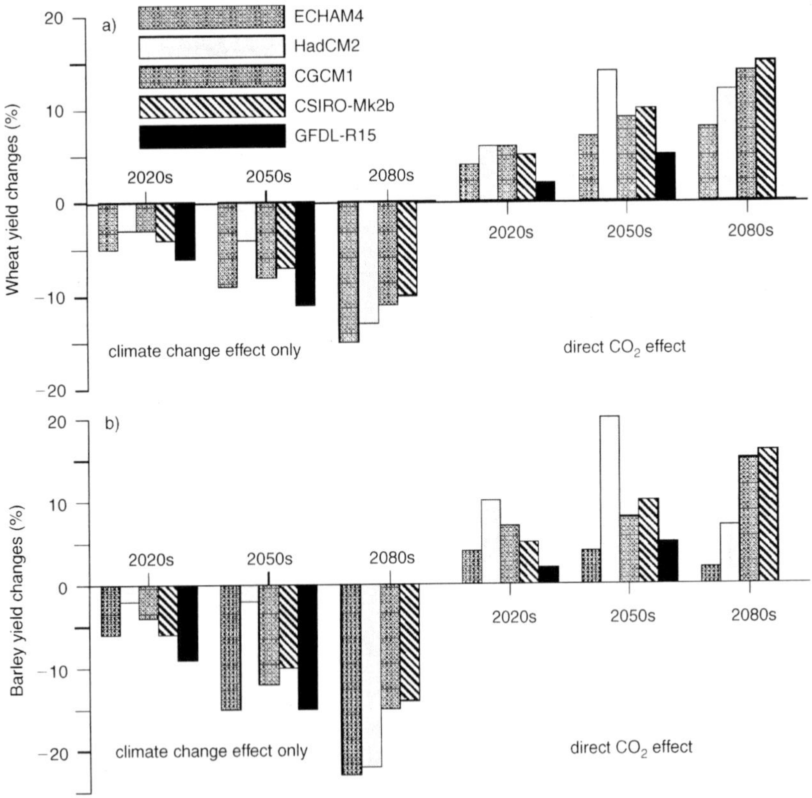

When the direct effect of an increased level of CO_2 was assumed in the study, most GCM climate change scenarios (except ECHAM4 in some locations) projected an increase in winter wheat and spring barley yield (Figures 7 and 8). The major reason of this change of impacts is that cereal crops, such as wheat and barley, belong to C3 crops, which are more sensitive to changes in CO_2 concentration than C4 crops. The CO_2 effect alone increased grain yield for wheat in the 2020s up to 8% above the baseline (1961-1990) in Upper Austria and Styria and up to 11-15% in Lower Austria. The simulated deviations of wheat yield increased in the 2050s up to 17-22% under the HadCM2 climate change scenario. Despite expected high air temperatures during the 2080s, projected increases in wheat yield were in more cases between 3 and 20% due to the fertilization impact of the increased CO_2 level. Similar findings were located for spring barley yield (Figure 8).

The GCM climate change scenarios, including the climate change effect only projected also a reduction in winter wheat yield across Bulgaria. For example, the projected yield reductions at the experimental station Radnevo (Southern Bulgaria) varied between 0 and 7% during the 2020s and 2050s, and between 4 and 20% in the 2080s. When the direct effect of higher CO_2 levels was assumed, all GCM climate change scenarios projected an increase in winter wheat yield (Tables 2, 3, and 4). The CO_2 effect alone caused an increase in wheat yield 10 to 20% above the baseline (1961-1990) for the 2020s. The simulated deviations of wheat yield increased in the 2050 by more than 20 to 25% for the ECHAM4, HadCM2, CGCM1, and CSIRO-Mk2b climate change scenarios. The increase in wheat yield varied from 14 to 37% for the GFDL-R15 scenario, depending on the location. Despite expected high air temperatures and precipitation reductions during the spring in the 2080s, projected increases in wheat yield varied between 12 and 49% due to the fertilization impact of the increased CO_2 level. The above results are consistent with results obtained in previous European simulation studies (e.g., Downing et al., 2000; Harrison et al., 1995; Nonhebel, 1996; Parry, 2000; Semenov et al., 1996) and most of the CO_2 enrichment experiments (e.g., Idso and Idso, 1994; Parry, 2000).

Soybean in Austria

The increase in simulated soybean seed yield for the next century was caused primarily by the positive impact of warming and especially by the beneficial direct CO_2 effect. The simulated low increase in rain-fed soybean yield in Hohenau (heavy to medium soil) under the CGCM1 climate change scenario for the 2080s (climate change effect only) was due to the reduced summer precipitation projected by this GCM model (Figure 9a). In some medium and especially light soils in some locations, there was even a simulated decrease in rain-fed soybean yield in the 2080s as a result of the GCCM1-pro-

FIGURE 8. Range of winter wheat and spring barley yield changes in the selected Austrian regions, relative to the current yields, under all applied GCM scenarios for the 2020s, 2050s and 2080s assuming direct CO_2 effect; CERES model.

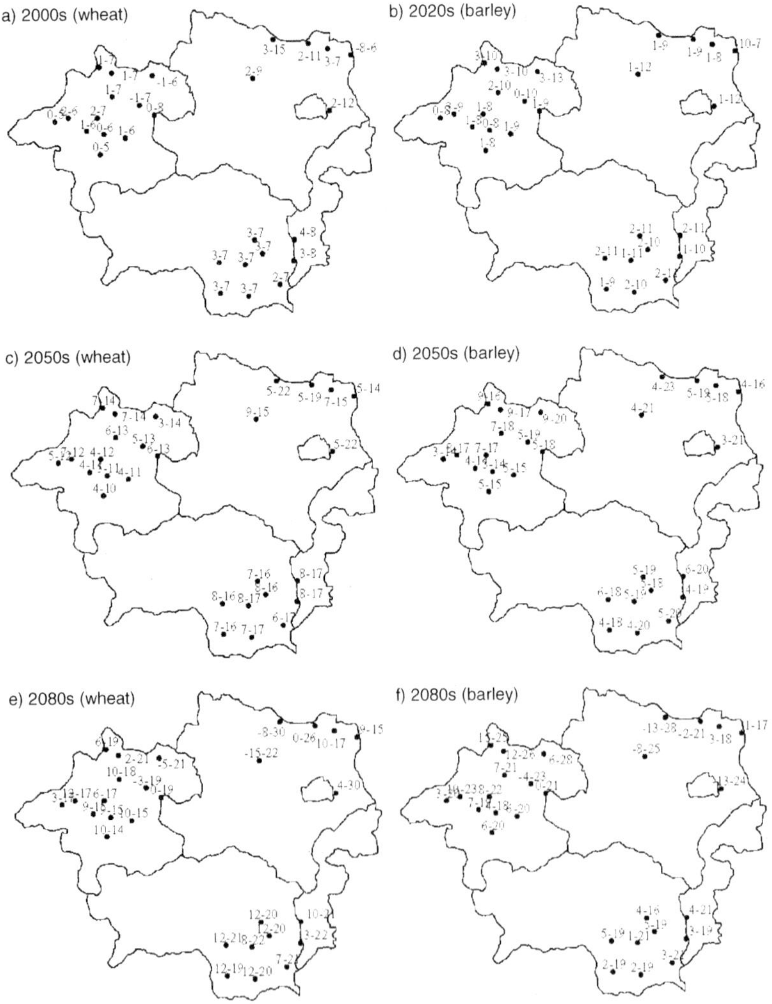

TABLE 2. Departures of grain yield (%) for maize and winter wheat in Bulgaria under transient global climate model scenarios for the 2020s (CO_2 = 447 ppm), relative to the current climate.

Station in Bulgaria	Region	ECHAM4		HadCM2		CGCM1		CSIRO-Mk2b		GFDL-R15	
		maize	wheat	maize	wheat	maize	wheat	maize	wheat	maize	wheat
Kapitanovci	NW	−7	16	4	10	−5	18	−8	17	−13	16
Medkovec	NW	−7	18	4	14	−7	22	−7	19	−11	16
Selanovci	NW	−6	18	9	16	−6	30	−6	24	−11	18
Kojnare	N	−5	15	12	9	−8	18	−8	16	−12	14
Pordim	N	−3	15	10	9	−3	20	−4	16	−8	15
Pavlikeni	N	−3	15	10	15	−4	27	−4	22	−8	14
Svetlen	N	−5	15	12	9	−8	18	−8	16	−12	14
Brashljan	N	−8	10	5	12	−6	18	−8	17	−14	8
Kubrat	N	−6	16	10	9	−6	20	−6	15	−11	15
Sitovo	NE	−5	12	5	10	−6	16	−6	13	−12	8
Carev brod	NE	−5	16	12	10	−8	20	−8	17	−12	17
Dobrich	NE	−5	12	5	10	−6	16	−6	13	−12	8
Jitnica	NE	−4	20	11	14	−6	25	−6	21	−12	21
Burgas	SE	−4	14	11	8	−6	12	−7	12	−11	13
Zimnica	S	−5	13	11	9	−7	13	−6	13	−13	15
Bojanovo	S	−6	14	12	8	−7	15	−7	14	−12	14
Radnevo	S	−6	14	11	8	−7	14	−7	12	−12	12
Ljubimec	S	−8	13	6	12	−7	17	−8	15	−14	10
Gorski izvor	S	−5	14	12	8	−6	14	−7	12	−11	12
Benkovski	S	−7	12	5	12	−6	17	−8	16	−14	9
Ognjanovo	S	−7	12	5	12	−6	17	−8	16	−14	9

TABLE 3. Departures of grain yield (%) for maize and winter wheat in Bulgaria under transient global climate model scenarios for the 2050s (CO_2 = 554 ppm), relative to the current climate.

Station in Bulgaria	Region	ECHAM4		HadCM2		CGCM1		CSIRO-Mk2b		GFDL-R15	
		maize	wheat	maize	wheat	maize	wheat	maize	wheat	maize	wheat
Kapitanovci	NW	−15	28	−7	30	−9	25	−10	26	−21	24
Medkovec	NW	−14	27	−8	32	−10	29	−9	27	−19	24
Selanovci	NW	−12	42	−6	43	−11	45	−9	40	−16	36
Kojnare	N	−12	27	1	30	−11	31	−10	29	−19	29
Pordim	N	−10	28	−4	32	−7	31	−6	29	−15	29
Pavlikeni	N	−11	35	−1	40	−7	41	−7	37	−17	30
Svetlen	N	−12	27	1	30	−11	31	−10	29	−19	29
Brashljan	N	−15	26	−6	29	−11	24	−12	23	−21	17
Kubrat	N	−11	29	−4	32	−10	31	−9	30	−17	31
Sitovo	NE	−10	23	1	29	−10	23	−9	23	−18	18
Carev brod	NE	−12	29	2	32	−11	32	−10	31	−19	32
Dobrich	NE	−10	23	1	29	−10	23	−9	23	−18	18
Jitnica	NE	−11	34	5	38	−10	40	−9	37	−18	37
Burgas	SE	−8	22	6	22	−10	23	−9	22	−18	22
Zimnica	S	−10	28	4	27	−11	29	−10	28	−19	29
Bojanovo	S	−9	25	5	25	−11	26	−10	25	−18	26
Radnevo	S	−12	24	0	25	−11	25	−11	24	−19	25
Ljubimec	S	−13	20	−7	27	−11	23	−11	20	−21	14
Gorski izvor	S	−11	24	1	25	−10	25	−10	24	−19	25
Benkovski	S	−12	20	−6	28	−11	23	−11	19	−20	14
Ognjanovo	S	−12	20	−6	28	−11	23	−11	19	−20	14

TABLE 4. Departures of grain yield (%) for maize and winter wheat in Bulgaria under transient global climate model scenarios for the 2080s (CO_2 = 697 ppm), relative to the current climate.

Station in Bulgaria	Region	ECHAM4		HadCM2		CGCM1		CSIRO-Mk2b		GFDL-R15	
		maize	wheat	maize	wheat	maize	wheat	maize	wheat	maize	wheat
Kapitanovci	NW	−19	24	−24	24	−22	27	−13	31	NA	NA
Medkovec	NW	−17	27	−28	29	−20	28	−10	34	NA	NA
Selanovci	NW	−14	39	−27	41	−18	48	−9	57	NA	NA
Kojnare	N	−19	40	−14	24	−22	40	−12	41	NA	NA
Pordim	N	−12	36	−22	27	−18	40	−7	42	NA	NA
Pavlikeni	N	−15	30	−20	39	−18	38	−8	49	NA	NA
Svetlen	N	−19	40	−14	24	−22	40	−12	41	NA	NA
Brashljan	N	−19	17	−21	29	−21	20	−14	30	NA	NA
Kubrat	N	−14	44	−23	25	−18	44	−10	46	NA	NA
Sitovo	NE	−16	23	−10	32	−18	28	−11	36	NA	NA
Carev brod	NE	−18	42	−14	24	−22	42	−12	43	NA	NA
Dobrich	NE	−16	23	−10	32	−18	28	−11	36	NA	NA
Jitnica	NE	−16	47	−12	33	−20	46	−12	49	NA	NA
Burgas	SE	−10	31	−9	16	−14	28	−10	30	NA	NA
Zimnica	S	−19	40	−14	24	−22	40	−12	41	NA	NA
Bojanovo	S	−10	35	−11	19	−15	32	−11	34	NA	NA
Radnevo	S	−13	33	−20	18	−19	32	−13	34	NA	NA
Ljubimec	S	−18	11	−21	23	−21	14	−13	21	NA	NA
Gorski izvor	S	−15	33	−15	18	−20	32	−13	34	NA	NA
Benkovski	S	−19	12	−20	23	−20	14	−14	20	NA	NA
Ognjanovo	S	−19	12	−20	23	−20	14	−14	20	NA	NA

jected water stress (Figure 10c). Under the impact of climate change effect only, soil water storage capacity between 200 and 400 mm was projected to have a linear impact on soybean seed yield changes. The highest deviations from the linear fits in Figure 10 were obtained in soils with very low or very high soil water storage capacities. There was a strong correlation between soil water storage capacity ($r^2 = 0.54$) and soybean seed changes under GCMs simulating summer precipitation deficit (i.e., the CGCM1 model). This behaviour is consistent with the sensitivity analysis on the basis of soil storage capacity mentioned earlier (Table 1).

When the direct effect of increased CO_2 was assumed, all GCM climate change scenarios projected a strong increase in soybean yield (Figure 9). The

FIGURE 9. Soybean (a–rainfed; b–irrigated) yield changes, relative to the current yields, under GCM climate change scenarios for the 2020s, 2050s and 2080s in Hohenau (Lower Austria, heavy to medium soil) assuming climate change effect only and direct CO_2 effect.

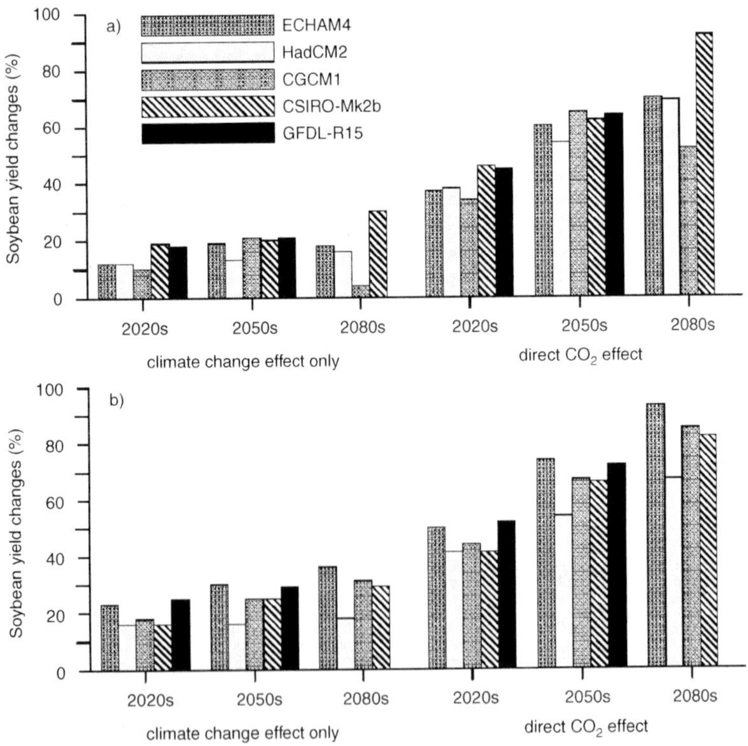

FIGURE 10. Relations between soil water storage capacity over the whole soil profile and soybean seed yield change under climate change effect only and direct CO_2 effect for the 2020s (a) and 2080s (b, c) in Lower Austria; CROPGRO model.

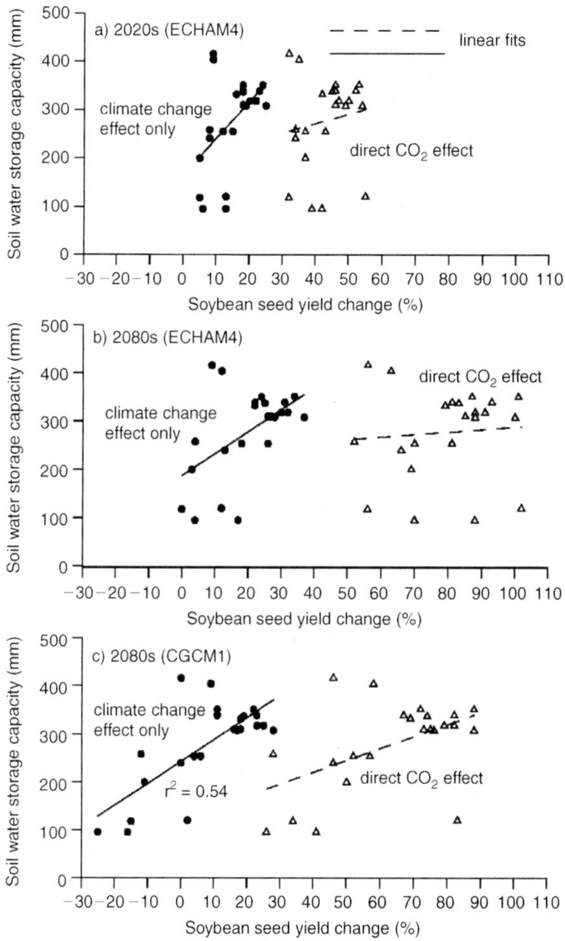

simulated increases started from 30 to 40% in the 2020s and rose to between 50 and 90% in the 2080s. These results are consistent with other simulation and experiment results (e.g., Downing et al., 2000; Parry, 2000). Yield changes for irrigated soybean simulations were higher than rain-fed simulations, especially with the climate change effect only (Figure 9b). The rates of yield increase simulated by different GCM outputs, changed when irrigation was applied to overcome water stress.

Most transient GCM climate change scenarios for the 21st century, including the climate change effect only, projected increases in water demand for soybean irrigation (Table 5). Because of the relatively low warming and increased precipitation in May, June and July, minor negative changes in irrigation were simulated for some soil types only under the CSIRO-Mk2b scenario for the 2020s. The higher warming projected by the ECHAM4 GCM model for the 2050s and 2080s led to higher water demands for irrigation at the middle (9-23%) and at the end (24-48%) of the 21st century. Higher CO_2 levels in general reduces stomatal opening of plant leaves. By so doing, CO_2 reduces transpiration per unit leaf area while enhancing photosynthesis. This often improves water-use efficiency (Raschi et al., 1997; Schulze and Caldwell, 1994), which explains the reduction in the above-mentioned water irrigation requirements (Table 5). The direct CO_2 effect also decreased the influence of soil water storage capacity on soybean yield changes, as shown in Figure 10 for the 2080s.

Maize in BG

The decrease in simulated maize yield for the next century was primarily caused by a shorter growing season duration and reductions in precipitation. All GCMs simulated a decrease in precipitation from March to June for the 2080s, which affected soil moisture recharge during the spring and the early developmental stages of maize. The simulated increase in maize grain yield for the HadCM2 climate change scenario for the 2020s was due to a relatively low projected increase in air temperature, as well as a predicted increase in precipitation in July. Because maize is a C_4 crop, an increased level of CO_2 alone had no significant impact on either maize crop growth, and development or final yield (Tables 2, 3, and 4). Maize yield decreased by 3 to 8% in the 2020s for the ECHAM4, CGCM1, and CSIRO-Mk2b model scenarios. The projected decrease was highest for the GFDL-R15 model, e.g., between 8 and 14%, while the HadCM2 scenario projected an increase from 4 to 12% for the next decades. A slight increase at the most experimental stations in Northeastern and Southern Bulgaria is even projected under the HadCM2 climate change scenario for the 2050s (Table 3). The decrease in simulated maize yield for the 2050s ranged for most stations from 10 to 20% for the ECHAM4, CGCM1, CSIRO-Mk2b, and GFDL-R15 GCM scenarios. The largest decrease in maize yield is expected to occur at the end of the century.

Results from Changes in Climate Variability and Model Comparison

The use of observed (1961-1990) climate data to create climate change scenarios provides spatial and temporal variability. However, it assumes that this

TABLE 5. Changes in total irrigation water amount during the soybean crop-growing season in Lower Austria.

Location in Lower Austria	Soil Type	Irrigation amount changes with climate change effect only (%)														
		2020s					2050s					2080s				
		1	2	3	4	5	1	2	3	4	5	1	2	3	4	5
Grossenzersdorf	l	12	7	10	0	12	18	10	10	10	19	41	14	40	7	NA
	m	18	6	12	1	13	23	10	10	11	23	48	18	44	12	NA
	h	11	3	9	−2	9	18	4	5	6	16	44	11	44	3	NA
	l*	17	8	12	3	13	22	10	15	15	23	52	19	46	13	NA
	m*	12	4	10	1	11	14	8	9	9	19	37	12	32	10	NA
	h*	14	7	13	2	11	19	10	9	11	19	44	16	41	9	NA
Langenlois	m	14	4	6	0	4	19	9	6	7	14	45	15	44	2	NA
	mh	14	8	11	1	11	18	15	9	12	18	43	16	48	4	NA
	l	11	5	7	−1	7	16	8	7	8	14	36	12	29	7	NA
	m	11	4	8	−3	8	12	6	6	6	15	34	11	33	6	NA
Retz	m	14	3	9	−2	10	19	8	7	7	17	45	14	40	6	NA
	mh	12	0	6	−4	7	15	4	4	5	13	37	13	36	1	NA
	l	14	7	8	3	10	20	8	9	10	18	36	14	32	11	NA
	m	12	5	9	0	11	15	8	7	8	17	34	11	35	6	NA
Poysdorf	l	8	3	6	0	7	10	6	7	5	11	26	8	25	4	NA
	m	9	4	5	−3	11	14	8	6	5	16	25	11	29	3	NA
	hm	12	7	10	−1	10	17	8	10	9	17	34	12	34	7	NA
Hohenau	l	4	2	4	−1	5	9	4	4	3	9	26	7	24	2	NA
	m	8	5	6	−1	7	14	10	6	4	14	30	14	33	4	NA
	hm	15	6	7	−2	13	15	8	9	10	18	34	12	34	7	NA
Laa/Thaya	mh	13	9	14	1	16	20	10	8	10	22	39	14	36	7	NA
	mh	8	3	7	0	6	11	4	7	6	12	27	7	24	6	NA
	m	9	4	10	−1	8	14	9	8	6	14	33	10	36	6	NA

TABLE 5 (continued)

Location in Lower Austria	Soil Type	Irrigation amount changes with direct CO_2 effect only (%)														
		2020s					2050s					2080s				
		1	2	3	4	5	1	2	3	4	5	1	2	3	4	5
Grossenzersdorf	l	9	0	3	−5	6	10	−1	1	2	11	29	0	28	−2	NA
	m	13	3	5	−3	8	16	1	2	3	15	36	3	31	0	NA
	h	7	−1	2	−10	4	10	0	−1	−1	9	30	−2	28	−7	NA
	1*	13	4	9	−2	8	18	3	6	6	15	39	5	34	1	NA
	m*	7	−1	4	−3	6	8	1	1	2	12	27	1	22	−3	NA
	h*	11	3	8	−3	7	11	1	3	3	13	30	3	28	−3	NA
Langenlois	m	7	1	2	−7	1	10	0	−2	−2	4	29	−3	26	−9	NA
	mh	11	1	5	−4	4	10	1	−1	−2	8	28	1	27	−6	NA
	l	8	−1	3	−5	3	10	−3	0	−2	8	24	1	20	−4	NA
	m	6	−3	1	−7	4	4	0	−2	−2	6	20	0	20	−9	NA
Retz	m	7	0	3	−4	6	10	0	1	0	10	29	0	27	−6	NA
	mh	7	−4	−1	−8	2	7	−5	−4	−5	5	21	−5	21	−10	NA
	l	9	2	4	−3	6	11	1	4	3	11	26	4	23	0	NA
	m	6	−2	3	−7	5	6	0	−1	−2	7	23	0	21	−7	NA
Poysdorf	l	5	−1	2	−5	2	6	−3	0	−1	4	16	−1	12	−6	NA
	m	4	−2	0	−8	2	−1	−5	−5	−6	4	16	−5	17	−12	NA
	hm	9	−1	5	−5	8	10	1	−2	0	12	25	1	23	−4	NA
Hohenau	l	1	−2	1	−6	0	2	−4	−3	−3	3	16	−2	10	−7	NA
	m	2	−2	0	−8	4	2	−3	−6	−6	4	16	−1	16	−12	NA
	hm	7	3	3	−4	7	10	2	0	−1	9	23	2	19	−3	NA
Laa/Thaya	mh	8	−1	7	−3	6	8	2	1	−3	12	26	1	24	−7	NA
	mh	7	3	7	0	6	10	2	6	6	10	24	2	22	2	NA
	m	6	−1	6	−8	2	8	−1	−1	−1	6	22	−1	20	−10	NA

Legend: 1–ECHAM4; 2–HadCM2; 3–CGCM1; 4–CSIRO; 5–GFDL; l–light, m–medium, h–heavy, mh–medium to heavy, hm–heavy to medium soils; *–different field capacity; NA -not available data will not change from current conditions. Some investigations have looked not only at changes in averages of climate parameters but also at inter-annual and intra-annual temperature and precipitation variability (e.g., Dubrovsky et al., 2000; Riha et al., 1996; Semenov and Porter, 1995; Wilks and Riha, 1996). Changes in climate variability might be an important factor in crop production. For this reason, some numerical experiments in which weather elements and parameters were varied were performed in the study. Changing the variability of maximum air temperature by increasing its standard deviation resulted in a decrease in both mean winter wheat and soybean yield in Hohenau (Lower Austria, light soils) (Table 6). Changes in the extremes and an increase in the standard deviation and the coefficient of variation of winter wheat yield were also obtained (Table 6). Changes in solar radiation variability resulted in a decrease in mean wheat yield, but there were no significant changes in mean soybean yield. Changes in precipitation characteristics also led to changes in both winter wheat and soybean mean yields. Doubling the precipitation distribution scale parameter α decreased both winter wheat and soybean yield. A decrease in average monthly rainy days N resulted in a lower projected winter

TABLE 6. Winter wheat and soybean yield statistical characteristics under the HadCM2 climate change scenario for the 2080s (CO_2 = 697 ppm) in Hohenau (Lower Austria, light soils) without and with changes in current maximum air temperature, solar radiation and precipitation variability.

Climate Variability		Mean (kg/ha)	Max (kg/ha)	Min (kg/ha)	σ (kg/ha)	As	Cv	Ex
Winter wheat								
No changes		6452	8074	4932	794	0.155	0.123	−0.678
Max temperature	2σ	6044	8755	3455	1088	−0.104	0.180	0.192
Solar radiation	2σ	6204	7442	4245	740	−0.585	0.119	0.257
	2α	6258	7561	4558	760	−0.355	0.122	−0.283
Precipitation	N/2	6280	7826	5129	603	0.830	0.096	−0.004
Soybean								
Climate Variability		Mean (kg/ha)	Max (kg/ha)	Min (kg/ha)	σ (kg/ha)	As	Cv	Ex
No changes		4587	5490	3066	524	−0.907	0.114	0.800
Max temperature	2σ	3944	4765	2517	476	−1.043	0.121	1.161
Solar radiation	2σ	4504	5216	2779	510	−1.461	0.113	2.680
	2α	4490	5132	2522	532	−1.645	0.119	3.831
Precipitation	N/2	4826	5427	3461	428	−1.371	0.089	2.002

Legend: σ–standard deviation; As, Cv, Ex–coefficients of asymmetry, variation and excess; α–precipitation distribution scale parameter, N–average monthly rainy days wheat yield. In this case, however, the minimum soybean yield increased by 400 kg/ha. A more comprehensive study of the influence of precipitation variability on crops, especially the impact of drought frequency and duration, is required.

The sensitivity of the CERES and WOFOST models in respect to the growing season duration and grain yield was compared under the projected climate change for some locations in Austria and soil profiles with different available soil water storage capacity (Tables 7 and 8). The CERES and WOFOST models are simulating the shortening of spring barley growing season in a similar way under warming. However, there are significant differences between the changes of the growing season of winter wheat under climate change, simulated by the two crop growth models. The WOFOST model simulates significantly higher decreases of the duration of winter wheat growing season under projected warming than the simulated one by the CERES model. It seems there is a model difference between the CERES and WOFOST models regarding the process and duration of vernalization of winter wheat. In general, the WOFOST model was found to be more sensitive to climate change effect only and especially direct CO_2 effects as well as available soil water storage capacity in comparison to the CERES model when the nitrogen stress was not assumed in the study.

Adaptation Measures

Several economic and agronomic adaptation strategies can be suggested with a view to avoiding or reducing negative effects and exploiting possible beneficial options. The agronomic strategies available include both short-term adjustments and long-term adaptation. The short-term adjustments for dealing with the effect of climate change include changes in planting dates and

TABLE 7. Growing season duration of winter wheat and spring barley, relative to the 1961-1990 duration, under HadCM1 GCM climate change scenarios for the 2020s, 2050s and 2080s in Linz (Upper Austria) and Graz (Styria).

Station	Changes from 1961-1990 (%)											
	Winter wheat						Spring barley					
	2020s		2050s		2080s		2020s		2050s		2080s	
	CER	WOF	CER	WOF	CER	WOF	CER	WOF	CER	WOF	CER	WOF
Linz	−7	−10	−9	−19	−16	−34	−5	−5	−6	−5	−9	−10
Graz	−6	−10	−9	−17	−18	−32	−5	−4	−6	−4	−11	−10

Legend: CER–CERES; WOF–WOFOST

TABLE 8. Winter wheat and spring barley yield changes, relative to the 1961-1990 yields, under CGCM1 GCM climate change scenarios for the 2020s, 2050s and 2080s in some locations in Upper Austria and Styria assuming climate change effect only and direct CO_2 effect.

Winter wheat															
				Changes from 1961-1990 (%)											
Station	SW (mm)	CERES yield 1961-90 (kg/ha)	WOFOST yield 1961-90 (kg/ha)	Climate change effect only						Direct CO_2 effect					
				2020s		2050s		2080s		2020s		2050s		2080s	
				CER	WOF	CER	WOF	CER	WOF	CER	WOF	CER	WOF	CER	WOF
	96	6289	5063	−7	−10	−12	−16	−22	−28	4	11	9	22	9	29
Linz	279	7150	7036	−2	−1	−7	−6	−9	−8	7	20	11	31	16	52
	249	7150	6874	−2	−2	−7	−7	−9	−9	7	19	11	31	16	51
	284	8668	8603	−4	−7	−8	−14	−10	−19	6	17	10	26	17	40
Graz	185	8398	5579	−4	−19	−9	−31	−14	−54	7	8	11	11	17	−10
	286	8668	8604	−4	−7	−8	−14	−10	−19	6	17	10	26	17	40
	198	8454	8125	−4	−9	−8	−17	−13	−26	7	15	11	22	17	31

Spring barley															
				Changes from 1961-1990 (%)											
Station	SW (mm)	CERES yield 1961-90 (kg/ha)	WOFOST yield 1961-90 (kg/ha)	Climate change effect only						Direct CO_2 effect					
				2020s		2050s		2080s		2020s		2050 s		2080s	
				CER	WOF	CER	WOF	CER	WOF	CER	WOF	CER	WOF	CER	WOF
	289	5037	4946	−2	−6	−10	−10	−13	−17	9	13	12	23	18	32
Enns	119	4831	4396	−5	−10	−12	−14	−19	−32	8	9	12	22	16	22
	249	5037	4963	−2	−5	−10	−10	−13	−16	9	14	12	23	18	33
	192	5968	6411	−4	−5	−11	−11	−15	−17	7	14	10	20	16	29
Weiz	236	5976	6429	−4	−5	−11	−11	−14	−16	6	14	9	20	15	29
	225	5976	6434	−4	−5	−11	−11	−14	−16	6	14	9	20	15	29

Legend: SW–available soil water storage capacity; CER–CERES; WOF–WOFOST cultivars; changes in external input such as irrigation; techniques to conserve soil water. Long-term adaptations refer to major structural changes to overcome disadvantages caused by climate change. These may include changes in land use; biotechnology applications; crop substitution; microclimate modification; and changes in farming systems (IPCC WG2 TAR, 2001).

Changes in Sowing Date

Crops in Austria, and especially spring crops could be sown earlier under the GCM climate change scenarios in order to reduce the yield loss or to further increase the projected gain resulting from an increase in temperature. A related problem is the effective use of soil water accumulated during the winter period by the young crops to avoid drought impact in spring. If plant photoperiod reaction were adapted to a shorter day length, this alteration would allow the spring crops to develop during an earlier period of the year with ample soil water, thereby increasing the growth duration, especially the grain filling period.

The sowing dates of spring crops in Bulgaria could also shift under projected climate change. The selection of an earlier sowing date for maize will probably be the appropriate response to offset the negative effect of a potential increase in temperature (Figure 11). The results depicted in Figure 11 show that the sowing date of maize for the experimental station Carev brod (Northeast Bulgaria) should occur at least two weeks earlier in the 2080s under the ECHAM4 scenario, relative to the current climate conditions. It should be noted, however, that although changes in sowing date are a non-cost decision that can be taken at the farm level, a large shift in sowing dates could interfere with the agrotechnological management of other crops.

Changes in Crop Cultivars

Crop diversification allows farmers to cope with climate variation from year to year and to reduce the risk in crop production. This type of adaptation is likely to occur at the farm system level. Switching from monocultures, which are more vulnerable to climate change, pest and diseases, to more diversified agricultural production systems such as appropriate crop rotation systems will also help farmers to cope with changing climatic conditions.

Another option for adaptation to climate change is to use different and adapted hybrids and cultivars. Plant breeding could provide more productive, later/earlier-maturing, disease- and pest-tolerant hybrids and cultivars. This study suggested that winter wheat cultivars with a shorter vegetative growth period than the "Perlo" cultivar (sowing-flowering) would be beneficial for

FIGURE 11. Cumulative probability function of maize yield in Carev brod (Northeastern Bulgaria) under the ECHAM4 climate change scenarios for the 2080s and different sowing dates (CO_2 = 697 ppm); CERES model.

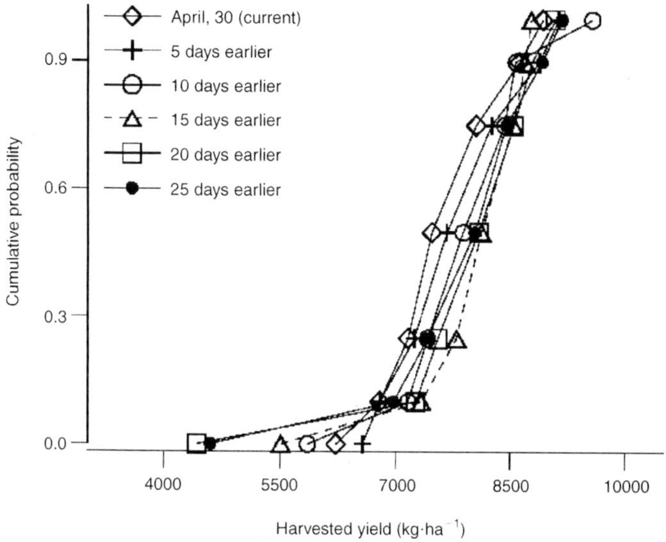

wheat productivity in Austria under expected climate change. Using spring barley cultivars with longer reproductive (flowering-maturity) period would be beneficial for wheat productivity in Austria under expected climate change (Figure 12). Switching to soybean cultivars of maturity groups 000 and 00 (shorter growing season) or to maturity groups 4 and 5 (very long growing season) would probably result in a decrease in final yield in a warmer-climate scenario. However, uing soybean cultivars of maturity groups 1 and 2 (with slightly longer growth duration than the 0 maturity group of the current cultivar "Apache") could increase soybean productivity in Austria under a warmer climate.

Switching from maize hybrids with a long to a short or very short growing season projected an additional decrease of final yield under a potential warming in Bulgaria. However, using hybrids with a medium growing season, would be beneficial for maize productivity (Figure 13). Technological innovations, including the development of new crop hybrids and cultivars that may be bred to better match the changing climate, are considered as a promising adaptation strategy. However, the cost of these innovations is still unclear.

FIGURE 12. Simulated spring barley yield in Linz (Upper Austria) for the CGCM1 scenario in the 2080s under different reproductive period duration; WOFOST model.

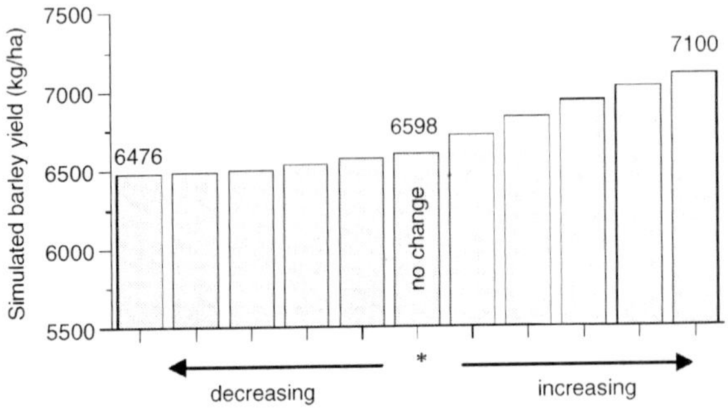

FIGURE 13. Simulated maize yield in Ognjanovo (Southern Bulgaria) under the HadCM2 transient climate change scenarios for the 2020s, 2050s, and 2080s and hybrids with different growing duration; CERES model.

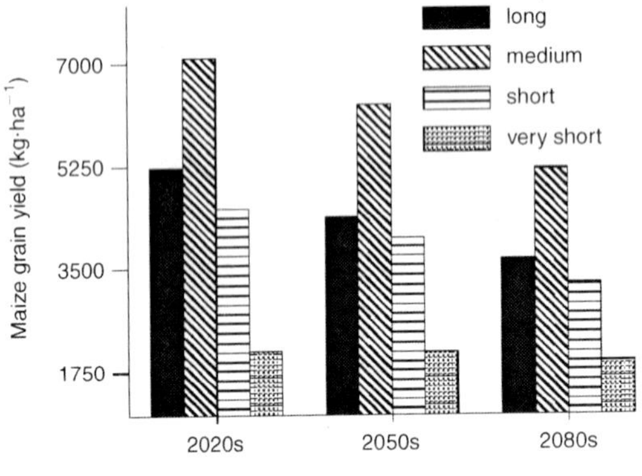

CONCLUDING REMARKS

It is important to emphasize that the GCM climate scenarios are not predictions like weather forecasts. The transient GCM climatic scenarios created in the study are a plausible indication of what the future climate could be like over the 21st century globally and in the selected agricultural regions in Austria and Bulgaria given a specific set of assumptions.

As the results show, warming will generally shorten the growing period of the selected crops in Austria and Bulgaria because of enhanced phenological development. Generally, for winter wheat and spring barley an increase of precipitation in Austria did not have a positive effect on yield since temperature is the dominant yield-limiting factor in the given climatic conditions as it reduces the length of growing season. However, for some light soils with low soil water storage capacity, an increase of precipitation has a positive effect on final grain yield. Soybean in Austria, especially on soils with low soil water storage capacity, showed the greatest sensitivity to changes in precipitation, in contrast to heavy soils.

All transient GCM climate change scenarios for the 21st century, including the adjustment for only air temperature, precipitation and solar radiation, projected reductions in winter wheat yield both in Austria and Bulgaria and spring barley yield in Austria, mainly caused by a shorter vegetation period. However, when the direct effect of an increased CO_2 level was assumed, most GCM climate change scenarios projected an increase in wheat and barley yield and especially in soybean yield. Because maize is a C_4 crop, an increased level of CO_2 alone had no significant impact on final grain yield.

Farmers and agricultural systems will, in fact, try to adjust to changing environmental conditions. Results from the adaptation assessments suggest that possible changes in sowing date as well as hybrid and cultivar selection could be beneficial for agricultural crop production in the countries in Central and Southeastern Europe during the 21st century. The results of the study are generally consistent with those of other studies investigating potential effects of climate change in Europe (e.g., Downing et al., 2000; Harrison et al., 1995; Parry, 2000; Nonhebel, 1996; Semenov et al., 1996).

It should be emphasized that the DSSAT CERES and WOFOST models embody a number of simplifications and limitations. For example, many physiological processes and their response to diurnal weather conditions are not simulated on an hourly base but using daily incremental weather data. The impact of weeds, diseases and insect pests on crop growth, development and final yield formation were assumed to be controlled. There were also no limitations from soil conditions (e.g., salinity or acidity) and no extreme weather events, such as heavy rains, hailstorms and dry winds, or changes in the frequency of drought occurrence. Advances in cultivars through plant breeding were not considered. Technology and land use were assumed to be constant for the vulnerability assessments, even though it is certain that they will change in the future. Thus the effects of climatic change on yields in farmers' fields may be different from those simulated by the DSSAT CERES and CROPGRO as well as WOFOST crop growth models.

There are many sources of potential uncertainties in climate change impact studies–uncertainties resulting from both unknown future developments and biases or variability in databases and methods. Some of the major uncertainties can be attributed to the impact crop models. Most crop growth models have been validated using data from across a wide range of currently observed climates and may therefore be seen as being validated in an extrapolated sense as well. Some crop model comparisons have been performed with respect to their sensitivity to climate changes (e.g., Wolf et al., 1996) and response to selected climate change scenarios (e.g., Semenov et al., 1996). There are sometimes manifest differences between models and their sensitivity (e.g., Carter et al., 1999) as shown also in this study. One of the major uncertainty factors is, for example, how the models simulate water stress or the response to increased CO_2 under different environmental conditions. These and other factors related to the differing sensitivities of impact models should be taken into account when estimating uncertainties in the near future. For example, further comparison between the DSSAT and WOFOST model results and analysis of uncertainties need to be done.

REFERENCES

Alexandrov, V. (1999). Vulnerability and adaptation of agronomic systems in Bulgaria. *Climate Research* 12(2-3):161-173.

Alexandrov, V. and G. Hoogenboom. (2000). The impact of climate variability and change on crop yield in Bulgaria. *Agricultural and Forest Meteorology*, 104(4): 315-327.

Alexandrov, V., J. Eitzinger, V. Caijc and M. Oberfoster. (2002). Potential climate change impacts on selected agricultural crops in northeastern Austria. *Global Change Biology* 8(4):372-389.

Alexandrov, V., J. Eitzinger and M. Oberforster. (2001). Adaptation of crop-weather models in Austria and Bulgaria. *Proceedings of the European Conference on Applied Meteorology (ECAM)*, Budapest, Hungary, 22 pp. (CD version).

Argonne National Laboratory (ANL). (1994). *Guidance for Vulnerability and Adaptation assessments.* U.S. Country Studies Program, Washington DC, USA, 205 pp.

Boogaard, H., C. Van Diepen, R. Roetter, J. Cabrera and H. Van Laar. (1998). *WOFOST 7.1 User's Guide for the WOFOST 7.1 Crop Growth Simulation Model and WOFOST Control Center 1.5.* SC-DLO, IRRI, Wageningen, 144 pp.

Carter, T.R., M. Hulme and D. Viner. (1999) *Representing Uncertainty in Climate Change Scenarios and Impact Studies.* ECLAT-2 Report No.1, Helsinki Workshop, 1999, CRU, Norwich, UK, 130 pp.

Cramer, W., R. Doherty, M. Hulme and D. Viner. (eds.) (2000). *Climate Scenarios for Agricultural, Forest and Ecosystem Impacts.* ECLAT-2 Report No.2, Potsdam Workshop, 1999, CRU, Norwich, UK, 120 pp.

Cure, J.D. and B. Acock. (1986). Crop responses to carbon dioxide doubling: a literature survey. *Agricultural and Forest Meteorology* 38:127-145.

Downing, T.E. P.A. Harrison, R.E. Butterfield and K.G. Lonsdale. (eds.) (2000). *Climate Change, Climatic Variability and Agriculture in Europe. An Integrated Assessment,* Research Report No. 21, Brussels, Belgium: Commission of the European Union, Contract ENV4-CT95-0154, 445 pp.

Dubrovsky, M., Z. Zalud and M. Stastna. (2000). Sensitivity of CERES-Maize yields to statistical structure of daily weather series. *Climatic Change* 46:447-472.

Eitzinger, J, V. Alexandrov, V. Cajic and H. Formater. (2000). A site specific study on the potential range of climatic change impact on soil water balance and crop production under consideration of various models. *Proceedings of the 3rd European Conference on Applied Climatology "Tools for the Environment and Man of the Year 2000"* Pisa, Italy, 6 pp. (CD version).

Eitzinger, J., Z. Zalud, V. Alexandrov, C.A. Diepen, M. Trnka, M., Dubrovsky, D. Semeradova and M. Oberfoster. (2001). A local simulation study on the impact of climate change on winter wheat production in north-eastern Austria. *Austrian Journal of Agricultural Research* 52(4):199-212.

FAOSTAT (1998). *On-Line and Multilingual Database*, FAO, Rome.

Goudriaan, J. and M.H. Unsworth. (1990). Implications of increasing carbon dioxide and climate change for agricultural productivity and water resources. In *Impact of Carbon Dioxide, Trace Gases and Climate Change on Global Agriculture*. ASA Spec. Publ, No.53, American Society of Agronomy, Madison, USA, pp. 111-120.

Harrison, P. and R. Butterfield. (1995). Effects of climate change on Europe-wide winter wheat and sunflower productivity. *Climate Research* 7:225-241.

Harrison, P., R. Butterfield and T. Downing. (eds.) (1995). *Climate Change and Agriculture in Europe-Assessment of Impacts and Adaptation*, University of Oxford, UK, 411 pp.

Hoogenboom, G., P.W. Wilkens and G.Y. Tsuji. (1999). *DSSAT v3*, Vol. 4. University of Hawaii, Honolulu, Hawaii.

Houghton, J., L. Meira Filho, B. Callander, N. Harris, A. Kattenberg and K. Maskell (eds.) (1996). *Climate Change 1995–The Science of Climate Change*. Contribution of WG I to the Second Assessment Report of the IPCC, 570 pp.

Idso, K.E. and S. Idso. (1994). Plant responses to atmospheric CO_2 enrichment in the face of environmental constraints: a review of the past 10 years research. *Agricultural and Forest Meteorology* 69:154-203.

Intergovernmental Panel on Climate Change (IPCC) (1997). *The Regional Impacts of Climate Change: An Assessments of Vulnerability*, IPCC, 22 pp.

Intergovernmental Panel on Climate Change Data Distribution Centre (IPCC DDC). (1999). Data/Information Supplied by the IPCC Data Distribution Centre for Climate Change and Related Scenarios for Impact Assessments, Version 1.0, April 1999.

Intergovernmental Panel on Climate Change–Task Group on Scenarios for Climate Impact Assessment (IPCC-TGCIA). (1999). *Guidelines on the Use of Scenario Data for Climate Impact and Adaptation Assessment*, IPCC, 69 p.

Intergovernmental Panel on Climate Change Working Group I Third Assessment Report (IPCC WG2 TAR) (2001). *Climate Change 2001: Impacts, adaptation and Vulnerability*. http://www.ipcc.ch/pub/tar/wg1/index.htm (accessed on 4 October).

Intergovernmental Panel on Climate Change Working Group II Third Assessment Report (IPCC WG2 TAR) (2001). *Climate Change 2001: Impacts, Adaptation and Vulnerability*. http://www.ipcc.ch/pub/tar/wg2/index.htm (accessed on 4 October 2002).

International Benchmark Sites Network for Agrotechnology Transfer (IBSNAT). (1993). *The IBSNAT Decade*. University of Hawaii, Honolulu.
Li, A., Y. Hou and A. Trent. (2001). Effects of elevated atmospheric CO_2 and drought stress on individual grain filling rates and durations of the main stem in spring wheat. *Agricultural and Forest Meteorology* 106:289-301.
Nonhebel, S. (1996). Effects of temperature rise and increase in CO_2 concentration on simulated wheat yields in Europe. *Climatic Change* 34:73-90.
Parry, M. (ed.) (2000). *Assessment of Potential Effects and Adaptations for Climate Change in Europe. The Europe Acacia Project*. Jackson Environment Institute, University of East Anglia, Norwich, UK, 320 pp.
Pickering, N., J. Hansen, J. Jones, C. Wells, V. Chan and D. Godwin. (1994). Weatherman: A utility for managing and generating weather data. *Agronomy Journal* 86: 332-337.
Raschi, A., F. Miglietta, R. Tognetti and P.R. Van Gardingen. (eds.) (1997). *Plant Responses to Elevated CO_2*. Cambridge University Press, UK, 272 pp.
Reilly, J. (1996). Agriculture in a changing climate: Impacts and adaptation. In *Climate Change 1995–Impacts, Adaptation and Mitigation of Climate Change*, eds. Watson, R., M. Zinyowera and R. Moss. Contribution of WG II to the Second Assessment Report of the IPCC, Cambridge Univ. Press.
Riha, S., D.S. Wilks and P. Simoens. (1996). Impact of temperature and precipitation variability on crop model predictions. *Climatic Change* 32:293-311.
Rosenzweig, C. and A. Iglesias. (eds.) (1994). *Implication of Climate Change for International Agriculture: Crop Modeling Study*. US EPA.
Sathaye, J.A., R.K. Dixon and C. Rosenzweig. (1997). Climate change country studies. *Applied Energy* 36(3-4):225-235.
Schulze, E. and M.M. Caldwell. (eds). (1994). *Ecophysiology of Photosynthesis*. Springer-Verlag, 576 pp.
Semenov, M.A. and J.R. Porter. (1995). Climatic variability and the modelling of crop yields. *Agricultural and Forest Meteorology* 73(3-4):265-284.
Semenov, M.A., J. Wolf, L.G. Evans, H. Eckersten and A. Iglesias. (1996). Comparison of wheat simulation models under climate change. II. Application of climate change scenarios, *Climate Research* 7:271-281.
Sirotenko, O.D., H.V. Abashina and V.N. Pavlova. (1997). Sensitivity of the Russian agriculture to changes in climate, CO_2 and tropospheric ozone concentrations and soil fertility. *Climatic Change* 36:217-232.
Slavov, N., and G. Georgiev. (1985). Estimating photosynthetic active radiation (PhAR) by sunshine duration. *Crop Physiology* 11(4):65-71 (In Bulgarian).
Tegart, W., G. Sheldon and G. Griffiths. (1990). *Climate Change–The IPCC Impacts Assessment*. WMO/UNEP Intergovernmental Panel on Climate Change. Australian Government Publishing Service, Canbera, Australia.
Thornton, P. and G. Hoogenboom. (1994). A computer program to analyze single-season crop model outputs. *Agronomy Journal* 86:860-868.
Tsuji, G., G. Hoogenboom and P. Thornton. (1998). *Understanding Options for Agricultural Production*. Kluwer Acad. Publ., 399 pp.
Tsuji, G., G. Uehara and S. Balas. (eds.) (1994). *DSSAT version 3*, Vol. 1-3, IBSNAT, University of Hawaii, Honolulu.

Van Diepen, C., J. Wolf, H. Van Keulen and R. Rappolt. (1989). WOFOST: a simulation model of crop production. *Soil Use Manage* 5:16-24.

Watson, R. M. Zinyowera and R. Moss. (eds.) (1996). *Climate Change 1995–Impacts, Adaptation and Mitigation of Climate Change.* Contribution of WG II to the Second Assessment Report of the IPCC, Cambridge Univ. Press.

Wilks, D.S. and S.J. Riha. (1996). High-frequency climatic variability and crop yields. *Climatic Change*, 32:231-235.

Wolf, J., and C.A. Van Diepen. (1995). Effects of climate change on grain maize yield potential in the European Community. *Climatic Change* 29(3):299-331.

Wolf, J., L.G. Evans, M.A. Semenov, H. Eckersten and A. Iglesias. (1996). Comparison of wheat simulation models under climate change. I. Model calibration and sensitivity analyses. *Climate Research* 7:253-270.

World Meteorological Organization (WMO). (2000). *WMO Statement on the Status of the Global Climate in 1999, World Meteorological Organization*, WMO-No. 913, Geneva, 12 pp.

Production Ecophysiology of Hungarian Green Pepper Under Elevated Air CO_2 Concentration

Lajos Helyes
Zoltán Tuba
János Balogh
Katalin Réti

SUMMARY. Production, dry matter (including reproductive) allocation, photosynthesis, transpiration, water use efficiency and carbon and nitrogen responses of a Hungarian sweet pepper (*Capsicum annum* L.) under continuous elevated CO_2 concentrations are reported. Plants were grown in open top chambers under a temperate-continental climate in Hungary from plantation at ambient (350 μmol mol^{-1}) and elevated

Lajos Helyes is affiliated with the Department of Horticulture, Faculty of Agricultural and Environmental Sciences, Szent István University, Gödölõ, H-2103, Páter K. u.1., Hungary.

Zoltán Tuba and János Balogh are affiliated with Plant Ecology Departmental Research Group of Hungarian Academy of Sciences and the Department of Botany and Plant Physiology, Faculty of Agricultural and Environmental Sciences, Szent István University, Gödöllõ, H-2103, Páter K. u.1., Hungary.

Katalin Réti is PhD Student, Department of Botany and Plant Physiology, Faculty of Agricultural and Environmental Sciences, Szent István University, Gödöllõ, H-2103, Páter K. u.1., Hungary.

The present work was supported by the Hungarian Széchenyi Agroecology (OM-3B/0057/2002) project.

[Haworth co-indexing entry note]: "Production Ecophysiology of Hungarian Green Pepper Under Elevated Air CO_2 Concentration." Helyes, Lajos, et al. Co-published simultaneously in *Journal of Crop Improvement* (Food Products Press, an imprint of The Haworth Press, Inc.) Vol. 13, No. 1/2 (#25/26), 2005, pp. 333-344; and: *Ecological Responses and Adaptations of Crops to Rising Atmospheric Carbon Dioxide* (ed: Zoltán Tuba) Food Products Press, an imprint of The Haworth Press, Inc., 2005, pp. 333-344. Single or multiple copies of this article are available for a fee from The Haworth Document Delivery Service [1-800-HAWORTH, 9:00 a.m. - 5:00 p.m. (EST). E-mail address: docdelivery@haworthpress.com].

http://www.haworthpress.com/web/JCRIP

Digital Object Identifier: 10.1300/J411v13n01_15

(700 μmol mol^{-1}) CO_2 concentrations. The CO_2 assimilation responses of the green pepper plants grown in high CO_2 from plantation until final harvest reflected down-regulation of their photosynthesis. The integrated and combined effect of the increased net CO_2 assimilation rate and the unchanged rate of transpiration resulted in higher WUE at elevated CO_2 concentrations in the high CO_2 plants than in the control ones grown at ambientCO_2. However, the improved water use efficiency in the high CO_2 plants was not followed by an acclimation in C-translocation and C-allocation to the reproductive organs in the required degree. This was reflected in a slightly increased overall plant production and higher reproductive allocation, but was not accompanied by an increased fresh or dry berry mass production. The acclimation discussed may be of advantage for plant growth in a high CO_2 environment with restricted water availability. We did not find worthy statistical difference between the yield mass of the control and elevated CO_2, although the dry matter production parameters of the high CO_2 plants had statistically not significantly higher values.

KEYWORDS. Antioxidants, CO_2 assimilation, dry matter and reproductive allocation, stomatal conductance, transpiration, water use efficiency, yield

ABBREVIATIONS. A = CO_2 assimilation rate; Ca = ambient CO_2 concentration; Ci = intercellular CO_2 concentration; dm = dry mass; E = transpiration rate; gs = stomatal conductance; RA = reproductive allocation; Rubisco = RuBP carboxylase/oxygenase; WUE = water use efficiency

INTRODUCTION

There have been a large number of reports on plant responses to increased concentrations of atmospheric CO_2; however, no definitive picture can be drawn up on the impacts of long-term exposure to high CO_2 as yet. Reports on the changes in plant growth and production have shown different degrees of responsiveness and have sometimes been conflicting (e.g., reviews by Jarvis, 1993 and Dahlman, 1993). These differences may have been due to the differences in the protocols used, differences in the length of period of exposure and CO_2 concentration. C3 plants generally have been shown to respond to high CO_2 with increased growth and production, sometimes dramatically so (Kramer, 1981; Dahlman, 1993).

Exposure to high CO_2 caused increased CO_2 assimilation and decreased the rate of transpiration with a resultant increase in water use efficiency (Maidner and Mansfeld, 1965; Farquhar and Sharkey, 1982; Sage and Sharkey, 1987). In long-term studies sometimes the CO_2 assimilation rate in high CO_2 plants was lower than those grown at ambient concentration (Mauney et al., 1979; DeLucia et al., 1985; Sasek et al., 1985). This has been explained by a down-regulation of photosynthesis during acclimation (e.g., Eamus and Jarvis, 1989). An accumulation of starch seems to be characteristic during acclimation and an inverse correlation was found between photosynthesis rate and starch accumulation (e.g., Cave et al., 1981; Peet et al., 1986). Stomatal conductance and transpiration rates are reported to be decreased in the high CO_2 grown plants while their water use efficiency of photosynthesis is considerably increased (Radoglou et al., 1992). Another feature of high CO_2 plants is the higher allocation of carbon to roots and the development of larger root systems (Rogers and Dahlman, 1993). This has been interpreted as the plants response to maintain an equilibrium C:N ratio in the leaves (Jarvis, 1993). The above changes largely apply to the various crops (Sionit et al., 1981; Havelka et al., 1984; Du Cloux et al., 1989). Flowers and vegetables show very positive effects from CO_2 enrichment by increased dry weight, plant height, number of leaves and lateral branching (Mortensen 1982). If the CO_2 content of the air increases by about 300 μmol mol^{-1}, an increase in crop productivity of about 30% is forecast, with varied manifestations in different species (Kimball, 1983).

According to Vijverberg and van Uffelen (1977), sweet pepper yield was 33% higher when plants were grown under CO_2 concentration of 1000 ppm.

In the meantime, there is a dearth of information on long-term production responses of sweet pepper plants to high CO_2 and there are caveats in our understanding of how elevated air CO_2 affects its production, C-allocation, and more importantly reproductive allocation.

We report here the results of a high CO_2 exposure study on a Hungarian green pepper (*Capsicum annuum* L.). Dry matter production, dry matter allocation (including reproductive allocation), photosynthesis, transpiration and water use efficiency responses were compared in paprika plants grown from the plantation continuously at ambient (350 μmol mol^{-1}) and high (700 μmol mol^{-1}) CO_2 concentrations.

MATERIALS AND METHODS

Species Examined, Growth Conditions, and CO_2 Fumigation

The experiment was carried out in the "Global Climate Change and Plants" Long-term Experimental Ecological Research Station in the Botanical Garden of the Department of Botany and Plant Physiology at Szent István University, Gödöllõ, Hungary. The climate of the region is temperate-continental with hot dry summers; annual mean temperature is 11°C; annual precipitation is 500 mm. The soil is light moderately calcareous sandy soil. NPK fertilisers were applied at rates of 105, 85 and 175 kg ha^{-1} at the beginning and 0.01% nutrient solution was given at bi-weekly intervals. Plantation bed was prepared with compost mixed into the topsoil. A hybrid cultivar, HRF (F_1) green pepper (*Capsicum annum*) was used. The plants were sown on 2 April 2001 and planted at a spacing of 7.8 plant m^{-2} on 25 May 2001. Perspex open top chambers with a diameter of 128 cm and height of 130 cm were set up at the beginning of April and CO_2 concentrations were kept at ambient (350 μmol mol^{-1}) for the control and at 700 μmol mol^{-1} for the high CO_2 treatment. An Infralyt-4 IRGA (VEB Junkalor, Dessau, Germany) controlled a magnetic valve and regulated the rate of CO_2 flow from the supply cylinders to the chambers. Axial ventilators ensured the mixing of CO_2 with air before it entered the chambers through four inlets. Identical ventilation was used in the control chambers. Weed control was by mechanical means and the chambers were occasionally irrigated with the amount of 10-20 l water during the growing period.

The fruit-production, physiological and biochemical measurements were carried out at the Department of Botany and Plant Physiology (Szent István University, Gödöllõ, Hungary). The harvesting period started on 4th July 2001 and there were 9 harvests during the growing period. The last harvest was completed on 9th October 2001, when final production and allocation parameters were also measured.

CO_2 Assimilation and Transpiration Rate Measurements

Light saturated net CO_2 assimilation rate were measured on attached leaves using an LCA2-type IRGA system (ADC Co. Ltd., Hoddesdon, UK), operated in differential mode and Parkinson LC-N leaf chamber with an air flow of 300 ml min^{-1} at ambient and 700 μmol mol^{-1} CO_2 concentrations, which were produced by a gas diluter (GD 600, ADC Co. Ltd., Hoddesdon, UK). Photosynthetic active radiation (PAR) and leaf temperature were kept constant at 800 μmol m^{-2} s^{-1} and 20 ± 0.5°C, respectively, by using a light and temperature-controlled air-conditioned chamber (Tuba, 1987) and videolamp (55060 × 6.35-25 1000 W Tungsram, Hungary). Prior to measurements leaves were acclimated to the measurement conditions for at least 20 minutes. Gas-exchange parameters (A, E, Ci, gs) were calculated according to von Caemmerer and Farquhar (1981).

Water Use Efficiency of Photosynthesis

Water use efficiency of photosynthesis (WUE) is expressed as the ratio of net CO_2 assimilation rate to transpiration rate. Leaf area was measured by a leaf area meter (LAM 001, Delta T Devices, Cambridge, and UK).

Carbohydrate Analysis

The soluble sugar content was measured using the phenol-sulphuric acid reaction (5% phenol (v/v) and concentrated sulphuric acid) according to Dubois et al. (1956). The starch content was determined by the method of McCready et al. (1950), using of the anthrone reaction after hydrolytic digestion of the starch with 35% perchloric acid (v/v).

Dry Matter Production and Reproductive Allocation Measurements

Roots, shoots, stems, and reproductive organs were air dried at 90°C until constant mass was achieved. Reproductive allocation (RA), in the present study is given as the ratio of total vegetative biomass: to total reproductive biomass on a dry mass basis (Fekete et al., 1988).

Yield and Growth Measurements

After each harvest the berries were weighed individually and net yield weight was calculated (fresh weight). At the end of the harvesting periods average fresh and dry mass were calculated for plant individuals. Number of buds and flowers and growing fruits were recorded at each harvest, on every plant. These data were used for analysing fruit formation and fruit growth parameters. Leaves were also weighed after cutting and their weight was measured fresh and dry. Specific leaf weight was calculated by using leaf area data.

Statistical Analysis

All measurements were done using 5-10 replicates. F and t-statistics were calculated to determine significant differences (Sváb, 1981). Linear regression and other statistical analyses were done using GB-STAT (Friedman, 1988).

RESULTS AND DISCUSSION

Net CO_2 Assimilation, Transpiration Rate, and Water Use Efficiency

When measured at 350 μmol mol^{-1} CO_2 air concentration the CO_2 assimilation rate of the green pepper grown at elevated CO_2 was similar to those grown under present-day CO_2 concentration (Table1). Net CO_2 assimilation measured at 700 μmol mol^{-1} CO_2 concentrations the plants grown at high CO_2 had a slightly (statistically not significant) lower rate (Table 1). These responses indicate the downward acclimation of photosynthesis in the high CO_2 grown paprika plants. However, this downward acclimation of CO_2 assimilation was not associated with accumulation of starch in leaves (see below), which is characteristic during the downward acclimation.

Exposure of plants to elevated CO_2 generally resulted in positive effects on biomass production due to an increased photosynthesis (reviewed by Ceulemans and Mousseau, 1994). Allocation of assimilates and initial growth stimulation by CO_2 has been shown to be strongly dependent on the level of mineral nutrition (Stulen et al., 1993).

The reduction in Rubisco capacity on exposure to high CO_2 has been reported many times (e.g., Caemmerer and Farquhar, 1981; Sage et al., 1989) and has been explained by a decrease in enzyme activation state or in the amount of Rubisco or both (Sharkey, 1985). As a large proportion of leaf protein-nitrogen of the C_3 plants is found in Rubisco (Evans, 1989), the reduced leaf nitrogen content might have been an indication of a decrease in the amount of Rubisco.

In the investigated plants there was no change in the rate of transpiration due to the high CO_2 environment (Table 1). The acclimation of the stomatal regulation in plants under elevated CO_2 mostly appears to have been downward with the consequence of decreased conductivity (g_s) and transpiration rate (*E*) (Drake et al., 1997). It has long been established that transpiration is directly related to stomatal conductance (Meidner and Mansfield, 1965). Stomatal conductance and transpiration are usually reduced by long-term CO_2 exposure (Rogers et al., 1983; Jarvis, 1993). However in some cases no change or even an increase in transpiration rate was reported (Eamus and Jarvis,

TABLE 1. Net CO_2 assimilation rate (A), transpiration rate (E) and water use efficiency (WUE) in sweet pepper plants (*Capsicum annum*) grown at present (350 μmol mol^{-1}) and elevated (700 μmol mol^{-1}) CO_2 concentrations, measured both at present and elevated CO_2.

	A measured at		E measured at		WUE measured at	
	350 μmol mol^{-1} CO_2	700 μmol mol^{-1} CO_2	350 μmol mol^{-1} CO_2	700 μmol mol^{-1} CO_2	350 μmol mol^{-1} CO_2	700 μmol mol^{-1} CO_2
Grown at present CO_2	13.77 ± 2.4	28.6 ± 3.3	3.8 ± 0.2	3.8 ± 0.5	3.6 ± 0.6	7.6 ± 1.2
Grown at elevated CO_2	12.89 ± 1.6	28.0 ± 6.6	3.6 ± 0.5	3.5 ± 0.5	3.6 ± 0.2	8.0 ± 2.0

1989), but the increased g_s and E similar to that in our study are less frequently documented even in herbaceous species.

The above changes in the net CO_2 assimilation and transpiration rates in the high CO_2 plants increased the WUE (Table 1). The increase in the WUE of these plants occurred primarily due to the increased net CO_2 assimilation rate and not to the decrease in transpiration rate as it is frequently reported (Jarvis, 1993). Thus the increased air CO_2 concentration environment seems to have favourably affected the WUE in the investigated sweet pepper. Increased WUE has long been recognised as a characteristic change in the plants exposed to long-term high CO_2 concentration (Morison, 1993), which can be of particular importance for crops in arid and semi-arid habitats (Sionit et al., 1980; Tuba et al., 1993) and in countries, like Hungary with frequent drought events.

Carbon, Nitrogen Content, C/N Ratio, and Carbohydrate Content

The total nitrogen content was lower in the leaves of the high CO_2 plants (Table 2), although this difference was not statistically significant.

Carbon content was unchanged in the leaves of the plants grown at present day and elevated CO_2 plants. The unchanged leaf carbon content supports the downward acclimation of photosynthesis in the elevated CO_2 sweet pepper plants. As a consequence the leaves C/N ratio was slightly higher in the elevated CO_2 plants. The above changes in leaf N content and C/N ratio reflect the nitrogen dilution effect observed frequently under elevated CO_2 (Körner, 1994). There were no any statistically significant differences between the carbohydrate (soluble sugars and starch) content in the leaves and roots of the plants growing under 350 $\mu mol\ mol^{-1}$ and 700 $\mu mol\ mol^{-1}$ CO_2 air concentra-

TABLE 2. Soluble sugar, starch, carbon and nitrogen contents and C/N ratio in the leaves and roots of sweet pepper plants (*Capsicum annum*) grown at present-day (350 μmol^{-1}) and elevated (700 μmol^{-1}) air CO_2 concentrations.

	Grown at present-day CO_2		Grown at elevated CO_2	
	Leaves	Roots	Leaves	Roots
C ($\mu g\ g^{-1}$)	374772.7 ± 12036.0	394678.7 ± 22794.0	370003.5 ± 4116.1	391300.7 ± 20194.4
N ($\mu g\ g^{-1}$)	39587.7 ± 784.8	30428 ± 3002.4	31700.7 ± 2404.9	29712.7 ± 2225.4
C/N ratio	9.5 ± 0.4	13.0 ± 1.2	11.6 ± 0.7	13.2 ± 0.4
Soluble sugars (mg g^{-1} dw)	86.9 ± 6.5	101.6 ± 19.1	85.6 ± 9.9	99.2 ± 4.1
Starch (mg g^{-1} dw)	113.1 ± 1.0	124.6 ± 4.6	118.9 ± 5.0	132.2 ± 5.6

tions, although the starch content in the roots of the high CO_2 plants was higher.

Dry Matter Production, Allocation, and Yield Pattern

Both the control and treated plant groups had similar shoot length. The number of leaves of the elevated CO_2 plants was seemingly higher but without statistically significant difference. The significantly increased number of buds is especially important from developmental and yield viewpoints (Table 3).

Under elevated CO_2, all vegetative dry matter production parameters (leaf, stem, shoot and root dry mass and whole plant dry mass) have shown an increased tendency. But we have experienced statistically significantly enhanced dry matter accumulation only in the reproductive organs, which led to a remarkably higher reproductive allocation in the elevated CO_2 plants. However, this increased reproductive effort of the high CO_2 plants was indicated only by the almost three times higher number of buds.

According to the individual plant fresh and dry berry yield mass, the increased reproductive allocation effect was not accompanied by an increased fresh or dry berry mass production (Table 4). Further this increased reproduc-

TABLE 3. Individual shoot length, number of leaves, number of buds, dry matter production, shoot to root dry mass ratio, reproductive allocation (RA) and specific leaf mass of sweet pepper plants (*Capsicum annum*) grown at present-day (350 μmol^{-1}) and elevated (700 μmol^{-1}) CO_2 concentrations. * Indicates significant difference at $P < 0.05$. Comparisons were made between the treatments (elevated versus present-day CO_2 concentrations).

	Grown at present-day CO_2	Grown at elevated CO_2
Shoot length (cm)	56.4 ± 10.5	52.4 ± 11.8
Number of leaves	240.8 ± 97.4	357.6 ± 191.0
Number of buds	23.0 ± 24.0	67.8 ± 62.9*
Leaf dry mass (g)	12.1 ± 6.9	14.0 ± 9.6
Stem dry mass (g)	13.7 ± 8.4	15.7 ± 11.3
Shoot dry mass (g)	26.3 ± 15.2	29.7 ± 20.8
Root dry mass (g)	5.6 ± 2.7	6.7 ± 3.6
Whole plant dry mass (g)	31.9 ± 17.8	36.4 ± 24.3
Reproductive organs dry mass (g)	0.1 ± 0.2	0.4 ± 0.4*
Shoot/root dry mass ratio (g)	4.5 ± 0.8	4.0 ± 1.0
Reproductive allocation	229.1 ± 100.7	89.7 ± 59.6*
Specific leaf mass (mg w cm^2)	4.0 ± 6.2	4.6 ± 8.3

TABLE 4. Number of berries, average fresh and dry weight yield of sweet pepper plants (*Capsicum annum*) grown at present-day (350 μmol^{-1}) and elevated (700 μmol^{-1}) CO_2 concentrations.

	Grown at present-day CO_2	Grown at elevated CO_2
Number of berries	15.5 ± 8.0	15.0 ± 10.5
Fresh mass (g)	41.3 ± 6.7	38.5 ± 7.1
Dry mass (g)	2.3 ± 0.8	2.3 ± 0.4

tive allocation was even not expressed in the higher number of berries per plant (Table 4).

The shoot to root dry mass ratio is an indicator of the rate of dry matter allocation from the aboveground parts to roots and this ratio was increased in the sweet pepper plants growing under elevated air CO_2.

The lower shoot/root ratio with the higher root mass and root starch content indicates an enhanced carbon allocation to the roots in the high CO_2 treated plants (Table 1). The explanation offered by Jarvis (1993) for this typical high CO_2 growth response (e.g., Rogers and Dahlman, 1993) is to increase both the C-sink capacity and nitrogen uptake in high CO_2 exposed plants to counter the shift in the C/N ratio and keep it at a constant level. In the sweet pepper this can be seen in the unchanged C/N ratio of the roots (Table 2).

The lower yield mass showed in Table 3 may have been the result of the inhibition of C-allocation to the developing fruits. The mere 13% increase in total plant dry mass production in the high CO_2 plants and the lower economic yield mass conflicted with the results of a large number of high CO_2 exposure studies (e.g., reviewed by Eamus and Jarvis, 1989 and Rogers and Dahlman, 1993). This can probably be explained by the different concentrations of applied CO_2, the different lengths of period to that plant were exposed to high CO_2–in a number of cases CO_2 exposition was not sustained during the whole growing season and restricted only to a certain phenophase–and to a lesser extent by other differences in the protocols used in the various studies.

CONCLUSIONS

The photosynthesis responses in Hungarian green pepper grown in high CO_2 from plantation until final harvest reflected both down-regulation and stimulation effects. The integrated and combined effect of the increased net CO_2 assimilation rate and the unchanged rate of transpiration resulted in higher WUE at elevated CO_2 concentrations in the high CO_2 plants than in the control ones grown at ambient CO_2. However, the higher C-assimilation in high CO_2 plants was followed by an increased C-translocation and C-allocation to the reproductive organs, but this was not accompanied by an increased fresh or dry berry mass production. This was reflected in a statistically not significant increase in the overall plant dry mass production and unchanged sweet pepper berry mass yield. The acclimation discussed may be of advantage for plant growth in a high CO_2 environment with restricted water availability.

REFERENCES

Caemmerer, S. von and G. D. Farquhar. (1981). Some relationship between the biochemistry of photosynthesis and the gas exchange of leaves. *Planta* 153: 376-387.

Cave, G., L. C. Tolley and B. R. Srain. (1981). Effect of carbon dioxide enrichment on chlorophyll content, starch content and starch grain structure in *Trifolium subterraneum* leaves. *Phys. Plant.* 51: 171-174.

Ceulemans, R. and M. Mousseau. (1994). Effects of elevated atmospheric CO_2 on woody plants (Tansley Review No. 71). *New Phytologist* 127:425-446.

Dahlman, R. C. (1993). CO_2 and plants: revisited. *Vegetatio* 104/105: 339-355.

DeLucia, E. H., T. W. Sasek and B. R. Strain. (1985). Photosynthetic inhibition after long-term exposure to elevated levels of atmospheric carbon dioxide. *Photos. Res.* 7: 175-184.

Drake, B. G., M. Gonzalez-Meler and S. P. Long. (1997). More efficient plants: A consequence of rising atmospheric CO_2. *Annual Reviews of Plant Physiology and Plant Molecular Biology* 48: 607-637.

Du Cloux, H., M. Andre, A. Gerbaud and A. Daguenet. (1989). Wheat response to CO_2 enrichment: effect on photosynthetic and photorespiratory characteristics. *Photosynthetica* 23: 145-153.

Dubois, M., K. A. Giles, J. K. Hamilton, P. A. Rebers and F. Smith. (1956). Colorimetric method for determination of sugar and related substances. *Anal. Chem.* 28: 350-356.

Eaumus, D. and P. G. Jarvis. (1989). Direct effects of increase in the global atmospheric CO_2 concentration on natural and commercial trees and forests. *Advances in Ecology Res.* 19: 1-55.

Evans, J. R. (1989). Photosynthesis and nitrogen relationships in leaves of C3 plants. *Oecologia* 78: 9.

Farquhar, G. H. and T. D. Sharkey. (1982). Stomatal conductance and photosynthesis. *Ann. Rev. Plant Physiol.* 33: 317-345.

Fekete, G., Z. Tuba and E. Melkó. (1988). Background processes at the population level during succession in grasslands on sand. *Vegetatio* 77: 33-41.

Friedman, P. (1988). GB-STAT Professional Statistics and Graphics (Version 1.5), Silver Spring, MD.

Havelka, U. D., V. A. Wittenbach and M. G. Boyle. (1984). CO_2-enrichment effects on wheat yield and physiology. *Crop Sci.* 24: 1163-1168.

Jarvis, P. G. (1993). Global change and plant water relations. In: *Water transport in plants under climatic stress*, eds. Borghetti, M., J. Grace, and A. Raschi, Cambridge University Press, pp. 1-13.
Körner, Ch. (1994). Biomass fractionation in plants: a reconsideration of definitions based on plant functions. In: *A whole plant perspective on carbon-nitrogen interactions*, eds. J. Roy and E. Garnier, SPB Acad Publishing, The Hague, The Netherlands, pp. 173-185.
Kramer, P. J. (1981). Carbon dioxide concentration, photosynthesis, and dry matter production. *BioScience* 31: 29-33.
Mauney, J. R., G. Guinn, K. E. Fry and J. D. Hesketh. (1979). Correlation of photosynthetic carbon dioxide uptake and carbohydrate accumulation in cotton, soybean, sunflower and sorghum. *Photosynthetica* 13: 260-266.
McCready, R. M., J. Guggolz, V. Silviera and H. S. Owens. (1950). Determination of starch and amylase in vegetables. *Anal. Chem.* 22: 1156-1158.
Meidner, H. and T. A. Mansfield. (1965). *Physiology of stomata*. McGraw-Hill, London, pp. 179.
Morison, J. I. L. (1993). Response of plants to CO_2 under water limited conditions. *Vegetatio* 104/105: 193-209.
Mortensen, L. M. (1982): Charcoal as a carbon dioxide source in greenhouses. *Gartenbauwissenschaft* 47:14-18.
Peet, M. M., S. C. Huber and D. T. Patterson. (1986). Acclimation to high CO_2 in monoccious cucumbers. II. Carbon exchange rates, enzyme activities, and starch and nutrient concentrations. *Plant Physiol.* 80: 63-67.
Radoglou, K. M., P. Aphalo and P. G. Jarvis. (1992). Response of photosynthesis, stomatal conductance and water use efficiency to elevated CO_2 and nutrient supply in acclimated seedlings of *Phaseolus vulgaris* L. *Annals of Botany* 70: 257-264.
Raschi, A. (ed.) *Water transport in plants under climatic stress*. Cambridge University Press, Cambridge, pp. 1-13.
Rogers, H. H. and R. C. Dahlman. (1993). Crop responses to CO_2 enrichment. *Vegetatio* 104/105: 117-131.
Rogers, H. H., G. E. Bingham, J. D. Cure, J. M. Smith and K. A. Surano. (1983). Responses of selected plant species to elevated carbon dioxide in the field. *J. Environ. Qual.* 12:569-574.
Sage, R. F. and T. D. Sharkey. (1987). The effect of temperature on the occurence of O_2 and CO_2 insensitive photosynthesis in field grown plants. *Plant Physiol.* 84: 658-664.
Sage, R. F., T. D. Sharkey, and J. R. Seeman. (1989). Acclimation of photosynthesis to elevated CO_2 in five C3 species. *Plant Physiol.* 89: 590-596.
Sasek, T. W., E. H. DeLucia and B. R. Strain. (1985). Reversibility of photosynthetic inhibition in cotton after long-term exposure to elevated CO_2 concentrations. *Plant Physiology* 78: 619-622.
Sharkey, T. D. (1985). O_2-insensitive photosynthesis in C_3 plants. *Plant Physiology* 78: 71-75.
Sionit, N., H. Hellmers and B. P. Strain. (1980). Growth and yield of wheat under carbon dioxid enrichment and water stress conditions. *Crop Science* 20: 687-690.
Sionit, N., B. P. Strain and H. Hellmers. (1981). Effects of different concentrations of atmospheric CO_2 on growth and yield components of wheat. *J. Agric. Sci.* 79: 335-339.

Stulen, I. and De Kok, L. J. (1993). Whole plant regulation of sulfur metabolism: a theoretical approach and comparison with current ideas on regulation of nitrogen metabolism. In: *Sulfur nutrition and assimilation in higher plants: regulatory, agricultural and environmental aspects*, eds. De Kok et al., SPB Ac. Publ., The Hague, pp. 77-91.

Sváb, J. (1981). *Biometriai módszerek a kutatásban.* (Biometric methods in research.) Mezõgazdasági Kiadó, Budapest, Hungary.

Tuba, Z. (1987). Light, temperature and desiccation responses of CO_2 exchange in desiccation tolerant moss, Tortula ruralis. In: *Proceedings of the IAB Conference of Bryoecology. Symp. Biol. Hung.*, eds. T. Pócs, T. Simon, Z. Tuba and J. Podani, Vol. 35: Part A. Akadémiai Kiadó, Budapest, pp. 137-150.

Tuba, Z., K. Szente, Z. Nagy, Zs. Csintalan, J. Koch, G.Kemeny, E. Laitat, E., Masarovicova and Z. Takács. (1993). The response of Hungarian loess grassland species to long-term elevated CO_2 (Ecophysiological responses to the first 9,5 months exposure). *Hung. Agric. Res.* 2: 37-40.

Vijverberg, A. J. and J. A. M. van Uffelen. (1977). The application of CO_2 in the culture of sweet peppers. *Acta Horticulturae* 58: 293-296.

Physiological and Yield Responses of Grapevine (*Vitis vinifera* L.) Exposed to Elevated CO_2 Concentrations in a Free Air CO_2 Enrichment (FACE)

Marco Bindi
Antonio Raschi
Mario Lanini
Francesco Miglietta
Roberto Tognetti

SUMMARY. Plants of *Vitis vinifera* (cv. Sangiovese) were submitted to elevated carbon dioxide concentrations in the field, using a FACE array; the experiment took place in a vineyard in Central Italy, throughout two growth seasons. The effects of fumigation on productivity, product quality and plant water relations were evaluated. Both vegetative growth and yield were increased, while no differences were evidenced in plant phenology. A significant decrease in stomatal conductance did not result in marked differences in leaf water potential, while osmotic adjustment

Marco Bindi is affiliated with the Dipartimento di Scienze Agronomiche e del Territorio, Università di Firenze, I-50144 Firenze, Italy.

Antonio Raschi, Mario Lanini, and Francesco Miglietta are affiliated with the Istituto di Biometeorologia, Consiglio Nazionale delle Ricerche, I-50145 Firenze, Italy.

Roberto Tognetti is affiliated with the Dipartimento di Scienze Animali, Vegetali e dell'Ambiente, Università del Molise, I-86100 Campobasso, Italy.

[Haworth co-indexing entry note]: "Physiological and Yield Responses of Grapevine (*Vitis vinifera* L.) Exposed to Elevated CO_2 Concentrations in a Free Air CO_2 Enrichment (FACE)." Bindi, Marco, et al. Co-published simultaneously in *Journal of Crop Improvement* (Food Products Press, an imprint of The Haworth Press, Inc.) Vol. 13, No. 1/2 (#25/26), 2005, pp. 345-359; and: *Ecological Responses and Adaptations of Crops to Rising Atmospheric Carbon Dioxide* (ed: Zoltán Tuba) Food Products Press, an imprint of The Haworth Press, Inc., 2005, pp. 345-359. Single or multiple copies of this article are available for a fee from The Haworth Document Delivery Service [1-800-HAWORTH, 9:00 a.m. - 5:00 p.m. (EST). E-mail address: docdelivery@haworthpress.com].

http://www.haworthpress.com/web/JCRIP

Digital Object Identifier: 10.1300/J411v13n01_16

induced by elevated CO_2 in leaves was evident at the final stage of the growing season, but such an observation was year dependent.

KEYWORDS. Vegetative growth, production, water relations

INTRODUCTION

The global change expected to take place in consequence of increasing atmospheric CO_2 concentration (Keeling and Whorf, 1994) has the potential to alter both the physiological functioning and growth of plants as well as the structure of plant communities. A large number of studies have been dealing with the effects of elevated CO_2 on agricultural crops (Cure and Acock, 1986), forest tree species (Ceulemans and Mousseau, 1994); and other plants types (Bazzaz, 1990; Poorter 1993; Idso and Idso, 1994). The great majority of these studies have concluded that increased atmospheric CO_2 enhanced net photosynthesis, biomass, seed yield, light use efficiency, nutrient use efficiency, plant water potential and water use efficiency. However, the interaction with other environmental factors (in particular, water and nutrient availability, soil and air temperature, light), the different fate of the extra carbon assimilated (different growth, respiration, addition to soil), the difficulty in discriminating long-term adaptive effects and short-term acclimation, often affecting the experiments, and finally the typically species-specific response evidenced in many experimental studies, make any predictive summarization extremely difficult (Amthor, 1995), in particular when data from single species are used to infer the behavior of whole systems.

The studies documenting the effect of elevated CO_2 on fruit trees are relatively limited in number (Idso and Kimball, 1997; Ito et al., 1999). This may be attributed, at least in part, to the practical problems arising when exposing adult trees in the traditional enrichment tools, such as greenhouses and open top chambers. In seedlings of fruit tree species, large photosynthetic and growth responses have been evidenced (Centritto et al., 1999; Pan et al., 1998; Keutgen and Chen, 2001), and the presence of large sinks, such as those represented by growing fruits, is expected to stimulate a strong response (Farrar, 1996). On the other hand, the large effects on fruit production obtained on vegetable species in enriched glasshouses might suggest a similar effect in field fumigated fruit trees, under optimal growth conditions. In the Mediterranean area, present rainfall barely meets potential evapotranspiration, and as a result of the predicted climate change, plant communities may have to face longer drought periods, characterized by high temperatures (Wigley et al., 1984). These climatic effects may thereafter counteract the direct positive effects of elevated CO_2 on photosynthesis and growth, while the direct effects on plant water relations may be critical in determining the overall response of plants.

The variegated nature of European agricultural communities, climate conditions, landforms and soil types (as well as social environments influencing land use and infrastructures) is expected to greatly affect the responsiveness to climate change (Olesen and Bindi, 2002). Mediterranean agro-ecosystems have, however, received less attention compared to temperate agricultural systems, despite this region, which is warm, drought stressed, moderately fertile and densely populated, should host systems fairly responsive to global warming (Moreno and Oechel, 1995). The aim of the present study was to assess the effect of elevated CO_2 on some water relations and growth characters of adult *Vitis vinifera* L. plants fumigated in the vineyard by means of a the Free Air CO_2 Enrichment (FACE) set-up. Since growth and development of grapevine are influenced by environmental factors, such as temperature and radiation (Mullins et al., 1992), fruit and total biomass accumulation may be affected by increased CO_2 through the stimulation of photosynthesis and plant growth. Changes in climatic and atmospheric conditions will probably have a significant impact on grapevine growth.

MATERIALS AND METHODS

Site Details and Experimental Design

The experiment was done on a vineyard of 20-year-old-grapevines (cultivar Sangiovese) in the Chianti Region in Italy (Santa Cecilia, Rapolano Terme, Siena, Lat. 43.25°N, Long. 11.35°E), throughout the growing seasons 1995 and 1996. Grapevines were grown at a spacing of 3 m × 1.5 m, cordon trained (vertical trellised vines) and spur pruned. Six FACE arrays, enclosing 7 plants each, were installed on three rows in March 1994 and fumigation started at bud break in both years. Two CO_2 exposure levels (ambient, and 700 μmol mol^{-1}) were maintained. Each emission array was formed by a rectangular plenum (10 m × 2 m) positioned at the soil surface, connected with 22 vertical pipes. The whole system was assembled from polyvinyl chloride (PVC), using pipes with an internal diameter of 20 cm for the plenum and 2 m long pipes with an internal diameter of 4 cm for the vertical vents. Each vent pipe has 4 emission holes with 2 cm diameter, placed in two groups of two holes each, approximately at 120 and 190 cm above the soil surface. High volume blowers (APEM, Firenze, Italy) were used to blow air into the plenum, through flexible pipes; pure carbon dioxide was mixed with ambient air by injecting it in the flexible pipes. The CO_2 injection rates were regulated using the Proportional Integral Differential (PID) algorithm described by Lewin et al. (1994) and were controlled by motorised metering valves (Zonemaster, Satchwell Control System). The algorithm makes use of both horizontal mass flow based on wind velocity and a PID component based on CO_2 concentrations read in the centre of the array to calculate output voltage used to control the metering valves. The CO_2 concentration in the FACE array is measured by an infra-red gas analyser (WMA-2, PPS, Hitchin, UK) having a 0-2000 μmol mol^{-1} range, with an accuracy of 20 μmol mol^{-1} full scale and a temporal resolution of 1 s. The FACE system used geologic CO_2 obtained from a local company (Geogas S.p.A.) that extracts CO_2 for industrial and alimentary purposes. CO_2 fumigation was done only during daytime hours since natural CO_2 concentrations during nightime were very often above 400 μmol mol^{-1}. Distances between arrays ranged from 10 to 20 m to avoid unintended CO_2 enrichment of the controls. Further details on the performance of the fumigation system are given elsewhere (Bindi et al., 2001a).

Plant Growth and Productivity

At harvest time, destructive biomass samples were collected from two shoots per plant. These shoots were first used to determine shoot parameters (number of nodes, lateral branches, etc.), and were subsequently separated into stems, fruit and leaf components, dried in a ventilated oven at 70°C till constant weight. A multiple correlation analysis was carried out to calculate the relationship between non-destructive and destructive measurements, and was used to infer biomass parameters on the other shoots, on which only non-destructive measurements were performed (Williams et al., 1985). The specific leaf mass (SLA) was determined through the 1995 growing season, by collecting samples of known area from fully developed leaves. Plant phenology was also observed, and the dates of flowering and maturity recorded.

Water Relations

Plant water relations were monitored measuring the leaf osmotic potential at full turgor (ψ_π^{100}) and stem hydraulic parameters throughout 1995 and 1996 growing season, and following diurnal trends of leaf water potential (ψ) and stomatal conductance (g_s).

The osmotic potential at full turgor was determined on 6 samples per treatment by the analysis of pressure-volume curves (Tyree and Hammel, 1972), to estimate the occurrence of osmotic adjustment mechanisms under elevated CO_2. Each sample consisted of one, healthy, sun exposed, fully developed leaf. Each sample was weighted after cutting, then it was quickly placed inside a polyethylene sheath that was heat sealed leaving approximately 3 cm of stem protruding from this enclosure, thus providing a firm but not airtight seal. The protruding stem was fitted with a pre-weighted sap collection assembly. The base of the pressure chamber (Giudice, Pisa, Italy) was lined with moist filter paper to minimize tissue water loss and leaf temperature fluctuations during pressure changes. The flow of nitrogen into the chamber was set at approximately 0.004 to 0.005 MPa s^{-1}, until fluid appeared on the cut surface of the stem. This initial water potential (balance pressure) was recorded and the exudation of fluid from the cut surface absorbed by the sap collection assembly. This pressure was maintained for 2 min; after this exchange time, chamber pressure was slowly reduced to 0.1 to 0.2 MPa below the balance pressure, the sap collection assembly was removed and sealed, and immediately weighted. The stem was then fitted with a new pre-weighted sap collection assembly before the chamber was slowly pressurized to pre-determinate over-pressure of 0.4 MPa, still maintained for 2 min. This procedure was repeated 8 to 10 times, typically providing data points in the range of positive turgor and in the region of zero turgor. The pressure-volume curves were then analysed by transforming the potential isotherm to $1/\psi$ versus the relative water content, RWC (Tyree and Jarvis, 1982). This approach provides a framework for analysing tissue water relations in terms of the relationships between ψ, its component potentials ψ_π (osmotic potential) and ψ_p (turgor potential), and RWC. Extrapolation of the linear portion of RWC = 100% provides an estimation of ψ_π^{100}.

The daily trend of stomatal conductance was determined on the abaxial surface of 6 healthy, sun exposed leaves per treatment by using a steady-state porometer (LI-COR 1600, Lincoln, NE, USA). The natural inclination and azimuth of leaves were maintained during the measurements, each measurement was completed within 30 s and humidity inside the porometer chamber was kept near ambient values. Changes in leaf conductance were considered to reflect variation in stomatal conductance, assuming that boundary layer conductance inside the cuvette was constant and large. In parallel to stomatal conductance measurements, water potential was determined in the field on 6 leaves per treatment by using a portable pressure chamber (PMS 1000, Corvallis, OR, USA); again, sun exposed, healthy leaves were chosen.

Quality Analysis

Grape quality was assessed as total sugar (g l^{-1}) and acid (g l^{-1}) concentrations were estimated at harvest by using two samples of 100 berries per plot. A series of analyses on the principal chemical compounds that determine the basic red wine quality were also performed. Specifically, two sub-samples per array were collected at the final harvest and put in fermentation vats for the micro-vinification. After the first run off (about six months after the harvest) total poly-phenols, total flavonoids, total and free anthocyanin, non-anthocyanin flavonoids, colour intensity and tonality, alcohol, total and volatile acidity, total and free SO_2, and pH were determined. All the grape and wine analyses were made accordingly to the O.I.V. methods (O.I.V., 1990).

RESULTS

Fruit and total biomass accumulation for the two CO_2 treatments in 1995 and 1996 seasons are shown in Table 1. Although no significant differences were detected between the two years, yet vegetative production was generally higher in the year 1996, while yield showed an opposite trend. The effect of fumigation was, however, significant in both seasons, either on vegetative growth and on yield ($P < 0.05$), with increases ranging between 37 and 44% for fruit dry mass and 57 and 37% for vegetative biomass. Specific leaf area decreased with the season in both treatments (Table 2). Differences in specific leaf area between the two treatments were particularly evident in early season, values being lower under elevated CO_2 ($P < 0.05$). Plant phenology did not differ between treatments, day of flowering being 173 and 165 since 01 January, and day of maturity 291 and 274, respectively, in 1995 and 1996.

Daily trends of stomatal conductance usually showed a maximum in morning hours, followed by a midday depression, which continued till afternoon (Figure 1). Differences in stomatal conductance between the two treatments, both 1995 and 1996 growing seasons, were particularly evident in mid-summer ($P < 0.05$). The relationship between maximum stomatal conductance and vapour pressure deficit was not affected by CO_2 fumigation (data not shown).

TABLE 1. Shoot biomass accumulation and yield at final harvest, and plant phenology of grapevine grown at different CO_2 concentrations (ambient and 700 ppm). Mean values are reported ± SE, as well as the ratio FACE/control. See text for significance.

	Shoot + leaf dm g m^{-2}		
Time	Control	FACE	Relative response
1995	408.9 ± 61.3	645.8 ± 79.3	1.57
1996	487.3 ± 48.0	666.4 ± 64.6	1.36
	Berry dm g m^{-2}		
1995	309.4 ± 51.2	424.95 ± 75.3	1.37
1996	252.6 ± 27.1	364.41 ± 31.5	1.44

TABLE 2. Specific leaf area (SLA) in leaves of grapevine grown at different CO_2 concentrations (ambient and 700 ppm). Mean values are reported ± SE. See text for significance.

1995 Julian day	SLA (cm^2 g^{-1} DW)	
	Control	FACE
151	160.28 ± 0.53	136.76 ± 0.59
180	155.59 ± 0.64	135.96 ± 0.41
215	133.93 ± 0.37	128.27 ± 0.30
242	127.12 ± 0.44	122.57 ± 0.68

A decrease in predawn water potential was evidenced at the end of the growing season in 1995, while no differences were recorded for seasonal trend in 1996 (Figure 2). Midday water potential displayed lower values in August in 1995, then recovered in September; in 1996, midday water potential values were constantly low. No significant differences in predawn and midday water potential where evidenced between the two treatments, suggesting a scarce effect of the reduced transpiration (due to elevated CO_2) on plant water status, in absence of strong water stress.

In 1995, CO_2 fumigation consistently affected osmotic potential at saturation in the last part of the growing season, but this effect was not evident in 1996 (Figure 3).

Although the concentrations of acids and sugars were affected by CO_2 concentration throughout the vegetative seasons (data not reported), yet, as the grapes reached the maturity stage, acid and sugar concentrations in the control treatment became closer to those in the enriched treatment, and, at maturity, the CO_2 effect on both quality parameters had almost completely disappeared. The chemical analyses of the wine showed, with a few exceptions, no significant differences among the compounds of wine obtained from control and enriched areas in both seasons (Table 3). These exceptions regarded some pigments (such as total flavonoids, total anthocyanin, non-anthocyanin flavonoids) that in 1996 were statistically ($P < 0.05$) stimulated under elevated CO_2, and resulted in different colour intensity.

DISCUSSION AND CONCLUSIONS

Elevated CO_2 enhances the photosynthesis rates in the leaves of almost all C_3 species which are well supplied with nutrients (Long and Drake, 1992), and an increase in the net CO_2 exchange rate has been evidenced in plants of the

FIGURE 1. Diurnal trends in stomatal conductance (g_s) in leaves of grapevine grown at two CO_2 concentrations (ambient and 700 ppm). Mean values are reported ± SE. See text for significance. Sampling date and treatments are referred to by symbols in the legend.

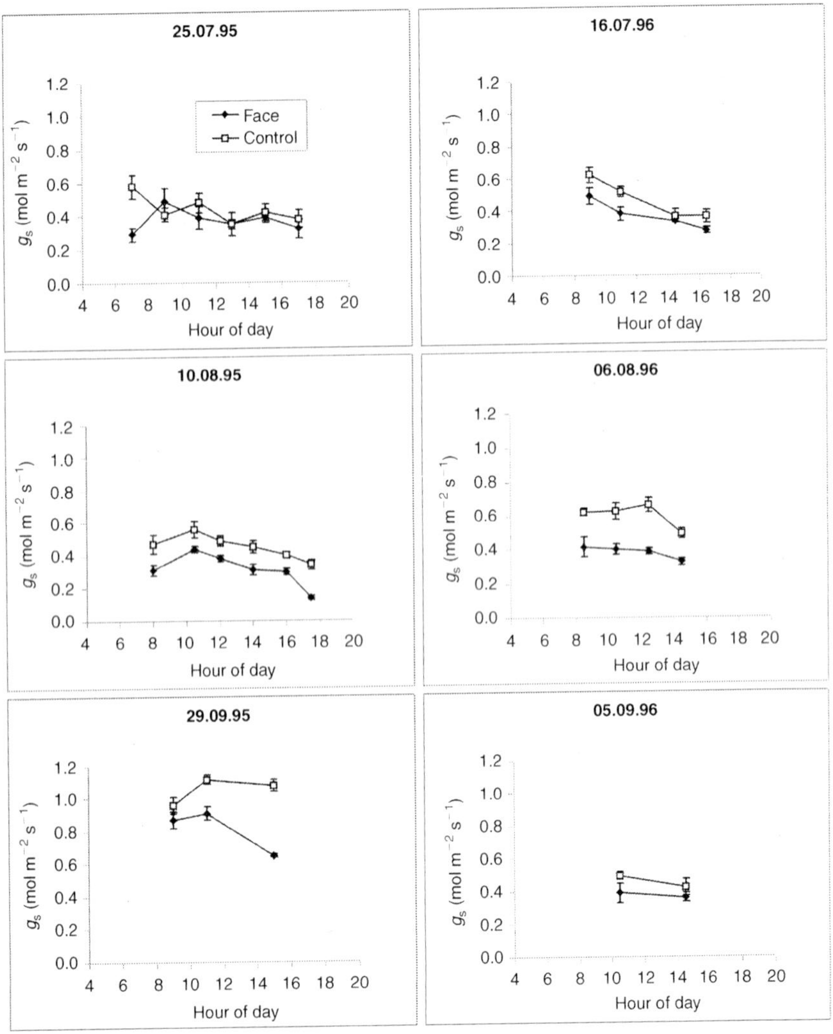

FIGURE 2. Diurnal trends in leaf water potential (Ψ) in grapevine grown at two CO_2 concentrations (ambient and 700 ppm). Mean values are reported ± SE. See text for significance. Sampling date and treatments are referred to by symbols in the legend.

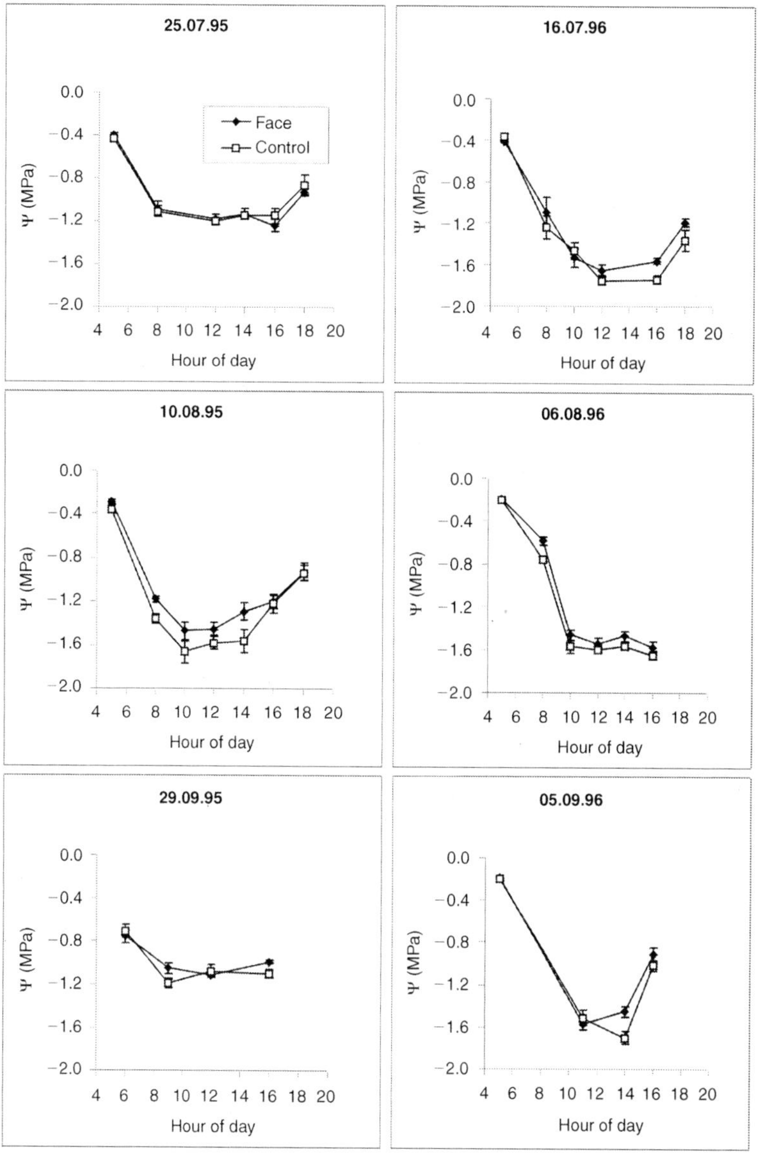

FIGURE 3. Osmotic potential at saturation (Ψ_{π}^{100}) in leaves of grapevine grown at two CO_2 concentrations (ambient and 700 ppm). Mean values are reported ± SE; two asterisks indicate differences between treatments at P<0.01. See Figures 1 and 2 for symbols.

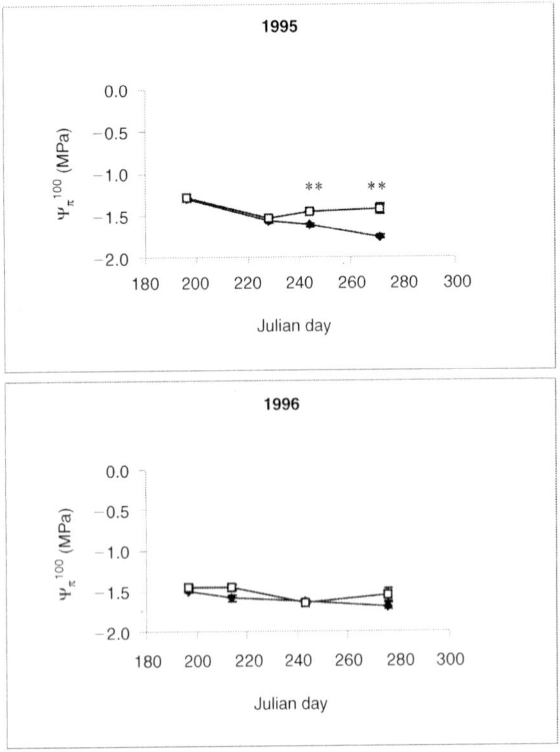

Vitis genus by several authors (Pearcy and Gamon, 1989; Lakso, 1986). It is thereafter not surprising that elevated CO_2 concentrations affected both vegetative growth and fruit production, but it is worth noting that the harvest index did not show relevant variations, as total biomass increased by 45 to 50 percent and harvestable yield increased by 40 to 45 percent. It is well understood that carbon accumulation, at the whole plant level, depends upon the co-ordination between the activity of source organs, transport and sink activity (Farrar, 1996), and, thereafter, transport capacity and sinks relative dimension for assimilates, can affect plant's investment of the extra fixed carbon into structural growth or long-term storage (Schnyder and Baum, 1992; Körner and Miglietta, 1994). On the other hand, to evaluate the patterns of extra car-

TABLE 3. Quality analysis of grapes and wine at final harvest of grapevine grown at different CO_2 concentrations (ambient and 700 ppm). Mean values are reported ± SE, as well as the ratio FACE/control. See text for significance.

Yield quality					
Time	GRAPES	Julian day	Control	FACE	Relative response
1996	Acidity g/l	275	10.65 ± 0.37	9.9 ± 0.22	0.93
1996	Sugar %	275	17.3 ± 0.26	17.6 ± 0.14	1.01
1996	pH	275	3.198 ± 0.090	3.045 ± 0.010	0.95
1995	Harvest index	291	0.43 ± 0.07	0.39 ± 0.05	0.92
1996	Harvest index	275	0.34 ± 0.08	0.35 ± 0.08	1.03
Yield quality					
Time	WINE	Time span	Control	FACE	Relative response
1996	Total polyphenol mg/l	after first fermentation	1286.0 ± 108.0	1398.0 ± 27.0	1.087
1996	Free anthocyanin mg/l	after first fermentation	100.5 ± 10.5	111.0 ± 3.0	1.104
1996	Total flavonoids mg/l	after first fermentation	581.0 ± 50.0	786.5 ± 5.5	1.35
1996	Total anthocyanin mg/l	after first fermentation	145.0 ± 17.0	178.0 ± 16.0	1.23
1996	Flavonoids non-anthocyanin mg/l	after first fermentation	410.0 ± 30.0	576.5 ± 13.5	1.41
1996	Colour intensity	after first fermentation	1.859 ± 0.006	2.656 ± 0.285	1.43
1996	Colour tonality	after first fermentation	0.850 ± 0.070	0.775 ± 0.015	0.91
1996	Alcohol %	after first fermentation	9.560 ± 0.240	10.405 ± 0.255	1.09
1996	Total acidity g/l	after first fermentation	7.100 ± 0.850	8.075 ± 0.075	1.14
1996	Volatile acidity g/l	after first fermentation	0.290 ± 0.010	0.300 ± 0.000	1.03
1996	Free SO_2 mg/l	after first fermentation	39.000 ± 3.000	39.000 ± 1.000	1.00
1996	Total SO_2 mg/l	after first fermentation	72.000 ± 0.000	68.000 ± 8.000	0.94
1996	pH	after first fermentation	3.415 ± 0.085	3.275 ± 0.005	0.96

bon allocation, hidden sinks, should also be considered, such as volatile compounds emission (Veteli et al., 2002; Tognetti et al., 1998), root exudation and turnover (Ponti et al., 2001), whose determinations falls out of the scope of this work. Although both acid and sugar quality components were positively affected by increases in CO_2 concentration, up to a maximum increase at the middle of the ripening season, yet, at maturity, the CO_2 effect on both quality parameters had almost completely disappeared. This was reflected in the results of wine analyses: wine quality was almost unchanged, apart for the colour, in consequence of the increase in the concentration of some compounds; this could result in an increased quality of the final product; yet, this result is difficult to extrapolate to other red cultivars, while an increase in colour might not be appreciated in white cultivars. Anthocyans and poliphenols content have been shown to vary in relation to the plant water status, and the response has been proved to be cultivar-specific (Escalona et al., 1999). These parameters might be thereafter affected by both the climatic aspect of global change, and by elevated carbon dioxide, both through its effect on assimilation, and its effect on plant water relations.

The specific leaf area may provide an estimate of the dry weight of mechanical and conducting tissues (which overwhelms that of parenchyma cells). Extra carbon available under elevated CO_2 might result in additional cell layer or starch accumulation. Indeed, the Specific leaf area was reduced under CO_2 enriched atmosphere; this finding agrees with most studies (Ceulemans and Mousseau, 1994), but not with results obtained by other authors for Mediterranean sclerophyllous species growing nearby moffettes (Jones et al., 1995; Tognetti et al., 1998). It is worth noting that the differences are evident in spring-formed leaves, but tend to be lower in summer season. A higher specific leaf area may contribute to improve drought resistance and to better resist herbivores. No differences were evidenced in phenology, while a shortening of the growth cycle was somewhat observed. A stronger effect of elevated CO_2 on phenology may, however, be expected if changes in temperatures are added to experimental designs aimed to study global warming.

Indication of osmotic adjustment induced by elevated CO_2 in leaves of grapevine was evident at the final stage of the growing season, but such an observation was year dependent. The present study confirms that stomatal conductance generally decreases under elevated CO_2 and that the effect is not transitory but persists over the long term (Morison, 1998). The reductions in stomatal conductance of grapevine fumigated with CO_2 observed in mid-summer did not result in marked differences in leaf water potential between the two treatments. In general, it is assumed that a large genetic range exists in stomatal responsiveness to environmental factors (Chaves et al., 1987; Schultz et al., 1999), and that cultivars adapted to stressful conditions usually display a higher stomatal sensitivity; extrapolation of our results to other cultivars, typical of different pedoclimatic environments, may thereafter be misleading.

In conclusion, our findings suggest that the rising in CO_2 concentration may strongly stimulate grapevine production without having any negative or positive repercussion on grapes and wine quality. However, these positive CO_2-fertilisation effects on grapevine yield may be reduced or completely cancelled by the negative effects determined by the warmer temperatures that are expected to be associated with elevated CO_2 concentration under future climates (decrease in the duration of active photosynthesis, higher respiration, shorter duration of development stages, etc.). In this sense year-to-year variations in physiological response to elevated CO_2 must be taken in account, particularly at Mediterranean latitudes with climate characterised by strong variability. Thus, the studies for predicting the likely impact of climate change should be carried out using tools like crop growth simulation models that allow to couple the hypotheses generated by the output of the General Circulation Models (GCMs) with those obtained from experimental studies (Bindi et al., 1996, 1998a, b).

REFERENCES

Amthor, J.S. 1995. Terrestrial higher-plant response to increasing atmospheric [CO_2] in relation to the global carbon cycle. Global Change Biology, 1: 243-274.

Bazzaz, F.A. 1990. The response of natural ecosystems to the rising global CO_2 levels. Annual Review of Ecology and Systematic, 21: 167-196.

Bindi, M., Fibbi, L., Gozzini, B., Orlandini, S., Miglietta, F. 1996. Modelling the impact of climate scenarios on yield and yield variability of grapevine. Climate Research, 7: 213-224.

Bindi, M., Fibbi, L. 1998. Modelling climate change impacts at the site scale on grapevine. In: Climate Change, Climatic Variability and Agriculture in Europe, eds. Butterfield, R.E., Lonsdale, K.G., Downing, T.E., Final Report (1996-1998) to the Commission of the European Union, Contract ENV4-CT95-0154, pp. 117-134.

Bindi, M., Fibbi, L., Maselli, F., Miglietta, F. 1998. Modelling climate change impacts on grapevine in Tuscany. In: Climate Change, Climatic Variability and Agriculture in Europe, eds. Butterfield, R.E., Lonsdale, K.G., Downing, T.E., Final Report (1996-1998) to the Commission of the European Union, Contract ENV4-CT95-0154, pp. 191-216.

Bindi, M., Fibbi, L., Lanini M., Miglietta F. 2001. Free Air CO_2 Enrichment (FACE) of grapevine (*Vitis vinifera* L.): I. Development and testing of the system for CO_2 enrichment. European Journal of Agronomy, 14: 135-143.

Chaves, M.M., Harley, P.C., Tenhunen, J.D., Lange, O.L. 1987. Gas exchange studies in two Portuguese grapevine cultivars. Physiologia Plantarum, 70: 639-647.

Centritto, M., Magnani, F., Lee, H.S.J. and Jarvis, P.G. 1999. Interactive effects of elevated [CO_2] and drought on cherry (*Prunus avium*) seedlings. II. Photosynthetic capacity and water relations. New Phytologist 141: 141-153.

Ceulemans, R., Mousseau, M. 1994. Effects of elevated atmospheric CO_2 on woody plants. New Phytologist, 127: 425-446.

Cure, J.D., Acock, B. 1986. Crop responses to carbon dioxide doubling: a literature survey. Agricultural and Forest Meteorology, 38: 127-145.

Escalona, J.M., Flexas, J., Schultz, H.R., Medrano, H. 1999. Effect of moderate irrigation on aroma potential and other markers of grape quality. Acta Horticulturae, 493: 261-267.

Farrar, J.F. 1996. Sinks, integral parts of a whole plant. Journal of Experimental Botany, 47: 1273-1280.

Idso, K.E., Idso, S.B. 1994. Plant responses to atmospheric CO_2 enrichment in the face of environmental constraints: a review of the past 10 years' research. Agricultural and Forest Meteorology, 69: 153-203.

Idso, S.B., Kimball, B.A. 1997. Effects of long-term atmospheric CO_2 enrichment on the growth and fruit production of sour orange trees. Global Change Biology, 3: 89-96.

Ito, J., Hasegawa, S., Fujita, K., Ogasawara, S., Fujiwara, T. 1999. Effect of CO_2 enrichment on fruit growth and quality in Japanese pear (*Pyrus serotina reheder* cv. Kosui). Soil Science and Plant Nutrition, 45: 385-393.

Jones, M.B., Clifton-Brown, J., Raschi, A., Miglietta, F. 1995. The effects on *Arbutus unedo* L. of long-term exposure to elevated CO_2. Global Change Biology, 1: 295-302.

Keeling, C.D., Whorf, T.P. 1994. Atmospheric CO_2 records from sites in the SIO air sampling network. In: Trends '93: A Compendium of Data on Global Change, eds. Boden, T.A., Kaiser, D.P., Sepanski, R.J., Stoss, F.W., Oak Ridge, USA: Carbon Dioxide Information Analysis Center, Oak Ridge National Laboratory, pp. 16-26.

Keutgen, N., Chen, K. 2001. Responses of citrus leaf photosynthesis, chlorophyll fluorescence, macronutrient and carbohydrate contents to elevated CO_2. Journal of Plant Physiology, 158: 1307-1316.

Körner, Ch., Miglietta, F. 1994. Long term effects of naturally elevated CO_2 on Mediterranean grassland and forest trees. Oecologia, 99: 343-351.

Lakso, A.N., Reisch, B.I., Mortensen, J. and Roberts, M.H. 1986. Carbon dioxide enrichment for stimulation of growth of in vitro-propagated grapevines after transfer from culture. Journal of the American Society for Horticultural Science 111: 634-638.

Lewin, K.F., Hendrey, G., Nagy, J., LaMorte, R.L. 1994. Design and application of a free air carbon dioxide facility. Agricultural and Forest Meteorology, 70: 15-29.

Long, S.P., Drake, B.G. 1992. Photosynthetic CO_2 assimilation and rising atmospheric CO_2 concentrations. In: Crop Photosynthesis: Spatial and Temporal Determinants, eds. Baker, N.R., Thomas, H., Elsevier, Amsterdam, pp. 69-101.

Moreno, J.M., Oechel, W.C., eds. 1995. Global Change and Mediterranean-Type Ecosystems. Ecological Studies vol. 117, Springer-Verlag: Berlin, Germany, 527 p.

Morison, J.I.L. 1998. Stomatal response to increased CO_2 concentration. Journal of Experimental Botany, 49: 443-452.

Mullins, M.G., Bouquet, A., Williams, L.E. 1992. Biology of Grapevine. Cambridge University Press, New York.

O.I.V., 1990. Recueil des méthodes internationales d'analyse des vine et des mouts. Officielle, Paris.

Olesen, J.E., Bindi, M. 2002. Consequences of climate change for European agricultural productivity, land use and policy. European Journal of Agronomy, 16: 239-262.

Pan, Q., Wang, Z., Quebedeaux, B. 1998. Responses of the apple plant to CO_2 enrichment: changes in photosynthesis, sorbitol, other soluble sugars, and starch. Australian Journal of Plant Physiology, 25: 293-297.

Pearcy, R.W., Gamon, J.A. 1989. Leaf movement, stress avoidance and photosynthesis in *Vitis californica*. Oecologia 79: 475-481.

Ponti, F., Raschi, A., Fumagalli, I., Lanini, M., Minotta, G. 2001. Dynamics of root growth in a grassland submitted to CO_2 fumigation. In: Proceedings of the 6th Symposium of the International Society of Root Research. Nagoya, Japan, 11-15/11/2001.

Poorter, H. 1993. Interspecific variation in the growth response of plants to an elevated ambient CO_2 concentration. Vegetatio, 104/105: 77-97.

Schnyder, H., Baum, U. 1992. Growth of the grain of wheat (*Triticum aestivum* L.). The relationship between water content and dry matter accumulation. European Journal of Agronomy, 1: 51-57.

Schultz, H.R., Lebon, E., Rousseau, C. 1999. Suitability of the Ball, Woodrow, Berry'-Model for the description of stomatal coupling to photosynthesis of different *Vitis* species and *Vitis vinifera* cultivars in different climatic regions at various levels of water deficit. Acta Horticulturae, 493: 17-30.

Tognetti, R., Johnson, J.D., Michelozzi, M., Raschi, A. 1998. Foliage metabolism in *Quercus pubescens* and *Quercus ilex* mature trees with lifetime exposure to local greenhouse effect. Environmental and Experimental Botany, 39: 233-245.

Tyree, M.T., Hammel, H.T. 1972. The measurements of turgor pressure and the water relations of plants by the pressure-bomb technique. Journal of Experimental Botany, 23: 267-282.

Tyree, M.T., Jarvis, P.G. 1982. Water in tissue and cells. In: Encyclopedia of Plant Physiology, New Series, vol 12B. Physiological Plant Ecology II, eds. Lange, O.L., Nobel, P.S., Osmond, C.B., Ziegler, H., Springer-Verlag: Berlin, Germany, pp. 35-77.

Veteli T.O., Kuokkanen K., Julkunen-Tiitto R., Roininen, H., Tahvanainen J. 2002. Effects of elevated CO_2 and temperature on plant growth and herbivore defensive chemistry. Global Change Biology 8: 1240-1252.

Wigley, T.M., Briffa, K.R., Jones, P.D. 1984. Predicting plant productivity and water resources. Nature, 312: 102-103.

Williams, D.W., Williams, L.E., Barnett, W.W., Kelly, K.M., McKenry, M.V. 1985. Validation of a model for the growth and development of the thompson seedless grapevine. I. Vegetative growth and fruit yield. American Journal of Enology and Viticulture, 36: 275-282.

Photosynthesis, Growth, and Yield of Citrus at Elevated Atmospheric CO_2

Joseph C. V. Vu

SUMMARY. Agricultural productivity is expected to be affected by rising atmospheric carbon dioxide concentration ([CO_2]) and changing climate. Increase in [CO_2] *per se* may lead to higher yield for several crop species, including citrus. However, the interactions between elevated [CO_2] and other climatic change factors are complex, and producing crops under global climate changes will be a challenge for world agriculture. This review focuses on our current understanding of citrus photosynthesis, growth, and yield in response to rising atmospheric [CO_2]. There is indeed a critical need to unravel the mechanisms of citrus responses to atmospheric [CO_2] enrichment and climate changes, and to identify the targets for a genetic/breeding approach designed to enhance citrus tolerance/resistance to environmental stresses.

KEYWORDS. Rising atmospheric CO_2, global climate change, citrus, photosynthesis, yield

---

Joseph C. V. Vu is Research Plant Physiologist, Crop Genetics and Environmental Research Unit, U.S. Department of Agriculture-Agricultural Research Service, and Professor, Agronomy Department, Institute of Food and Agricultural Sciences, 304 Newell Hall, P.O. Box 110500, University of Florida, Gainesville, FL 32611-0500 USA (E-mail: jcvu@mail.ifas.ufl.edu).

[Haworth co-indexing entry note]: "Photosynthesis, Growth, and Yield of Citrus at Elevated Atmospheric CO_2." Vu, Joseph C. V. Co-published simultaneously in *Journal of Crop Improvement* (Food Products Press, an imprint of The Haworth Press, Inc.) Vol. 13, No. 1/2 (#25/26), 2005, pp. 361-376; and: *Ecological Responses and Adaptations of Crops to Rising Atmospheric Carbon Dioxide* (ed: Zoltán Tuba) Food Products Press, an imprint of The Haworth Press, Inc., 2005, pp. 361-376. Single or multiple copies of this article are available for a fee from The Haworth Document Delivery Service [1-800-HAWORTH, 9:00 a.m. - 5:00 p.m. (EST). E-mail address: docdelivery@haworthpress.com].

http://www.haworthpress.com/web/JCRIP

Digital Object Identifier: 10.1300/J411v13n01_17

INTRODUCTION

The earth's atmospheric carbon dioxide concentration ([CO_2]) has increased from a mean concentration of approximately 280 parts per million (ppm) since the start of the Industrial Revolution in Western Europe (1750-1800) to about 375 ppm at present. With the rapid increase in world population and economic activity, a doubling of the present atmospheric [CO_2] could be expected before the end of this century (Morison and Lawlor, 1999). As CO_2 is responsible for about 61% of global warming (Shine et al., 1990), a doubling of the atmospheric [CO_2] and a rise in other "greenhouse" gases, including methane, chlorofluorocarbons, nitrous oxide and ozone, would increase the mean global air temperature, possibly by as much as 6°C (Morison and Lawlor, 1999; Schneider, 2001), and cause shifts in regional scale precipitation patterns that would result in unpredictable weather in many areas of the world (Wigley and Raper, 1992; Keeling et al., 1995).

Atmospheric CO_2 is the basis for almost all life on earth. It is the primary raw material used by green plants to produce, through photosynthesis, the basic compounds out of which they construct their tissues and provide sustenance for other living things, ourselves included. Present understanding of photosynthetic carbon metabolism classifies terrestrial plants into three major photosynthetic categories: C_3, C_4, and Crassulacean acid metabolism (CAM). Approximately 95% of terrestrial plant species fix atmospheric CO_2 by the C_3 photosynthetic pathway, while 1% fix CO_2 by the C_4 pathway and 4% by CAM (Bowes, 1993). Current atmospheric [CO_2] restricts the photosynthetic performance, growth and yield of many crop plants, most of which are C_3. Therefore, they are expected to respond to a greater extent to rising atmospheric [CO_2] than C_4 and CAM species. The reason for such an expectation is that current atmospheric CO_2 and O_2 levels result in up to 40% photorespiratory losses in C_3 plants (Matsuoka et al., 2001). Therefore, the assumption that a rise in atmospheric [CO_2] will reduce the deleterious effect of O_2 on C_3 photosynthesis has been supported by experimental research. Exposure of C_3 plants to elevated growth [CO_2] generally results in stimulated photosynthesis and enhanced growth and yield (Kimball, 1993; Poorter et al., 1996; Drake et al., 1997). A compilation of the existing data available from the literature for C_3 agricultural crop species indicates that a doubling of the present atmospheric [CO_2] would increase net photosynthetic rate up to 63%, and growth and yield up to 58% (Kimball, 1983, 1993; Poorter 1993; Poorter et al., 1996; Norby et al., 1999).

In this review, the responses of citrus photosynthesis, growth, and yield to rising atmospheric [CO_2] will be discussed. Comparisons will be made to research conducted on other crop species. Growth and development of citrus, as well as other agricultural crop species, are the results of many interacting processes, including photosynthesis and its direct relationship to crop yield (Vu, 1999). In terms of photosynthesis, the CO_2 exchange rate (CER) is dependent on a multitude of reactions, each with a potentially unique response to environmental factors (Sage and Reid, 1994). Rising atmospheric [CO_2] *per se* could benefit citrus, as well as many other economically important C_3 species, but gains may or may not be realized in long-term growth due to the interaction of various adverse environmental factors that complicate the issue. As with other crop species, the ability of citrus to acclimate and/or compensate to rising atmospheric [CO_2] and other climate change factors is critical to its performance.

CITRUS PHOTOSYNTHESIS AT ELEVATED ATMOSPHERIC CO_2

Photosynthetic Rates

Leaf or whole-tree CER of a variety of citrus genotypes, when measured at the [CO_2] used for growth, is enhanced by elevated [CO_2] (Table 1, column 2) (Dowton et al., 1987; Idso and Kimball, 1992b; Brakke and Allen, 1995; Syvertsen and Graham, 1999; Syvertsen et al., 2000; Jifon et al., 2002; Morinaga, 2002; Vu et al., 2002). In Valencia sweet orange, CER of both pre-existing leaves and new leaves is 16 to 98% higher at 800 ppm than at 400 ppm growth [CO_2] (Downton et al., 1987). In Ambersweet orange, the percentage enhancement in CER by double-ambient growth [CO_2] (720 ppm) is about 44% for the pre-existing mature leaves and 36% for the new young leaves (Figure 1A). Data collected over a three-year period on sour orange trees also show that the mean daylight photosynthetic rate of the leaves under summer conditions in Phoenix, Arizona (USA) is about 2.2-fold greater for the elevated (700 ppm) CO_2 treatment in comparison with the near-ambient (400 ppm) CO_2 treatment (Idso et al., 1991). Furthermore, leaf CER of several citrus genotypes commonly used as rootstocks in commercial citriculture, including lemon, citrange and mandarin, is enhanced up to 145% under elevated [CO_2] (from 720 to 840 ppm) (Brakke and Allen, 1995; Syvertsen et al., 2000).

Although supraoptimum temperatures adversely affect photosynthesis of many crop plants (Morison and Lawlor, 1999), there has been an indication that elevated growth [CO_2] may partially alleviate the high temperature-stressed effects on leaf photosynthetic rate (Long, 1991). This arises from the fact that photorespiration rates increase at higher air temperatures under current atmospheric CO_2 levels, but significantly decrease at elevated [CO_2]. Theoretically, a 300-ppm increase in growth [CO_2] could raise the temperature optimum of light-saturated CER of C_3 plants by 5°C (Long, 1991). Photo-

TABLE 1. Percent enhancement in photosynthesis (CER), growth parameters and fruit yield of various citrus genotypes by double-ambient growth [CO_2].

Citrus Genotypes	% Enhancement in CER	% Enhancement in Growth/Fruit
'Ambersweet' orange (Allen and Vu, 2000; Vu et al., 2002)	36-44	17-49 (specific leaf weight) 15-37 (total tree biomass)
'Ridge Pineapple' sweet orange (Jifon et al., 2002)	55-75	45 (total tree biomass)
'Valencia' sweet orange (Downton et al., 1987)	16-98	25 (specific leaf weight) 70 (fruit)
Sour orange (Idso et al., 1991; Idso and Kimball, 1992,1997)	~120	79 (leaves) 56-240 (branches) 70-2400 (fruit)
Sour orange (Jifon et al., 2002)	36-102	58 (total tree biomass)
'Carrizo' citrange (Brakke and Allen, 1995 Koch et al., 1986,1987)	~100	67-120 (total tree biomass)
'Swingle' citrumelo (Koch et al., 1986)	~100	115 (total tree biomass)
'Cleopatra' mandarin (Syvertsen et al., 2000)	13-37	23-33 (specific leaf weight)
'Troyer' citrange (Syvertsen et al., 2000)	72-145	31-34 (specific leaf weight)
'Volkamer' lemon (Syvertsen et al., 2000)	31-44	13-52 (specific leaf weight)

synthesis measurements made during four consecutive summers for sour orange trees grown outdoors at Phoenix, Arizona show that there is a negative linear relationship between leaf CER and leaf temperature from 31 to 47°C for both ambient and elevated CO_2-grown trees, indicating that this temperature range is above the optimum for net photosynthesis of this citrus species (Idso et al., 1995). Under hot and dry summer conditions in Phoenix, leaf CER of sour orange trees is highest (about 6 and 11 $\mu mol\ m^{-2}\ s^{-1}$ for ambient and elevated CO_2-grown trees, respectively) at the morning's first measurements at 7:00 a.m. (31°C), and steadily declines thereafter. By 2:00 p.m. (47°C), leaf CER of the ambient CO_2-grown sour orange trees drops to near zero, whereas the CO_2-enriched trees still maintain their leaf photosynthetic rate at a sub-

FIGURE 1. Midday CO_2 exchange rate (CER) (**A**) and water-use efficiency (WUE) (**B**) of pre-existing mature leaves and new young leaves of Ambersweet orange trees grown for 29 months under 360 and 720 ppm CO_2. Each data value represents the mean (with SE bar) of 8 to10 determinations.

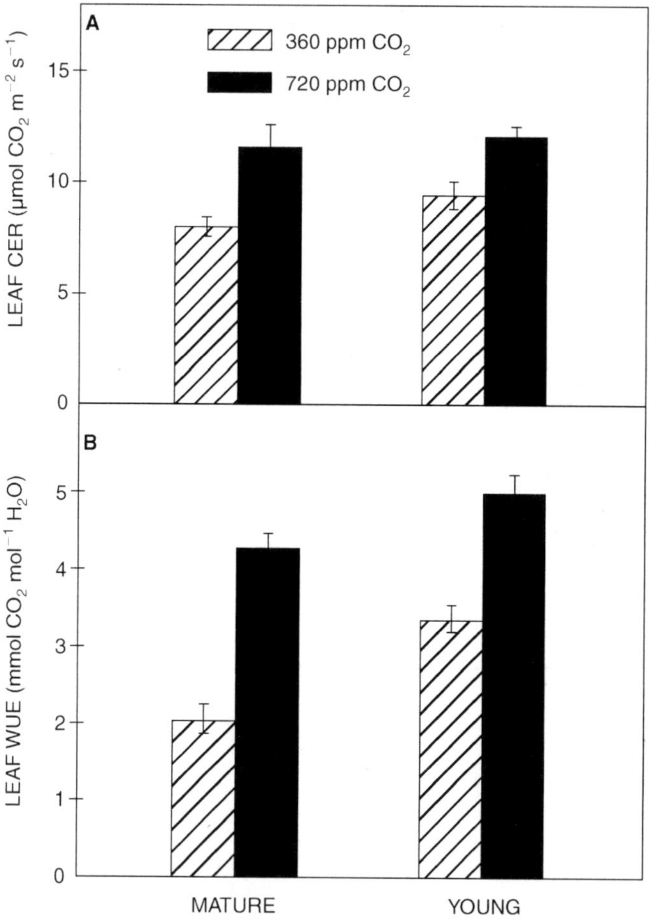

stantially high level (~4 µmol m^{-2} s^{-1}) (Idso et al., 1991, 1995). Nevertheless, the degree of CER enhancement in sour orange by CO_2 enrichment is 75% at a leaf temperature of 31°C, 100% at 35°C, and 200% at 42°C. Similar scenarios, although to a lesser extent, have been reported for soybean (Vu et al., 1997). The enhancements in sour orange leaf CER by elevated [CO_2] fall in the range of the predictions for an idealized C_3 plant, showing that a rise in temperature from 28 to 40°C increases CER enhancement from 66 to 190% when growth [CO_2] is raised from 350 to 650 ppm (Long, 1991). In Ambersweet orange, there is a small and consistent, although not significant, increase in CER at growth temperature of about 6°C above ambient temperature for both pre-existing mature and new young leaves at double-ambient growth [CO_2] (Vu et al., 2002). This indicates that for citrus there are genotype-specific differences in leaf CER responses to elevated growth CO_2 and temperature. In addition, differences in experimental and environmental growth conditions, CO_2 enrichment levels, leaf/tree growth stages used for the measurements, as well as time of the day when CER measurements were carried out, all of these would also contribute to the reported differences on CER and enhancements in CER by high [CO_2] and temperature.

Citrus responds, as do other reported herbaceous and woody species, to elevated atmospheric [CO_2] with a decrease in stomatal conductance, resulting in a reduction in leaf transpiration and consequently an improvement in leaf water-use efficiency (WUE), the ratio of leaf CER to leaf transpiration rate (Syvertsen and Graham, 1999; Syvertsen et al., 2000; Vu et al, 2002). However, the magnitude of stomatal response to elevated [CO_2] is species-specific and is generally smaller in trees than in annual crops (Field et al., 1995; Norby et al., 1999). In Ambersweet orange, stomatal conductance of pre-existing mature leaves of trees grown for more than two years at double-ambient [CO_2] is about 32% lower than their counterparts at ambient [CO_2] (Vu et al., 2002). WUE of the elevated CO_2-grown sweet orange trees is 105% greater for mature leaves and 47% higher for young leaves at near-ambient temperature (Figure 1B). Such enhancements in leaf WUE under elevated growth [CO_2] are even larger at growth temperature of 4.5°C above ambient: 150 and 94% for mature and young leaves, respectively (Vu et al., 2002).

In addition to a potential increase in mean global air temperature, shifts in regional precipitation patterns as a result of rising atmospheric [CO_2] will probably decrease soil water availability in many areas of the world. Soil water deficit, or drought, is the single most important factor limiting crop growth and yield (Hsiao, 1973). Production of agricultural crops, including citrus, in semiarid and arid areas of the world depends heavily on irrigation (Yelenosky, 1991; Vu, 1999). Even in normally humid areas, irrigation systems are installed to prevent yield reduction due to short dry periods. For citrus, as well as many other crops, an increase in drought stress is followed by decreasing CER, and such a reduction has been partially attributed to stomatal closure (Vu and Yelenosky, 1988; Vu et al., 1998; Vu, 1999). In Valencia sweet orange, drought stress also reduces photosynthetic carboxylating enzyme activity and protein concentration and shifts carbohydrate distribution in the leaves (Vu and Yelenosky, 1988, 1989). Unfortunately, not much is known about the interactive nature of rising atmospheric [CO_2] and drought on photosynthesis and growth of citrus.

Leaf Photosynthetic Biochemistry

Long-term exposure of C_3 plants to elevated [CO_2] leads to a variety of acclimation effects, including changes in leaf photosynthetic physiology and biochemistry, and alterations in morphology, anatomy, branching, tillering, timing of developmental events and life cycle completion (Bowes, 1993; Drake et al., 1997). A greater number of mesophyll cells and chloroplasts have been reported for a variety of plants grown at elevated [CO_2] (Thomas and Harvey, 1983; Vu et al., 1989). In terms of leaf photosynthetic physiology and biochemistry, acclimation occurs, ranging from species-specific changes in the CER versus intercellular CO_2 response curves (Radin et al., 1987; Campbell et al., 1988; Sage et al., 1989) to alterations in dark respiration (Drake et al., 1997) and enzyme biochemistry, especially ribulose bisphosphate carboxylase-oxygenase (Rubisco) (Bowes, 1996). Many C_3 species grown for long periods at elevated [CO_2] show a decrease in leaf photosynthesis (Sage et al., 1989; Besford et al., 1990). In addition, carbohydrate source-sink imbalance under growth at elevated [CO_2] is believed to have a major role in the regulation of photosynthesis through feedback inhibition (Arp, 1991; Stitt, 1991; Makino and Mae, 1999).

The photosynthetic carboxylating enzyme Rubisco "fixes" atmospheric CO_2 and thus plays a vital role in plant growth and productivity. Activity of Rubisco, concentration of the Rubisco protein, and the level of the Rubisco small subunit transcript (*rbcS*) have been frequently used as indicators/markers to evaluate the acclimation of leaf photosynthetic capacity at elevated growth [CO_2]. For many C_3 crops including sweet orange, longer exposure to elevated [CO_2] results in a down-regulation of Rubisco activity (Wong, 1979; Vu et al., 1983; Sage et al., 1989; Bowes, 1996; Drake et al., 1997; Vu et al., 1983, 1997, 2002). In addition, elevated [CO_2] affects the relative expression of the Rubisco protein concentration and its transcript levels in a variety of plant species (Gesch et al., 1998; Moore et al., 1998, 1999; Vu et al., 1999, 2001). Under long-term elevated growth [CO_2], some plants show reductions in both Rubisco protein content and *rbcS* abundance, whereas others show decreases in the Rubisco protein but not in the *rbcS* transcript (Moore et al., 1998). In Ambersweet orange, reductions in the Rubisco protein content at a double-ambient growth [CO_2] are substantial (Vu et al., 2002), but attempts to characterize the *rbcS* message levels were unsuccessful. Although no correlation between Rubisco protein content and *rbcS* mRNA abundance was made for sweet orange, total RNA levels in trees grown at double-ambient [CO_2]

were 38-49% less for pre-existing mature leaves, and 19-31% less for young leaves (Vu, unpublished data).

In Ambersweet orange, down-regulation of the Rubisco protein concentration under elevated growth [CO_2] does not entail a change in the level of total leaf soluble protein (Vu et al., 2002). New young leaves contain a similar amount of total soluble protein as pre-existing mature leaves, regardless of the growth [CO_2]. In sour orange, however, leaf soluble proteins show some decrease in concentration in new young leaves from trees grown at elevated [CO_2] (Weber et al., 1994). The decline in Rubisco protein concentration without a reduction in total soluble protein content in Ambersweet orange at elevated growth [CO_2], however, would allow an optimization of nitrogen use, either by reallocating the nitrogen resources away from Rubisco to other proteins within the leaves, or redistributing nitrogen from the photosynthetic proteins of source leaves to sink tissues (Stitt, 1991; Bowes, 1993).

For citrus, the reduction in activity and/or protein concentration of Rubisco at elevated growth [CO_2] may be genotype-specific, since Rubisco activity has been reported to be greater for Swingle citrumelo grown at twice ambient [CO_2], but not in leaf samples from Carrizo citrange (Koch et al., 1986). However, Rubisco activity for these two citrus rootstocks was expressed on a leaf chlorophyll basis, and not on a leaf area basis as for Rubisco activity of Ambersweet orange (Vu et al., 2002). Therefore, claims that Rubisco is modulated by elevated growth [CO_2] requires more careful evaluation, as the basis on which activity and/or concentration of the enzyme are expressed may vary or nullify observations (Bowes, 1993).

In addition to Rubisco, other proteins may be considered as photosynthetic genes, but available information on their responses to elevated growth [CO_2] is limited. In tomato, mRNA levels of Rubisco activase decline under elevated [CO_2] (van Oosten et al., 1994). Activity and mRNA levels of carbonic anhydrase, the enzyme facilitating diffusion of CO_2 from intercellular air spaces, are reduced in some plant species, but unchanged or even increased in the others (Moore et al., 1999). Activities of sucrose phosphate synthase (SPS) and acid invertase and adenosine 5′-diphosphoglucose pyrophosphorylase (ADGP), the enzymes involving in sucrose and starch metabolism, respectively, may be affected by elevated growth [CO_2]. However, the responses of these enzymes to elevated [CO_2] are also species-specific. In rice, leaf SPS activity is enhanced in CO_2-enriched plants, suggesting an acclimation response to optimize the capacity for carbon utilization and export for this crop (Hussain et al., 1999). On the other hand, activity of SPS is down-regulated by elevated [CO_2] in bean, cotton, cucumber, plantain and wheat, but up-regulated in pea, soybean, spinach, sunflower and tomato (Moore et al., 1998). Under elevated growth [CO_2], leaf acid invertase activity is down-regulated in cotton, cucumber, parsley, pea, radish, soybean, spinach, tobacco, and wheat, but up-regulated in bean, plantain, and sunflower (Moore et al., 1998). In Ambersweet orange grown under elevated [CO_2], activities of SPS and ADGP are down-regulated in the pre-existing mature leaves, but not in the new young leaves (Vu et al., 2002).

In sour orange grown at elevated [CO_2], concentrations of glucose, fructose, and sucrose in the new leaves are more than three-fold those of ambient-[CO_2] controls (Weber et al., 1994). In Ambersweet orange, the levels of foliar nonstructural carbohydrates are substantially higher in midday-sampled leaves of trees grown at double-ambient [CO_2] than their counterparts at ambient [CO_2]. At double-ambient [CO_2], glucose is increased up to 115%, fructose up to 164%, sucrose up to 73%, total soluble sugars up to 50%, starch up to 424%, and total nonstructural carbohydrates up to 166% (Vu et al., 2002). Diurnal leaf sampling analyses show that starch contents of both pre-existing mature and new young leaves of Ambersweet orange are several-fold higher under double-ambient growth [CO_2] (Figure 2). Particularly for new young leaves, starch content of the CO_2-enriched trees at late afternoon sampling is more than four-fold greater than their counterparts at ambient [CO_2]. For sour orange and sweet orange, growth at elevated [CO_2] (700 ppm) significantly increases starch content in leaves and fibrous roots (Jifon et al., 2002). Despite the large accumulation of leaf starch that often occurs at elevated growth [CO_2] for several crop plants, there is report that the levels of mRNA for ADPG shows little change (van Oosten et al., 1994).

CITRUS GROWTH AND YIELD AT ELEVATED ATMOSPHERIC CO_2

Growth of a variety of citrus genotypes, including sweet orange, sour orange, citrange and citrumelo, appears to be as responsive as that of several other annual crop species to elevated [CO_2] (Table 1, column 3) (Koch et al., 1986, 1987; Baker and Allen, 1993; Idso et al., 1991; Idso and Kimball, 1997, 2001; Syvertsen and Graham, 1999; Syvertsen et al., 2000; Jifon et al, 2002). Fourteen-week-old seedlings of Carrizo citrange and Swingle citrumelo grown for 5 months under double-ambient [CO_2] have 69 to 94% more new shoots, total shoot dry weight increased by over 100%, total leaf area increased by 85 to 124% due to greater leaf number and size, total root dry weight increased by 37 to 100%, and total dry matter accumulation increased by 111 to 115% (Koch et al., 1986). In a subsequent experiment, exposure of 9-week-old seedlings of Carrizo citrange to double-and triple-ambient [CO_2] for 17 weeks enhances total dry matter by 67 and 120%, respectively (Koch et al., 1987).

A long-term atmospheric CO_2 enrichment study on growth and fruit production of sour orange trees was conducted outdoors, in clear-plastic wall

FIGURE 2. Diurnal changes in starch levels of pre-existing mature leaves (M) and new young leaves (Y) of Ambersweet orange trees grown for 29 months under 360 and 720 ppm CO_2. Each data value represents the mean (with SE bar) of 4 determinations.

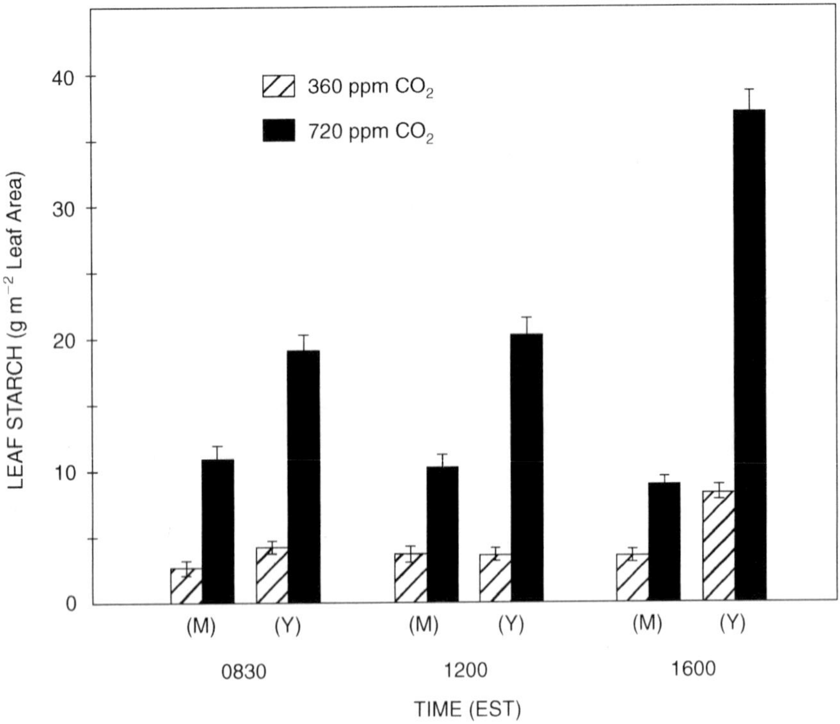

open-top chambers, at Phoenix, Arizona (Idso et al.,1991). Identical seedlings of sour orange, which had been grown outdoors and rooted in the ground, were half maintained at near-ambient [CO_2] (400 ppm) and half at 300 ppm above ambient (CO_2-enriched). In less than two years, the trunks of the CO_2-enriched trees are twice as large as their ambient-CO_2 treatment counterparts. By the end of the second year, the CO_2-enriched trees had 79% more leaves, 56% more primary branches, 70% more secondary branches, 240% more tertiary branches, and 279% more total trunk plus branch volume than the ambient-CO_2 trees (Idso et al., 1991; Idso and Kimball, 1992a). Three years later, the trunk plus branch volume factor of the CO_2-enriched trees was still about 200% that of the ambient-CO_2 trees (Idso and Kimball, 1997). These results, when compared with CO_2 enrichment studies on annual C_3 crops (Kimball,

1983), suggest that the perennial citrus species may respond proportionally more strongly to rising atmospheric [CO_2] than several other annual plant species. Fruit harvests during a six consecutive year period (years 3 to 8) for sour orange show that the average numbers of fruits produced per tree for the CO_2-enriched treatment are 25.0-, 6.9-, 3.6-, 2.3-, 2.0- and 1.7-fold those produced per tree for the ambient-CO_2 treatment (Idso and Kimball, 1997). Although not significantly different, the mean volume of the CO_2-enriched fruits is about 4% larger than that of the ambient-CO_2 treatment fruits. Over the first two years of this on-going study, elevated growth [CO_2] caused the CO_2-enriched/ambient-treatment ratio of the tree aboveground wood biomass to rise to a value in excess of 3.0, but gradually declined thereafter to level out at about 1.8 at the end of seven more years. Similarly, the ratio of yearly fruit biomass production of the CO_2-enriched/ambient-treatment over the last four years (years 10-13) also maintained the same average and essentially identical value as the aboveground wood biomass ratio (Idso and Kimball, 2001). This indicates that the CO_2-enriched sour orange trees may have reached by year 8-10 an equilibrium condition in terms of CO_2-induced enhancements in wood biomass and fruit production, and such steady-state responses likely will be maintained over the remainder of their lifespan (Idso and Kimball, 2001).

Downton et al. (1987) conducted experiments on yield response to elevated [CO_2] of 3-year-old Valencia sweet orange trees over a 12-month period. Trees grown under enriched (800 ppm) [CO_2] produce 70% more fruit, although similar in size and weight to that of control trees grown at 400 ppm [CO_2]. In their study, although leaf areas between treatments are similar, specific leaf dry weight is 25% greater for the CO_2-enriched trees. In Ambersweet orange, double-ambient growth [CO_2] treatment over a 29-month period does not significantly affect leaf area, but increases specific leaf weight 17 to 23% for pre-existing mature leaves, and 19 to 49% for young leaves (Vu et al., 2002). Dry weight response ratios to growth CO_2 (elevated/ambient) for the first year are 1.01 for leaves, 1.57 for shoot wood, 1.40 for roots, and 1.37 for total tree. At the end of the second year, the ratios are 1.05 for leaves, 1.20 for shoot wood, 1.15 for roots, and 1.15 for total tree (Allen and Vu, 2000). The large decrease in Ambersweet orange response to elevated [CO_2] in the second year is attributed to crowding of shoot and root space. As for the differences in CO_2 enhancements on citrus photosynthetic rate (Table 1, column 2), the variations in citrus growth and yield responses to elevated growth [CO_2] (Table 1, column 3) also indicate that, among citrus, genotype-specific differences will be encountered as a result of future increases in global [CO_2] and air temperatures. Besides, differences in growth and environment conditions as well as levels and durations of CO_2 treatments during the experimental periods would definitely contribute to reported variations in citrus growth and yield responses to CO_2 enrichment.

CONCLUDING REMARKS

Citrus, with its high value for nutritional and palatable qualities, enters the 21st century as one of the most important fruit crops in world trade (Vu, 1999). Sizable increases in yield of citrus could be anticipated in response to future rises in global atmospheric [CO_2], but expected climate changes as induced by rising [CO_2] and other greenhouse gases could have severe adverse impacts. Rising atmospheric [CO_2] and potential climate changes critically challenge our needs in understanding the mechanisms and processes directing citrus growth and yield. There is merit in the concept that "the superior genetically engineered citrus tree will ultimately provide the long-term solutions to environmental stresses" (Yelenosky, 1991). It is imperative therefore that citrus environmental stress research continues unabated with the eventual result of increased growth and yield under future global climate changes.

REFERENCES

Allen, L.H., Jr., and J.C.V. Vu. (2000). Carbon dioxide and temperature effects on young sweet orange trees. 2000 Annual Meetings Abstracts. American Society of Agronomy, November 5-9, Minneapolis, Minnesota, p. 34.

Arp, W.J. (1991). Effects of source-sink relations on photosynthetic acclimation to CO_2. *Plant, Cell and Environment* 14: 869-875.

Baker, J.T., and L.H. Allen, Jr. (1993). Contrasting crop species responses to CO_2 and temperature: rice, soybean and citrus. *Vegetatio* 104/105: 239-260.

Besford, R.T., L.J. Ludwig, and A.C. Withers. (1990). The greenhouse effect: acclimation of tomato plants growing in high CO_2, photosynthesis and ribulose-1,5-bisphosphate carboxylase protein. *Journal of Experimental Botany* 41: 925-931.

Bowes, G. (1993). Facing the inevitable: Plants and increasing atmospheric CO_2. *Annual Review of Plant Physiology and Plant Molecular Biology* 44: 309-332.

Bowes, G. (1996). Photosynthetic responses to changing atmospheric carbon dioxide concentration. In *Photosynthesis and the Environment*, ed. N.R. Baker. Dordrecht: Kluwer Academic Publishers, pp. 387-407.

Brakke, M., and L.H. Allen Jr. (1995). Gas exchange of *Citrus* seedlings at different temperatures, vapor-pressure deficits, and soil water contents. *Journal of the American Society for Horticultural Science* 120: 497-504.

Campbell, W.J., L.H. Allen Jr., and G. Bowes. (1988). Effects of CO_2 concentration on rubisco activity, amount, and photosynthesis in soybean leaves. *Plant Physiology* 88: 1310-1316.

Downton, W.J.S., W.J.R. Grant, and B.R. Loveys. (1987). Carbon dioxide enrichment increases yield of Valencia orange. *Australian Journal of Plant Physiology* 14: 493-501.

Drake, B.G., M.A. Gonzalez-Meler, and S.P. Long. (1997). More efficient plants: A consequence of rising atmospheric CO_2? *Annual Review of Plant Physiology and Plant Molecular Biology* 48: 609-639.

Field, C.B., R.B. Jackson, and H.A. Mooney. (1995). Stomatal responses to increased CO_2: implications from the plant to the global scale. *Plant, Cell and Environment* 18: 1214-1225.

Gesch, R.W., K.J. Boote, J.C.V. Vu, L.H. Allen Jr., and G. Bowes. (1998). Changes in growth CO_2 result in rapid adjustments of ribulose-1,5-bisphosphate carboxylase/oxygenase small subunit gene expression in expanding and mature leaves of rice. *Plant Physiology* 118: 521-529.

Hsiao, T.C. (1973). Plant responses to water stress. *Annual Review of Plant Physiology* 24: 519-570.

Hussain, M.W., L.H. Allen Jr., and G. Bowes. (1999). Up-regulation of sucrose phosphate synthase in rice grown under elevated CO_2 and temperature. *Photosynthesis Research* 60: 199-208.

Idso, S.B., B.A. Kimball, and S.G. Allen. (1991). CO_2 enrichment of sour orange trees: 2.5 years into a long-term experiment. *Plant, Cell and Environment* 14: 351-353.

Idso, S.B., and B.A. Kimball. (1992a). Aboveground inventory of sour orange trees exposed to different atmospheric CO_2 concentrations for 3 full years. *Agricultural and Forest Meteorology* 60: 145-151.

Idso, S.B., and B.A. Kimball. (1992b). Effects of atmospheric CO_2 enrichment on photosynthesis, respiration, and growth of sour orange. *Plant Physiology* 99: 341-343.

Idso, S.B., K.E. Idso, R.L. Garcia, B.A. Kimball, and J.K. Hoober. (1995). Effects of atmospheric CO_2 enrichment and foliar methanol application on net photosynthesis of sour orange tree (*Citrus aurantium*; Rutaceae) leaves. *American Journal of Botany* 82: 26-30.

Idso, S.B., and B.A. Kimball. (1997). Effects of long-term atmospheric CO_2 enrichment on the growth and fruit production of sour orange trees. *Global Change Biology* 3: 89-96.

Idso, S.B., and B.A. Kimball. (2001). CO_2 enrichment of sour orange trees: 13 years and counting. *Environmental and Experimental Botany* 46: 147-153.

Jifon, J.L., J.H. Graham, D.L. Drouillard, and J.P. Syvertsen. (2002). Growth depression of mycorrhizal *Citrus* seedlings grown at high phosphorus supply is mitigated by elevated CO_2. *New Phytologist* 153: 133-142.

Keeling, C.D., T.P. Whorf, M. Wahlen, and J. van der Plicht. (1995). Interannual extremes in the rate of rise of atmospheric carbon dioxide since 1980. *Nature* 375: 660-670.

Kimball, B.A. (1983). Carbon dioxide and agricultural yield: An assemblage and analysis of 430 prior observations. *Agronomy Journal* 75: 779-788.

Kimball, B.A. (1993). Effects of elevated CO_2 and climate variables on plants. *Journal of Soil and Water Conservation* 48: 9-14.

Koch, K.E., P.H. Jones, W.T. Avigne, and L.H. Allen Jr. (1986). Growth, dry matter partitioning, and diurnal activities of RuBP carboxylase in citrus seedlings maintained at two levels of CO_2. *Physiologia Plantarum* 67: 477-484.

Koch, K.E., L.H. Allen Jr., P. Jones, and W.T. Avigne. (1987). Growth of citrus rootstock (Carrizo Citrange) seedlings during and after long-term CO_2 enrichment. *Journal of the American Society for Horticultural Science* 112: 77-82.

Long, S.P. (1991). Modification of the response of photosynthetic productivity to rising temperature by atmospheric CO_2 concentrations: has its importance been underestimated? *Plant, Cell and Environment* 14: 729-739.
Makino, A., and T. Mae. (1999). Photosynthesis and plant growth at elevated levels of CO_2. *Plant, Cell and Environment* 40: 999-1006.
Matsuoka, M., R.T. Furbank, H. Fufayama, and M. Miyao. (2001). Molecular engineering of C_4 photosynthesis. *Annual Review of Plant Physiology and Plant Molecular Biology* 52: 297-314.
Moore, B.D., S-H. Cheng, and J.R. Seemann. (1998). Sucrose cycling, Rubisco expression, and prediction of photosynthetic acclimation to elevated atmospheric CO_2. *Plant, Cell and Environment* 21: 905-915.
Moore, B.D., S-H. Cheng, D. Sims, and J.R. Seemann. (1999). The biochemical and molecular basis for photosynthetic acclimation to elevated atmospheric CO_2. *Plant, Cell and Environment* 22: 567-582.
Morinaga, K. (2002). Effects of fruit load on photosynthesis and ribulose 1,5-bisphosphate carboxylase activity in satsuma mandarin (Citrus unshiu Marc.) trees under long-term elevated carbon dioxide. *Journal of the Japanese Society for Horticultural Science* 71: 311-316.
Morison, J.I.L., and D.W. Lawlor. (1999). Interactions between increasing CO_2 concentration and temperature on plant growth. *Plant, Cell and Environment* 22: 659-682.
Norby, R.J., S.D. Wullschleger, C.A. Gunderson, D.W. Johson, and R. Ceulemans. (1999). Tree responses to rising CO_2 in the field experiments: implications for the future forest. *Plant, Cell and Environment* 22: 683-714.
Poorter, H. (1993). Interspecific variation in the grown response of plants to an elevated ambient CO_2 concentration. *Vegetatio* 104/105: 77-97.
Poorter, H., C. Roumet, and B.D. Campbell. (1996). Interspecific variation in the growth response of plants to elevated CO_2: A search for functional types. In *Carbon Dioxide, Populations, and Communities*, ed. C. Korner, and F.A. Bazzaz. Academic Presss, New York, pp. 375-412.
Radin, J.W., B.A. Kimball, D.L. Hendrix, and J.R. Mauney. (1987). Photosynthesis of cotton plants exposed to elevated levels of carbon dioxide in the field. *Photosynthesis Research* 12: 191-203.
Sage, R.F., T.D. Sharkey, and J.R. Seemann. (1989). Acclimation of photosynthesis to elevated CO_2 in five C_3 species. *Plant Physiology* 89: 590-596.
Sage, R.F., and C.D. Reid. (1994). Photosynthetic response mechanisms to environmental change in C_3 plants. In *Plant-Environment Interactions*, ed. R.E. Wilkinson. Marcel Dekker, Inc., New york, pp. 413-499.
Schneider, S.H. (2001). What is "dangerous" climate change? *Nature* 411: 17-19.
Shine, K.P., R.G. Derwent, D.J. Wuebbles, and J.J. Morcrette. (1990). Radiative forcing of climate. In *Climatic Change. The IPCC Scientific Assessment*, ed. J.T. Houghton, G.J. Jenkins, and J.J. Ephraums. Cambridge University Press, Cambridge, pp. 41-68.
Stitt, M. (1991). Rising CO_2 levels and their potential significance for carbon flow in photosynthetic cells. *Plant, Cell and Environment* 14: 741-762.

Syvertsen, J.P., and J.H. Graham. (1999). Phosphorus supply and arbuscular mycorrhizas increase growth and net gas exchange responses of two *Citrus* spp. grown at elevated [CO_2]. *Plant and Soil* 208: 209-219.

Syvertsen, J.P., L.S. Lee, and J.W. Grosser. (2000). Limitations on growth and net gas exchange of diploid and tetraploid *Citrus* rootstock cultivars grown at elevated CO_2. *Journal of the American Society for Horticultural Science* 125: 228-234.

Thomas, J.F., and C.N. Harvey. (1983). Leaf anatomy of four species grown under continuous CO_2 enrichment. *Botanical Gazette* 144: 303-309.

van Oosten, J.-J, D. Wilkins, and R.T. Besford. (1994). Regulation of the expression of photosynthetic nuclear genes by high CO_2 is mimicked by carbohydrates: a mechanism for the acclimation of photosynthesis to high CO_2. *Plant, Cell and Environment* 17: 913-923.

Vu, C.V., L.H. Allen Jr., and G. Bowes. (1983). Effects of light and elevated atmospheric CO_2 on the ribulose bisphosphate carboxylase activity and ribulose bisphosphate level of soybean leaves. *Plant Physiology* 73: 729-734.

Vu, J.C.V., and G. Yelenosky. (1988). Water deficit and associated changes in some photosynthetic parameters in leaves of 'Valencia' orange (*Citrus sinensis* [L.] Osbeck). *Plant Physiology* 88: 375-378.

Vu, J.C.V., and G. Yelenosky. (1989). Non-structural carbohydrate concentrations on leaves of 'Valencia' orange subjected to water deficits. *Environmental and Experimental Botany* 29: 149-154.

Vu, J.C.V., L.H. Allen Jr., and G. Bowes. (1989). Leaf ultrastructure, carbohydrates and protein of soybeans grown under CO_2 enrichment. *Environmental and Experimental Botany* 29: 141-147.

Vu, J.C.V., L.H. Allen Jr., K.J. Boote, and G. Bowes. (1997). Effects of elevated CO_2 and temperature on photosynthesis and Rubisco in rice and soybean. *Plant, Cell and Environment* 20: 68-76.

Vu, J.C.V., J.T. Baker, A.H. Pennanen, L.H. Allen, Jr., G. Bowes, and K.J. Boote. (1998). Elevated CO_2 and water deficit effects on photosynthesis, ribulose bisphosphate carboxylase-oxygenase, and carbohydrate metabolism in rice. *Physiologia Plantarum* 103: 327-339.

Vu, J.C.V. (1999). Photosynthetic responses of citrus to environmental changes. In *Handbook of Plant and Crop Stress*, ed. M. Pessarakli, Marcel Dekker, Inc., New York, pp. 947-961.

Vu, J.C.V., R.W. Gesch, L.H. Allen, Jr., K.J. Boote, and G. Bowes. (1999). CO_2 enrichment delays a rapid, drought-induced decrease in Rubisco small subunit transcript abundance. *Journal of Plant Physiology* 155: 139-142.

Vu, J.C.V., R.W. Gesch, A.H. Pennanen, L.H. Allen, Jr., K.J. Boote, and G. Bowes. (2001). Soybean photosynthesis, Rubisco, and carbohydrate enzymes function at supraoptimal temperatures in elevated CO_2. *Journal of Plant Physiology* 158: 295-307.

Vu, J.C.V., Y.C. Newman, L.H. Allen Jr., M. Gallo-Meagher, and M-Q. Zhang. (2002). Photosynthetic acclimation of young sweet orange trees to elevated growth CO_2 and temperature. *Journal of Plant Physiology* 159: 147-157.

Webber, A.N., G-Y. Nie, and S.P. Long. (1994). Acclimation of photosynthetic proteins to rising atmospheric CO_2. *Photosynthesis Research* 39: 413-425.

Wigley, T.M.L., and S.C.B. Raper. (1992). Implications for climate and sea level of revised IPCC emissions scenarios. *Nature* 357: 293-300.

Wong, S.C. (1979). Elevated atmospheric partial pressure of CO_2 and plant growth. I. Interactions of nitrogen nutrition and photosynthetic capacity in C_3 and C_4 plants. *Oecologia* 44: 68-74.

Yelenosky, G. (1991). Responses and adaptations of citrus trees to environmental stresses. *Israel Journal of Botany* 40: 239-250.

The Effect of Elevated Atmospheric CO_2 Concentration on Two *Populus* Clones Grown in a FACE System

Roberto Tognetti
Antonio Raschi

SUMMARY. Two poplar clones, hybrid *Populus deltoides* Bartr. Ex Marsh × *Populus nigra* L. (*Populus ×euramericana*), clone I-214, and *Populus deltoides*, clone Lux, were grown from hardwood cuttings for one growing season in either ambient (360 μmol mol^{-1}) or elevated (560 μmol mol^{-1}) [CO_2] in FACE-system rings at Rapolano Terme (Siena, Italy). Both clones I-214 and Lux exhibited a higher aboveground biomass, photosynthesis at light saturation and instantaneous transpiration efficiency (ITE) in CO_2-enriched air. The elevated [CO_2]-induced responses of clone I-214 included increased investment in branch and leaf

---

Roberto Tognetti is affiliated with the Dipartimento di Scienze Animali, Vegetali e dell'Ambiente, Università del Molise, I-86100 Campobasso, Italy (E-mail: tognetti@unimol.it).

Antonio Raschi is affiliated with the Istituto di Biometeorologia-CNR, I-50145 Firenze, Italy

Address correpondence to: Roberto Tognetti at the above address.

The authors gratefully acknowledge M. Lanini and A. Longobucco for technical assistance in field measurements and experimental set up.

[Haworth co-indexing entry note]: "The Effect of Elevated Atmospheric CO_2 Concentration on Two *Populus* Clones Grown in a FACE System." Tognetti, Roberto, and Antonio Raschi. Co-published simultaneously in *Journal of Crop Improvement* (Food Products Press, an imprint of The Haworth Press, Inc.) Vol. 13, No. 1/2 (#25/26), 2005, pp. 377-392; and: *Ecological Responses and Adaptations of Crops to Rising Atmospheric Carbon Dioxide* (ed: Zoltán Tuba) Food Products Press, an imprint of The Haworth Press, Inc., 2005, pp. 377-392. Single or multiple copies of this article are available for a fee from The Haworth Document Delivery Service [1-800-HAWORTH, 9:00 a.m. - 5:00 p.m. (EST). E-mail address: docdelivery@haworthpress.com].

http://www.haworthpress.com/web/JCRIP

Digital Object Identifier: 10.1300/J411v13n01_18

biomass, and enhanced stem volume. The elevated [CO_2]-induced responses of clone Lux included an increase in the number of branches (and foliage area). Indication of photosynthetic acclimation under elevated [CO_2] was found during the early morning, but only in clone I-214. Stomatal conductance decreased under elevated [CO_2] particularly in clone Lux. Clone differences in response to elevated [CO_2] should be taken in account when planning future poplar plantations in forecast warmer and drier Mediterranean sites.

KEYWORDS. Elevated [CO_2], FACE-system, gas exchange, poplar

INTRODUCTION

The stimulation of tree growth by short-term exposure to elevated atmospheric CO_2 concentration ([CO_2]) (see Amthor 1995) is the result of the stimulation of a number of basic processes underlying overall plant growth and development. The primary effects of elevated [CO_2] on poplar generally include an increase in photosynthetic rates (e.g., Ceulemans et al. 1997), a reduction of stomatal conductance and decreased leaf transpiration rates (Radoglou and Jarvis 1990b). Trees grown in elevated [CO_2] can show evidence of downward acclimation of photosynthesis (e.g., Gunderson and Wullschleger 1994), i.e., a decrease in photosynthetic performance as compared with trees grown in ambient [CO_2], when measured under the same conditions, due to intrinsic changes in the photosynthetic machinery. Secondary effects may include growth traits, as well as morphological and developmental responses (Ceulemans et al. 1995, 1996).

The increasing demand for biomass (Hall 1997), has fostered the interest about the consequences of the increasing atmospheric [CO_2] on highly productive, short-rotation tree crops, such as *Populus* spp. and hybrids. Different tree genera, as well as different species and genotypes within the same genus and species, display considerably variable physiological and morphological responses to elevated [CO_2] (e.g., Ceulemans et al. 1995, 1996, 1997, Duff et al. 1994, El Kohen et al. 1993, Radoglou and Jarvis 1990a). Poplar species and hybrids generally show a large positive response to CO_2 enrichment (e.g., Gaudillère and Mousseau 1989, Radoglou and Jarvis 1990a, Brown 1991, Ceulemans et al. 1995, 1996) under various controlled environment conditions (glasshouse cabinets, growth chambers and open-top chambers). However, plant responses to elevated [CO_2] when grown in manipulated environments raise problems associated with interpreting results (Amthor 1995).

Currently, Free Air CO_2 Enrichment (FACE) studies on poplar species are being conducted in Wisconsin (USA, AspenFACE) and Italy (POPFACE) (Dicksn et al. 2000, Scarascia-Mugnozza et al. 2000). Poplar species showed to grow bigger and with increased leaf area index (LAI) under elevated [CO_2] than in control conditions in a FACE-system, during the first growing season (Calfapietra et al. 2001, Gielen et al. 2001). The concept of response specificity among poplar species to an increase in [CO_2] deserves to be extended to within species. The aim of this study was to examine the effects of an increase in [CO_2] on growth traits and gas exchange of two *Populus* clones, differing in crown architecture, leaf morphology, and resistance to climatic and biotic factors. We exposed clone cuttings to elevated [CO_2] for one growing season in the field by means of a FACE-system.

MATERIALS AND METHODS

Plant Materials and Planting Conditions

Two poplar clones, hybrid *Populus deltoides* Bartr. Ex Marsh × *Populus nigra* L. (*Populus ×euramericana*) clone I-214, and *Populus deltoides* clone Lux were selected. The former being relatively resistant to wind and susceptible to *Marssonina brunnea* (Ell. and Ev.) P. Magn., revealing a light crown, while the latter is moderately drought resistant and characterized by an open crown with large branches and leaves. Plants were grown from hardwood woody cuttings (25 cm long) in a FACE-system assemblage (two CO_2-enriched rings, 560 μmol mol^{-1}, and two ambient [CO_2] rings, 360 μmol mol^{-1}) at Rapolano Terme (Siena, Italy). Each ring was divided in four sectors. At the beginning of April (1997), the cuttings were planted at a distance of 1 m (1 × 1 m^2). The distance between rings was 30 m, and to reduce the boundary effect, several spare plants surrounded each ring. CO_2 enrichment started three weeks after planting at bud break. The rings were manually weeded, and all plants were regularly irrigated throughout the experiment. Because nutrient conditions were near optimal at the start of the experiment, fertilizer was only applied once during the spring.

FACE-System Design

The FACE-system consisted of a perforated circular annulus, CO_2 supply components, [CO_2] monitoring components and a PC-based control program. The circular array of multiple emitter port points was an 8 m-diameter toroidal distribution PVC plenum with internal diameter of 20 cm. A high volume blower injected air into the plenum. Pure CO_2 was mixed with ambient air by placing the outlet immediately after the blower at the level of a flexible pipe, which connected the blower to the plenum. A motorized metering valve controlled CO_2 injection rate (Zonemaster, Satchwell Control System, Milano, Italy). CO_2 was supplied 24 h per day. The height of the plenum was increased by means of extensive legs. This permitted to follow the growth of plants and allowed for CO_2 fumigation of the plant canopy up to 2 m height. A detailed description of the FACE-system can be found in Miglietta et al. (1997).

Growth Analysis

Total plant height (H), basal (D) and apical stem diameter, as well as total number of leaves and branches were monitored throughout the experiment. Stem volume index was estimated for each plant as $(\pi/3)H(R_1^2+R_1R_2+R_2^2)$, where R_1 and R_2 are the radii at the bottom and the top of the stem, respectively.

At the end of August, plants were harvested for analysis of aboveground biomass (stem, branches and leaves). All plant material (leaves, branches and stems) was oven dried at 70 °C until constant weight was reached. Leaf weight ratio (LWR) was calculated as the ratio of total leaf biomass to total plant biomass. Leaf area was determined using an area meter (Li-cor, Lincoln, NE, USA). Specific leaf area (SLA) was calculated as the ratio of total leaf area to total leaf dry mass.

Gas Exchange Measurements

Gas exchange measurements (light-saturated photosynthesis, stomatal conductance and leaf transpiration) were done using a portable, open-system gas analyzer (CIRAS, PP-systems, Hitchin, UK), on intact attached leaves at the same developmental stage. Mature, fully expanded leaves (fifth-sixth from the apex) of three replicate plants per treatment were sampled. Maximum photosynthetic rate, stomatal conductance and instantaneous transpiration efficiency (ITE, calculated as photosynthesis/transpiration) were measured at about two-week intervals on sunny days, under saturating PPFD conditions of about 1500 $\mu mol\ m^{-2}\ s^{-1}$. On several occasions, gas exchange was monitored along the plant profile; leaves were ranked from the bottom to the top of the stem. Two previously cross-calibrated open gas exchange systems were used for reciprocal photosynthetic rate determination, at the end of experiment (before harvest). The reference [CO_2] was set at 360 and 560 $\mu mol\ mol^{-1}$, and measurements performed in the two CO_2 treatments simultaneously (measuring the same leaf at both reference [CO_2], alternatively). The measurements were done on the labeled leaves; the measurements were first done setting the measurement [CO_2] at the plant growth [CO_2]; subsequently the measurement [CO_2] was switched from low to high in the case of plants grown at ambient [CO_2] or *vice versa* in the case of plants exposed to elevated [CO_2].

Statistical Analysis

Data of single leaves were pooled on a plant basis (physiological measurements were replicated twice). Results on a plant basis, at each sampling date, were subjected to either one-way or two-way analysis of variance (ANOVA) to statistically examine the effects of clone and CO_2 treatment.

RESULTS

Clone Lux showed a strong ($P < 0.05$) and positive effect of the elevated [CO_2] treatment on the number of branches produced during the growing season (Figure 1), and differences between clones were evident only under elevated [CO_2] (clone Lux had a higher number of branches than clone I-214). The number of leaves per plant did not differ between treatments throughout the study period, being higher ($P < 0.05$) in clone I-214 than in clone Lux (Figure 1). Stem volume index was constantly and significantly ($P < 0.05$) greater in the elevated [CO_2] treatment throughout the growing season only in clone I-214 (Figure 1). The increased stem volume production in the elevated [CO_2] treatment was explained by stimulated height growth but also by increased stem diameters. Clone I-214 showed overall significantly ($P < 0.05$) larger stem volume index than clone Lux, but only under elevated [CO_2].

At the end of the experiment there was a significant ($P < 0.05$) treatment difference in aboveground biomass (stem + branches + leaves) in both clones (Figure 2). The increase in aboveground biomass caused by the elevated [CO_2] treatment was proportionally larger for clone I-214. Clone Lux showed consistently ($P < 0.05$) greater total aboveground biomass than clone I-214, regardless of treatment, because of a much greater leaf dry mass. A significant ($P < 0.05$) and positive effect of the CO_2 enrichment was noticed on the biomass of all plant parts in clone I-214; the largest effect of elevated [CO_2] was found on branch biomass increase (despite branches did not change much in number). Biomass of branches of clone Lux did not increase significantly under elevated [CO_2] despite their increase in number. The relative stimulation in biomass of other plant parts of clone Lux was smaller compared to that observed for clone I-214.

LWR was not affected by elevated [CO_2] in the two clones (Figure 3), being generally higher ($P < 0.05$) for clone Lux than for clone I-214, regardless of the treatment. Elevated [CO_2] significantly ($P < 0.05$) decreased SLA in clone

FIGURE 1. Time course of average number of branches and leaves, and stem volume index for two poplar clones (I-214 and Lux) grown in a FACE-system under ambient or elevated [CO_2]. Vertical bars represent the standard error of the mean (12 replications), stem volume index data were fitted through power regression. Clones and treatments are referred to by symbols in the legend.

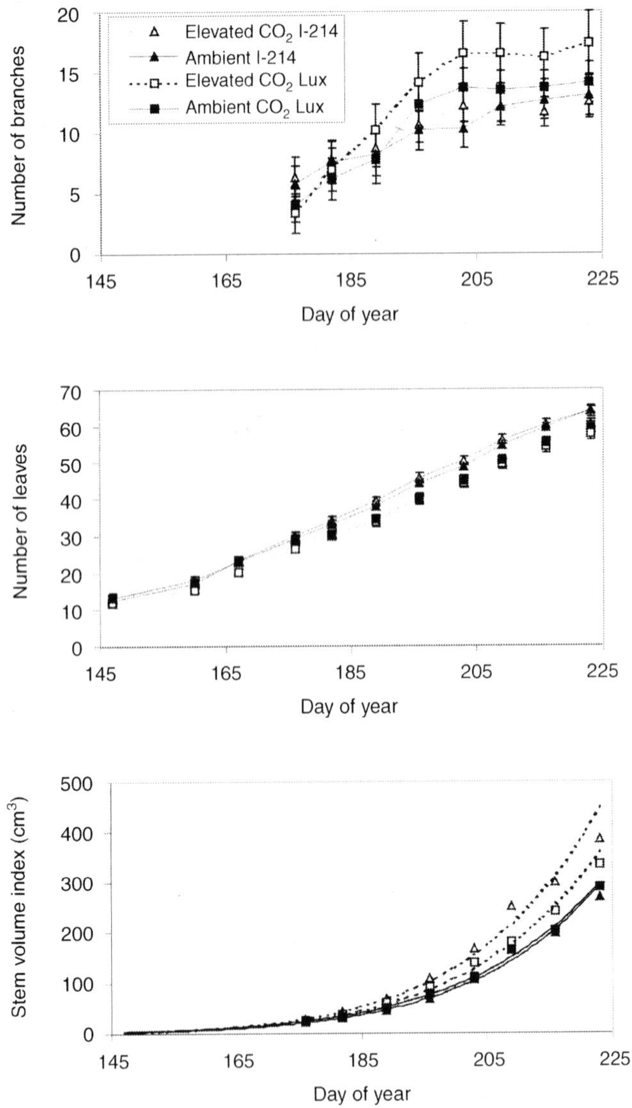

FIGURE 2. Plant biomass partitioning in two poplar clones (I-214 and Lux) grown in a FACE-system under ambient or elevated [CO_2], at final harvest. Vertical bars represent the standard error of the mean (12 replications).

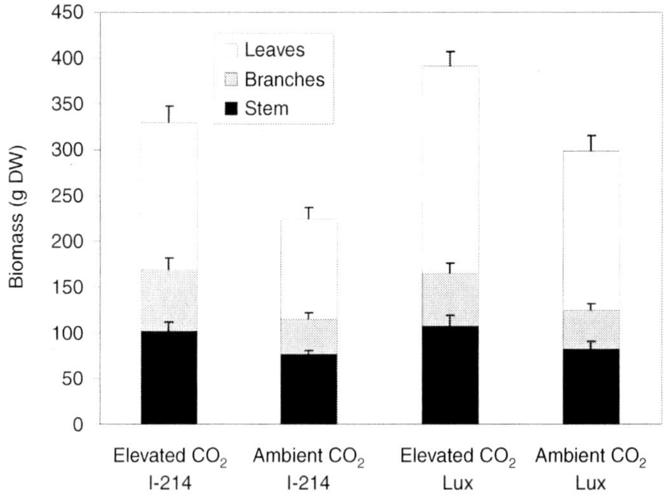

I-214 but not in clone Lux. Clonal differences were generally evident ($P <$ 0.05) in both treatments (relatively more pronounced in ambient [CO_2]).

Photosynthetic rates at light saturation were strongly ($P < 0.05$) and similarly enhanced by the elevated [CO_2] treatment in both clones (Figure 4). During the growing season, photosynthetic rates at light saturation remained stable in the elevated [CO_2] treatment, while at ambient [CO_2] there was a decrease towards the end of August. Stomatal conductance was generally lower in clone Lux, and significantly decreased in the elevated [CO_2] treatment ($P <$ 0.05) (Figure 4). Such a reduction was particularly evident in clone Lux (less in clone I-214) and at the end of the experiment. During the course of the summer, stomatal conductance decreased regardless of the treatment. As a result of the strong increase in photosynthetic rate and, secondarily, decrease in leaf transpiration, the elevated [CO_2] treatment significantly enhanced ITE ($P <$ 0.05) in both clones (Figure 5). The ratio of internal [CO_2] (Ci) to ambient (i.e., external) [CO_2] (Ca) did not change with CO_2 enrichment in both clones (Figure 5). Gas exchange measurements along the profile of plants resembled differences between treatments (at all leaf positions, $P < 0.05$) as previously described (Figure 6).

The reciprocal photosynthesis measurements at high [CO_2] (560 μmol mol^{-1}), were significantly ($P < 0.05$) lower for plants grown at elevated [CO_2]

FIGURE 3. Relationship between SLA and LWR (see text for abbreviations) in two poplar clones (I-214 and Lux) grown in a FACE-system under ambient or elevated [CO_2], at final harvest. Vertical and horizontal bars represent standard errors of the mean (12 replications). Clones are referred to by symbols in the legend.

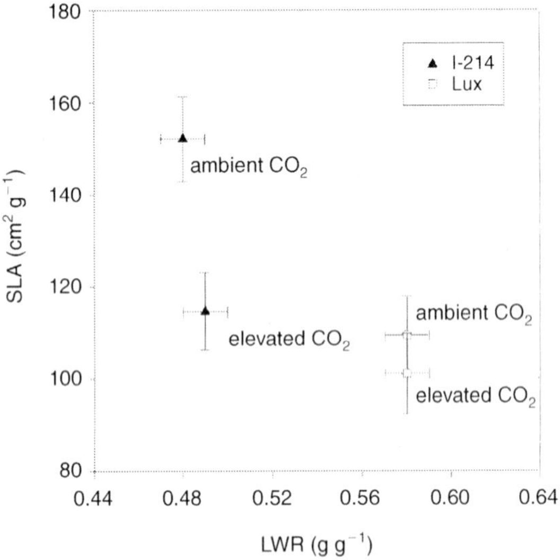

only in the early morning and for clone I-214 (Figure 7); the growth treatment had less than an effect, like for clone Lux, but the interaction between growth treatment and measurement [CO_2] was significant ($P < 0.05$). Photosynthesis measured under high [CO_2] was always found to be at least twice ($P < 0.05$) that measured under low [CO_2]. This was true for both growth treatments and clones.

DISCUSSION

Clone I-214 responded positively to high [CO_2] by increasing stem volume, but such an increase was much less evident in clone Lux. The elevated [CO_2]-induced stimulation on growth was affected by the genotype. Important differences in canopy characteristics among poplar species grown in a FACE facility have been observed by Gielen et al. (2002). The increase in stem volume in clone I-214 was primarily associated with increases in stem diameter and secondarily related to enhancements in stem height, which may result in different stem stiffness between poplar genotypes under elevated [CO_2]. The

FIGURE 4. Comparison of pooled seasonal photosynthesis and stomatal conductance (on a plant basis) for two poplar clones (I-214 and Lux) grown in a FACE-system under ambient or elevated [CO_2]. Clones are referred to by symbols in the legend.

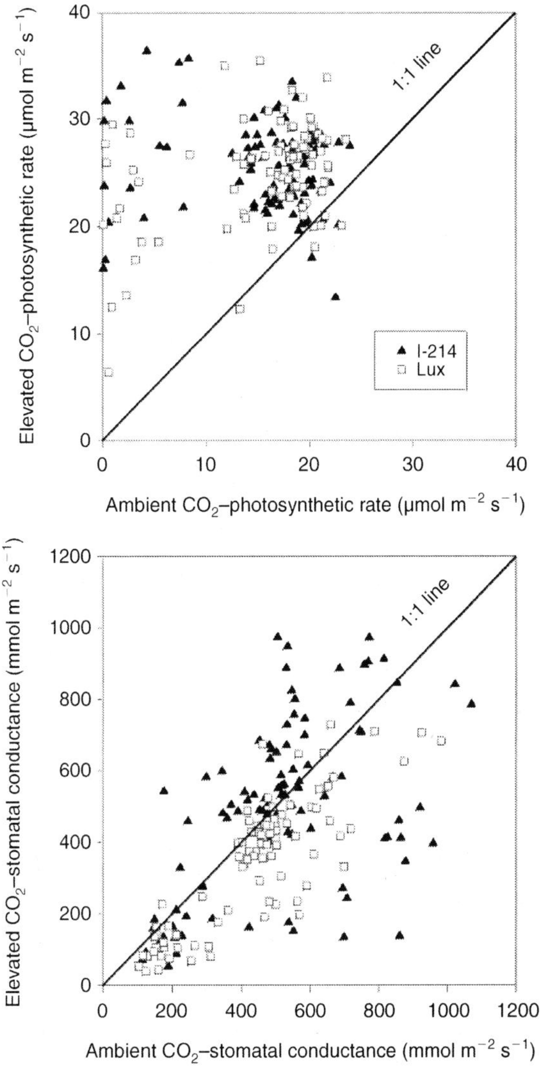

FIGURE 5. Relationship between averaged ITE and Ci/Ca (see text for abbreviations) in two poplar clones (I-214 and Lux) grown in a FACE-system under ambient or elevated [CO_2]. Vertical and horizontal bars represent standard errors of the mean (n = 24 plants). Clones are referred to by symbols in the legend.

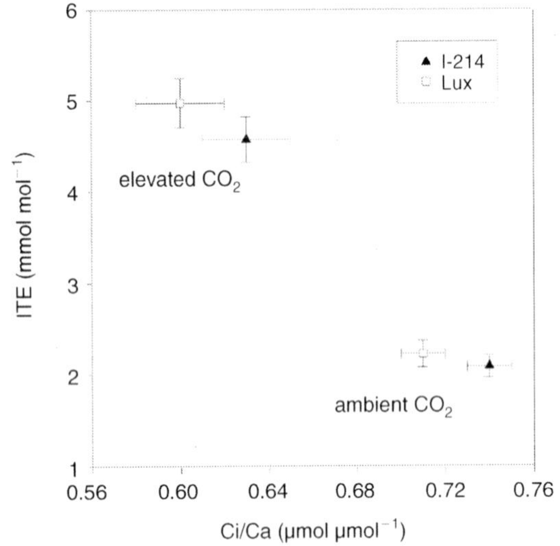

higher responsiveness to elevated [CO_2] of clone I-214 as compared to clone Lux was confirmed by the relatively larger increase in total aboveground, branch and leaf biomass, though clone Lux showed more branches in response to elevated [CO_2]. However, LWR did not vary much in response to elevated [CO_2] in both clones, so under elevated [CO_2] trees did not become more efficient in terms of the amount of biomass produced per unit leaf. Nevertheless, clone I-214 showed a pronounced and significant decrease in SLA under CO_2 enrichment, while in clone Lux the strong increase in foliage dry mass was accompanied by a significant increase in leaf area, which buffered such an effect. Gielen et al. (2001) did not observe FACE effects on SLA in three poplar genotypes. The total number of leaves per plant did not vary between treatments in both clones, though contrasting results are reported in the literature for poplar clones (Radoglou and Jarvis 1990a, Ceulemans et al. 1995, Gielen et al. 2001). In three different *Populus* species, Ferris et al. (2001) found an enhanced leaf size under elevated [CO_2], yet plant production and expansion of leaf cells had a different response, suggesting that mechanisms by which CO_2 promotes leaf growth might differ depending on the species, and are probably related to cell plasticity and cell wall elasticity. Although photosynthesis per

FIGURE 6. Leaf-level averaged photosynthetic rate along the plant profile (n = 4 plants, standard error within 5-10%, not reported for clarity) for two poplar clones (I-214 and Lux) grown in a FACE-system under ambient or elevated [CO_2]. Leaf number refers to rank position in the canopy profile. Clones and treatments are referred to by symbols in the legend.

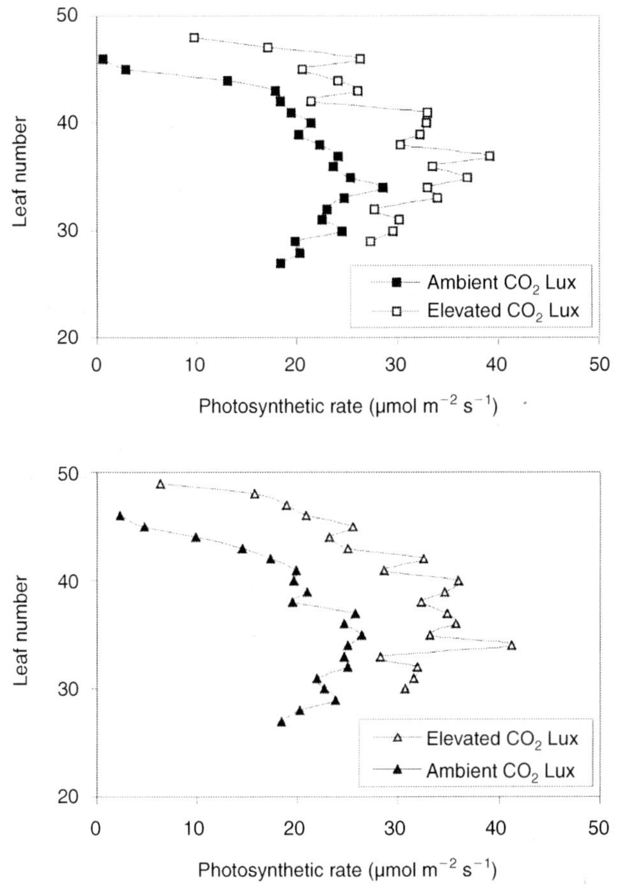

unit leaf area was similarly increased in both clones under elevated [CO_2], the different response to the CO_2 treatment between clones in terms of leaf area (Ceulemans et al. 1996) may result in different LAI of poplar plantations depending on the genotype used. Allometric LAI was found to increase in poplar genotypes grown in a FACE facility, as a result of increased tree dimension and larger individual leaf size.

FIGURE 7. Representative daily course of photosynthesis (mid-August) for two poplar clones (I-214 and Lux) grown in a FACE-system under ambient or elevated $[CO_2]$. Measurements were made in low $[CO_2]$ (360 μmol mol^{-1}) or high $[CO_2]$ (560 μmol mol^{-1}), see legend. Vertical bars represent the standard error of the mean (6-8 replications).

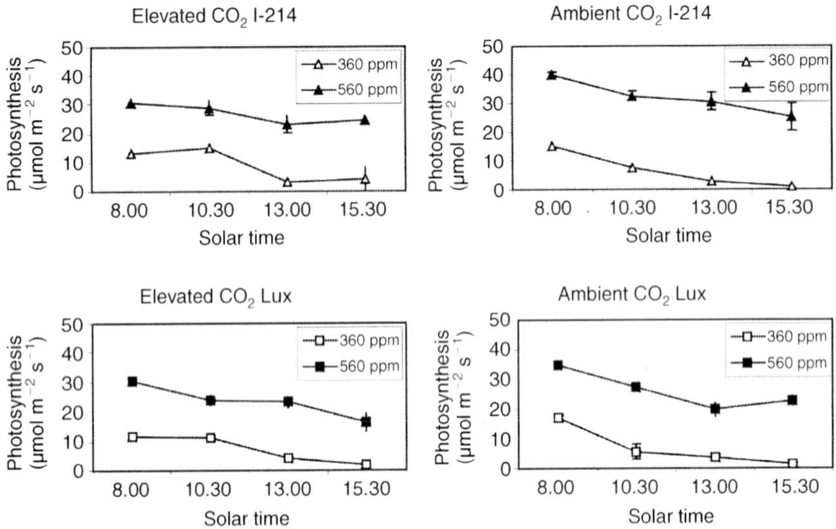

A generalized increase in the photosynthetic rate under elevated $[CO_2]$ (apparently not dependent on leaf position) has been often reported in poplar (e.g., Kalina and Ceulemans 1997), as well as a decrease in stomatal conductance (e.g., Radoglou and Jarvis 1990b). As a result of the strong increase in assimilation rate and decrease (less evident) in leaf transpiration, ITE of leaves increased under elevated $[CO_2]$ in both clones. This increase is a common response in woody species exposed to elevated $[CO_2]$ (see Ceulemans and Mousseau 1994, Saxe et al. 1998), but differences between genotypes can be important in planning future poplar plantations in Mediterranean environments. In particular, the relatively lower foliage transpiration of clone Lux under elevated $[CO_2]$ may allow this genotype to endure drought events by improving whole plant water use, though this advantage might be offset by an increased leaf area. The Ci/Ca ratio was not affected by CO_2 enrichment, despite at elevated $[CO_2]$, intercellular $[CO_2]$ should rise if stomata close consistently (Eamus and Jarvis 1989), suggesting that there was little or no stomatal acclimation to elevated $[CO_2]$ in these poplar clones. The stability of this ratio and the corresponding changes in gas exchange (changes in stomatal conductance appeared to be secondary with respect to photosynthesis, maintaining stable intercellular [CO_2] and indicating a close coupling between stomatal and mesophyll processes) suggest that variations in ITE in response to elevated [CO_2] may be primarily triggered by altered mesophyll processes in the two poplar clones.

It has been observed that the stomatal response to elevated [CO_2] is often stronger in growth cabinets than in open air fumigation systems, being linked to interactions between the photoenvironment and elevated [CO_2], as well as to phytohormone production (Saxe et al. 1998). As a matter of fact, as short-term experiments on potted plants are abandoned in favor of longer-term tests, the stomatal response to elevated carbon dioxide seems to decrease; this applies to several species, but not to others (e.g., Heath and Mansfield 1999). Although the stomatal response was limited, the results of this experiment suggest that different genotypes within the same species, may display different adaptive trends.

Gas exchange measurements along the profiles of plants showed similar differences between treatments at all leaf positions. Leaf response to elevated [CO_2] at low light intensities is still a matter of debate. Several authors report a higher photosynthetic stimulation at very low light intensities, than at full sunlight (e.g., Naumburg et al. 2001); yet, with a moderate light reduction, contrasting results are also reported (Herrick and Thomas 1999). In *Arbutus unedo* L., growing in a CO_2 spring, the higher photosynthetic enhancement evidenced in shade leaves was a consequence of the lack of midday stomatal closure (Bartak et al. 1999).

Several woody plants show evidence for acclimation (or down-regulation) of photosynthesis when grown for longer periods in elevated [CO_2] (e.g., Gunderson and Wullschleger 1994). In the present study, only an indication of acclimation was found and limited to the early morning hours for clone I-214. Gaudillère and Mousseau (1989) observed a lack of early acclimation in clone I-214, which was attributed to its high sink strength (i.e., roots). Similarly, Kalina and Ceulemans (1997) found no acclimation of photosynthesis in two hybrid poplar clones (Beauprè and Robusta). Ceulemans et al. (1997), studying poplar hybrids, observed some acclimation of photosynthesis in a glasshouse experiment but not in open-top chambers. Negative acclimation to elevated [CO_2] can hardly be observed in the field. Nevertheless, down-regulation of photosynthesis may sometimes be observed depending on the time of day and the genotype selected for measurements. There is a need for extreme caution in extrapolating results obtained in studies in controlled conditions to the real world.

In conclusion, the two clones responded positively to elevated [CO_2], both exhibiting a higher aboveground biomass, photosynthesis at light saturation and ITE in CO_2-enriched air, but the degree of the response varied with the genotype and the parameter considered. Indeed, absolute changes in stomatal conductance appeared to be secondary with respect to photosynthesis, and were relatively stronger in clone Lux than in clone I-214. The CO_2-induced responses of clone I-214 included increased investment in branch and leaf biomass, and enhanced stem volume. The CO_2-induced responses of clone Lux included an increase in the number of branches (and foliage area). We did not find an indication of strong photosynthetic acclimation under elevated [CO_2]. Growth stimulation and physiological changes have the potential to modify crown architecture and carbon allocation in response to FACE, but differences between poplar genotypes may become important in the long term (Gielen et al. 2002). Clone differences in response to elevated [CO_2] should be taken in account when planning future poplar plantations in warmer and drier Mediterranean sites as foreseen by the Global Circulation Models. The predicted responses of poplar species to rising [CO_2] have important implications for modeling the expected tree carbon sequestration (Gielen and Ceulemans 2001).

REFERENCES

Amthor J.S. 1995. Terrestrial higher-plant response to increasing atmospheric [CO_2] in relation to global carbon cycle. Global Change Biol. 1, 243-274.

Bartak M., Raschi A., Tognetti R. 1999. Photosynthetic characteristics of sun and shade leaves in the canopy of *Arbutus unedo* L. trees exposed to *in situ* long-term elevated CO_2. Photosynthetica 37, 1-16.

Brown K.R. 1991. Carbon dioxide enrichment accelerates the decline in nutrient status and relative growth rate of *Populus tremuloides* Michx. seedlings. Tree Physiol. 8, 161-173.

Calfapietra C., Gielen B., Sabatti M., De Angelis P., Scarascia Mugnozza G., Ceulemans R. 2001. Growth performance of *Populus* exposed to "Free Air Carbon dioxide Enrichment" during the first growing season in the POPFACE experiment. Ann. For. Sci. 58, 819-828.

Ceulemans R., Mousseau M. 1994. Effects of elevated atmospheric CO_2 on woody plants. New Phytol. 127, 425-446.

Ceulemans R., Jiang X.N., Shao B.Y. 1995. Effects of elevated atmospheric CO_2 on growth, biomass production and nitrogen allocation of two *Populus* clones. J. Biogeogr. 22, 261-268.

Ceulemans R., Shao B.Y., Jiang X.N., Kalina J. 1996. First-and second-year aboveground growth and productivity of two *Populus* hybrids grown at ambient and elevated CO_2. Tree Physiol. 16, 61-68.

Ceulemans R., Taylor G., Bosac C., Wilkins D., Besford R.T. 1997. Photosynthetic acclimation to elevated CO_2 in poplar grown in glasshouse cabinets or in open top chambers depends on duration of exposure. J. Exp. Bot. 48, 1681-1689.

Dickson R.E., Lewin K.F., Isebrands J.G., et al. 2000. Forest atmosphere carbon trasfer and storage-II (FACTS II) The Aspen free air CO_2 and O_3 enrichment (FACE) projects: An overview. USDA Forest Service North Central Forest Experiment Station, General Technical Report NC-214, 68 p.

Duff G.A., Berryman C.A., Eamus D. 1994. Growth, biomass allocation and foliar nutrient contents of two *Eucalyptus* species of the wet-dry tropics of Australia grown under CO_2 enrichment. Funct. Ecol. 8, 502-508.

Eamus D., Jarvis P.G. 1989. The direct effects of increase in the global atmospheric CO_2 concentration on natural and commercial temperate trees and forests. Adv. Ecol. Res. 19, 1-55.

El Kohen A., Venet L., Mousseau M. 1993. Growth and photosynthesis of two deciduous forest species at elevated carbon dioxide. Funct. Ecol. 7, 480-486.

Ferris R., Sabatti M., Miglietta F., Mills R.F., Taylor G. 2001. Leaf area is stimulated in *Populus* by free air CO_2 enrichment (POPFACE), through increased cell expansion and production. Plant Cell Environ. 24, 305-315.

Gaudillère J.P., Mousseau M. 1989. Short term effect of CO_2 enrichment on leaf development and gas exchange on young poplars (*Populus euramericana* cv. I-214). Oecol. Plant. 10, 95-105.

Gunderson C.A., Wullschleger S.D. 1994. Photosynthetic acclimation in trees to rising atmospheric CO_2: A broader perspective. Photosynth. Res. 39, 369-388.

Gielen B., Ceulemans R. 2001. Te likely impact of rising atmospheric CO_2 on natural and managed *Populus*: a literature review. Env. Pollution 115, 335-358.

Gielen B., Calfapietra C., Sabatti M., Ceulemans R. 2001. Leaf area dynamics in a poplar plantation under free-air carbon dioxide enrichment. Tree Physiol. 21, 1245-1255.

Gielen B., Calfapietra C., Claus A., Sabatti M., Ceulemans R. 2001. Crown architecture of *Populus* spp. is differently modified by free-air CO_2 enrichment (POPFACE). New Phytol. 153, 91-99.

Hall D.O. 1997. Biomass energy in industrialised countries-a view of the future. For. Ecol. Manage. 91, 17-45.

Heath J., Mansfield T.A. 1999. CO_2 enrichment of the atmosphere and the water economy of plants. In S.B. Agrawal, Agrawal, M. eds. Environmental Pollution and Plant Responses. Lewis Pub., Boca Raton, pp. 393.

Herrick J.D., Thomas R.B. 1999. Effects of CO_2 enrichment on the photosynthetic light response of sun and shade leaves of canopy sweetgum trees (*Liquidambar styraciflua*) in a forest ecosystem. Tree Physiol. 19, 779-786.

Kalina J., Ceulemans R. 1997. Clonal differences in the response of dark and light reactions of photosynthesis to elevated atmospheric CO_2 in poplar. Photosynthetica 33, 51-61.

Miglietta F., Lanini M., Bindi M., Magliulo V. 1997. Free air CO_2 enrichment of potato (*Solanum tuberosum*, L.): design and performance of the CO_2-fumigation system. Global Change Biol. 3, 417-427.

Naumburg E., Ellsworth D.S., Katul G.G. 2001. Modeling dynamic understory photosynthesis of contrasting species in ambient and elevated carbon dioxide. Oecologia 126, 487-499.

Radoglou K.M., Jarvis P.G. 1990a. Effects of elevated CO_2 enrichment on four poplar clones. I. Growth and leaf anatomy. Ann. Bot. 65, 617-626.

Radoglou K.M., Jarvis P.G. 1990b. Effects of elevated CO_2 enrichment on four poplar clones. II. Leaf surface properties. Ann. Bot. 65, 627-632.

Scarascia-Mugnozza G., De Angelis P., Sabatti M., Calfapietra C., Ceulemans R., Peressotti A., Miglietta F. 2000. A FACE experiment on short rotation, intensive poplar plantation: objective and experimental set up of POFACE. In M.A. Sutton, J.M. Moreno, W.H. van der Putten, S. Struwe eds. Terrestrial Ecosystem Research in Europe: Successes, Challenges and Policy. Office for Official Publications of the EU, Luxembourg, pp. 136-140.

Limited Increase of Agricultural Soil Carbon and Nitrogen Stocks Due to Increased Atmospheric CO_2 Concentrations

Pete Smith

SUMMARY. Increased atmospheric concentrations of CO_2 may lead to increases in agricultural soil carbon and nitrogen storage, but the impact is likely to be small and is uncertain due to limitations in other resources (e.g., nutrients, water) and interactions with climatic changes. Since only a small percentage of carbon added to the soil becomes stabilised, the impact of CO_2 fertilisation of crops is considered to be very small compared to deliberate efforts to increase soil carbon by improved agricultural management. Even if agricultural soil carbon stocks are increased, carbon credits cannot be claimed under the Kyoto Protocol since the increases are not directly human-induced, a condition which must be met in order for any carbon sink to be included in emission reduction targets. *[Article copies available for a fee from The Haworth Document Delivery Service: 1-800-HAWORTH. E-mail address: <docdelivery@haworthpress.com> Website: <http://www.HaworthPress.com> © 2005 by The Haworth Press, Inc. All rights reserved.]*

KEYWORDS. Agricultural soils, soil carbon, carbon sequestration, carbon storage, increased CO_2, Kyoto Protocol

---

Pete Smith is affiliated with the School of Biological Sciences, University of Aberdeen, St Machar Drive, Aberdeen, AB24 3UU, UK.

[Haworth co-indexing entry note]: "Limited Increase of Agricultural Soil Carbon and Nitrogen Stocks Due to Increased Atmospheric CO_2 Concentrations." Smith, Pete. Co-published simultaneously in *Journal of Crop Improvement* (Food Products Press, an imprint of The Haworth Press, Inc.) Vol. 13, No. 1/2 (#25/26), 2005, pp. 393-399; and: *Ecological Responses and Adaptations of Crops to Rising Atmospheric Carbon Dioxide* (ed: Zoltán Tuba) Food Products Press, an imprint of The Haworth Press, Inc., 2005, pp. 393-399. Single or multiple copies of this article are available for a fee from The Haworth Document Delivery Service [1-800-HAWORTH, 9:00 a.m. - 5:00 p.m. (EST). E-mail address: docdelivery@haworthpress.com].

http://www.haworthpress.com/web/JCRIP

Digital Object Identifier: 10.1300/J411v13n01_19

INTRODUCTION

In this article, the potential impact of increasing atmospheric CO_2 concentrations on agricultural soil carbon and nitrogen stocks is examined. Other global change drivers such as increasing global temperature, changing precipitation patterns and land-use change are beyond the scope of this paper. Further, it is not the purpose of this article to review in detail the many impacts of increased atmospheric CO_2 on crops since this is dealt with elsewhere in this special issue and in comprehensive reviews on this topic (e.g., Kimball 1983; Cure and Acock, 1986; Bazzaz, 1990; Drake et al., 1997). Instead, the impact on soil carbon and nitrogen stocks are examined. After reviewing these crop-mediated potential impacts, soil impacts are examined in the context of the Kyoto Protocol to the United Nations Framework Convention on Climate Change (UNFCCC).

Potential Effects of Increasing Atmospheric CO_2 Concentrations on Agricultural Soil Carbon and Nitrogen Pools

Soils are only one component of an ecosystem and often respond to global change only indirectly via physiological responses in plants. Soils are not directly affected by increases in atmospheric CO_2 since CO_2 concentrations in the air phase of soils can be 10-50 times higher than the atmospheric CO_2 concentration (Lamborg et al., 1983). In response to raised concentrations of CO_2 in the atmosphere, all ecosystem responses are mediated by direct effects on only three plant processes: photosynthesis, respiration and transpiration (Drake et al., 1997). Impacts on the soil are mediated through secondary responses in the plant such as changes in water use efficiency, specific leaf area, root: shoot ratio, C/N ratios and many others, so that impacts of raised CO_2 upon the soil can be regarded as tertiary in nature. Figure 1, developed by Smith et al. (2001), demonstrates schematically the complexity of the ecosystem response to raised CO_2.

Given that soil responses are far removed from the direct effects of increasing CO_2, it is not surprising that soil responses to global change are difficult to predict. Though the response of crops to raised CO_2 is highly species-specific, and dependent upon the conditions in which the plants are grown, there are two theoretical mechanisms by which soil organic carbon and nitrogen stocks are increased by raised atmospheric CO_2 concentrations (leading to the tentative arrows indicating an increase in soil carbon and nitrogen stocks under elevated CO_2 in Figure 1):

- *Increased carbon inputs to the soil:* Increased atmospheric CO_2 concentration leads to an increase in photosynthesis (and improved water use efficiency) which in turn leads to an increase in production. This increase in total dry matter production could increase the organic matter returns to the soil, via increased litter fall and higher root carbon and nitrogen inputs, thus increasing the stock of soil organic carbon and nitrogen.

- *Less decomposable carbon inputs to the soil:* Increased atmospheric CO_2 concentration leads to an increase in carbon assimilated by the plant. In the absence of extra mineral nitrogen, this could theoretically increase the overall C:N ratio of the plant tissue, which when it enters the soil as crop residue, may be more resistant to decomposition, since litters with a higher C:N ratio tend to be more recalcitrant. This may in turn lead to an increase in the rate of build up of soil organic carbon and nitrogen stocks though the effect may be short-lived.

There is conflicting evidence regarding an increase in carbon inputs to the soil due to increased atmospheric CO_2 concentration. In terms of increased production, a few studies show little impact, though most show some increase (Kimball, 1983; Drake et al., 1997). Results are highly species and condition specific and can even show negative impacts (Leadley and Stöcklin, 1996). Though there is much evidence of acclimation of plant responses to raised atmospheric CO_2 concentrations, whereby the effect diminishes with time (Cure

FIGURE 1. Schematic representation of the impacts of raised atmospheric CO_2 on agroecosystems (adapted from Smith et al., 2001). Arrows indicate likely direction of response. See text for further details.

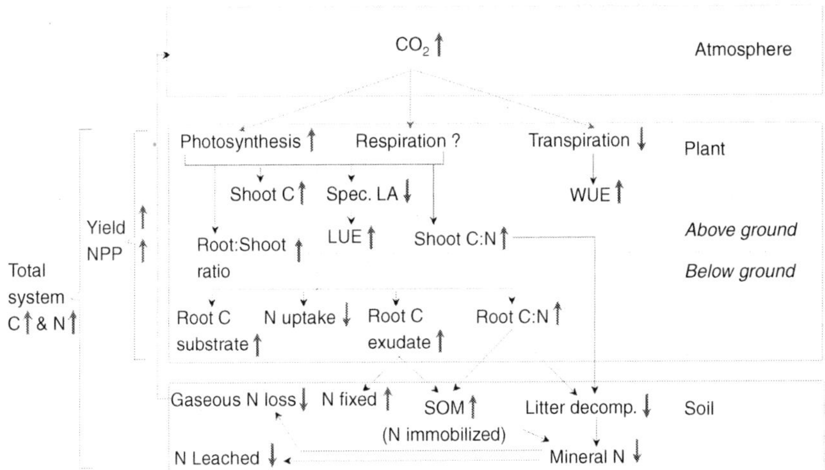

and Acock, 1986), the overall impact of increased CO_2 is likely to be that of increased photosynthesis and decreased transpiration (Drake et al., 1997) potentially leading to higher production. Fertilisation and irrigation may enhance this effect but the carbon costs of these activities may outweigh the carbon benefit in the soil (Schlesinger, 1999). Even if production is increased, however, soil carbon and nitrogen stocks will not increase if those extra carbon inputs do not reach the soil. If root systems become less well developed due to improvements in water use efficiency under raised CO_2, for example, then less carbon may be returned to the soil. Other factors may limit the potential for increased atmospheric CO_2 concentration to increase agricultural soil carbon stocks. These are described in the next section.

With respect to carbon inputs to the soil being less decomposable, Cotrufo et al. (1998) reviewed a large body of data on the nitrogen content of plant tissue grown under elevated CO_2. They found that, although the scatter of responses was large, there was, on average, a net reduction in tissue nitrogen content leading to a higher C:N ratio. A higher C:N ratio is associated with less decomposable organic matter which will tend to increase soil carbon stocks. Given that changes in the C:N ratio are small (Cotrufo et al., 1998), the effect may not be pronounced.

Limitations of, and Uncertainties About, the Impact of Increasing Atmospheric CO_2 Concentrations on Agricultural Soil Carbon and Nitrogen Pools

In addition to the potential limitations described above, other factors may limit an increase in carbon storage under increasing atmospheric CO_2 concentrations. Not least of these is water and nutrient limitation. Water and soil nutrients have been shown in many studies to limit production under raised atmospheric CO_2 (see Cure and Acock, 1986 for a thorough review).

Another important factor influencing carbon returns to the soil from crops over the coming decades is crop breeding/genetic modification. Amthor (1998) has shown that improvement in cultivars has had a far greater impact on yields in the past few decades than any other factor, and concludes that this will also be the case in the coming decades. With more of the photosynthetic product being directed toward harvestable yield, the dry matter (and hence carbon content) produced in other components of crop plants has decreased. This trend is likely to continue over the coming decades and may mean that less organic material (litter, straw, stubble, dead roots and root exudates) is returned to the soil each year. This may limit the impact of CO_2 on soil carbon levels. When compared to other global change drivers, modelling and experimental studies show that changes in temperature and precipitation pattern are likely to have much larger impacts upon the global carbon and nitrogen cycles in agricultural soils than increased atmospheric CO_2 concentration (e.g., Paustian et al., 1998).

Likely Overall Impact of Increasing Atmospheric CO_2 Concentrations on Agricultural Soil Carbon Sequestration

Despite the limitations described in the previous section, CO_2 is likely, if anything, to increase agricultural soil carbon and nitrogen stocks, though the effect may be very slight. Unpublished studies by the author using the Rothamsted carbon model (Coleman and Jenkinson, 1996), however, suggest that at least in some areas, this increase in soil carbon due to increased atmospheric CO_2 is likely to be outweighed by a loss in carbon due to climate change. Since only a small percentage of carbon added to the soil becomes stabilised, the impact of CO_2 fertilisation of crops is considered to be very small compared to deliberate efforts to increase soil carbon by changing tillage practice, agricultural extensification and use of organic amendments (e.g., Smith et al., 1997, 1998, 2000).

Implications for Policy

Article 3.4 of the Kyoto Protocol to the UNFCCC raised the possibility of soils being included in attempts to meet emission reduction targets if shown to be a sink for biospheric carbon. Following the sixth and seventh meetings of the Conference of Parties (COP6 and COP7 in The Hague/Bonn and Marrakech, respectively), this potential role for agricultural soils was confirmed for the first commitment period (2008-2012) with cropland management, grazing land management and revegetation allowable alongside forest management. The wording of Article 3.3 of the Kyoto Protocol (dealing with afforestation, deforestation and reforestation since 1990), however, suggests that not all increases in biospheric carbon stocks can be used. The phrase occurs "greenhouse gas emissions by sources and removals by sinks resulting from *direct human-induced* land-use change and forestry activities." The term direct human-induced in this context excludes increases in biospheric carbon stocks due to increasing atmospheric CO_2, which is considered to be an indirect effect. In article 3.4, dealing with agricultural soils, the word "direct" is not included, but the term *human induced* remains, and is mentioned twice within the article. This is also taken to mean that any increases in soil carbon stocks due to increasing atmospheric CO_2 are indirect and cannot be counted. Indeed, the Subsidiary Body for Scientific and Technical Advice (SBSTA) has asked the Intergovernmental Panel on Climate Change (IPCC) to prepare a report on methods to factor out direct human-induced changes in carbon stocks and greenhouse gas emissions by sources and removal by sinks, which was adopted at IPCC-XVIII in September 2001.

In short, the wording of the Kyoto Protocol and subsequent interpretations by IPCC, SBSTA and the COP mean that increases in soil carbon resulting from increased atmospheric concentrations of CO_2 cannot be included by parties to contribute towards carbon emission reduction targets. A corollary of this is that parties wishing to claim soil carbon credits due to improved management (i.e., a direct human induced effect), may need to factor out any increases in soil carbon stocks due to crop biomass increases resulting from elevated CO_2 concentrations (the CO_2 fertilisation effect), though this is not mentioned explicitly for cropland management increases in biospheric carbon in the Marrakech Accords. If required, such factoring out may be achieved by paired plot experiments (under raised and ambient CO_2) or through the use of well-calibrated, dynamic simulation models (IPCC, 2000). A group working under the auspices of the IPCC is currently (work began in 2002) preparing a report to advise on methods for factoring out indirect from direct human-induced changes in carbon stocks.

CONCLUSIONS

Increased atmospheric concentrations of CO_2 may lead to increases in soil carbon and nitrogen storage, but this is uncertain due to limitations in other resources (e.g., nutrients, water) and interactions with climatic changes. Since only a small percentage of carbon added to the soil becomes stabilised, the impact of CO_2 fertilisation of crops is considered to be very small compared to deliberate efforts to increase soil carbon through improved management. Even if soil carbon stocks are increased, carbon credits cannot be claimed under the Kyoto Protocol since the increases are not directly human-induced. In terms of the Kyoto Protocol then, CO_2 fertilisation yields no carbon benefit, and parties may even be inconvenienced if attempting to claim carbon credits in agricultural soils by having to factor out CO_2 fertilisation effects from the soil carbon stock increase attributable to improved management. In the context of the UNFCCC, the importance of CO_2 impacts on carbon storage in agricultural soils appears to be relatively limited.

REFERENCES

Amthor, J.S. (1998). Perspective on the relative insignificance of increasing atmospheric CO_2 concentration to crop yield. *Field Crops Research* 58: 109-127.

Bazzaz, F.A. (1990). The response of natural ecosystems to the rising global CO_2 levels. *Annual Review of Ecology and Systematics* 21: 167-196.

Coleman, K. and D.S. Jenkinson (1996). RothC-26.3–A Model for the turnover of carbon in soil. In: *Evaluation of Soil Organic Matter Models Using Existing, Long-Term Datasets* (eds. Powlson, D.S., P. Smith and J.U. Smith), pp. 237-246. NATO ASI Series I, Vol. 38, Springer-Verlag, Heidelberg.

Cotrufo, M.F., P. Ineson and A. Scott (1998). Elevated CO_2 reduces the nitrogen concentration of plant tissues. *Global Change Biology* 4: 43-54.

Cure, J.D. and B. Acock (1986). Crop responses to carbon dioxide doubling: a literature survey. *Agricultural and Forest Meteorology* 38: 127-145.

Drake, B.G., M.A. Gonzàlez-Meler and S.P. Long (1997). More efficient plants: A consequence of rising atmospheric CO_2? *Annual Review of Plant Physiology and Plant Molecular Biology* 48: 609-639.

IPCC (2000). Special Report on Land Use, Land-Use Change and Forestry (SR-LULUCF), Cambridge University Press, Cambridge, 377 p.

Kimball, B.A. (1983). Carbon dioxide and agricultural yield: An assemblage and analysis of 430 prior observations. *Agronomy Journal* 75: 779-788.

Lamborg, M.R., R.W.F. Hardy and E.A. Paul (1983). Microbial effects. In: *CO_2 and Plants: The Response of Plants to Rising Levels of Carbon Dioxide.* (ed: Lemon, E.R.) pp. 131-176. Westview Press, Boulder, CO.

Leadley, P.W. and J. Stöcklin (1996). Effects of elevated CO_2 on model calcereous grasslands: community, species, and genotype level responses. *Global Change Biology* 2: 389-397.

Paustian, K., Elliott, E.T. and Killian, K. (1998). Modeling soil carbon in relation to management and climate change in some agroecosystems in central North America. In: *Soil Processes and the Carbon Cycle* (eds. Lal, R. et al.) pp. 459-472. Advances in Soil Science, CRC Press, Boca Raton, FL.

Schlesinger, W.M. (1999). Carbon sequestration in soils. *Science* 284, 2095.

Smith, P., D.S. Powlson, M.J. Glendining and J.U. Smith (1997). Potential for carbon sequestration in European soils: preliminary estimates for five scenarios using results from long-term experiments. *Global Change Biology* 3: 67-79.

Smith, P., D.S. Powlson, M.J. Glendining and J.U. Smith (1998). Preliminary estimates of the potential for carbon mitigation in European soils through no-till farming. *Global Change Biology* 4: 679-685.

Smith, P., D.S. Powlson, J.U. Smith, P.D. Falloon and K. Coleman (2000). Meeting Europe's climate change commitments: Quantitative estimates of the potential for carbon mitigation by agriculture. *Global Change Biology* 6: 525-539.

Smith, P., P.D. Falloon, D.S. Powlson and J.U. Smith (2001). Carbon mitigation options in agriculture: Improving our estimates for Kyoto. In: *Sustainable Management of Soil Organic Matter* (eds. Rees, R.M., B.C. Ball, C.D. Campbell and C.A. Watson), pp. 324-329. CABI, Wallingford, UK.

Index

http://www.haworthpress.com/web/JCRIP